The Ecology of
Regulated Streams

Dam on the Middle Zambezi in Moçambique. Looking downstream,
January, 1974. Photo by B. R. Davies.

The Ecology of Regulated Streams

Edited by

JAMES V. WARD

Colorado State University
Fort Collins, Colorado

and

JACK A. STANFORD

North Texas State University
Denton, Texas

Springer Science+Business Media, LLC

Library of Congress Cataloging in Publication Data

International Symposium on Regulated Streams, 1st, Erie, Pa., 1979.
   The ecology of regulated streams.

   Includes index.
   1. Stream ecology—Congresses. 2. Rivers—Regulation—Environmental
aspects—Congresses. I. Ward, James V. II. Stanford, Jack Arthur, 1947-
   III. Title.
QH541.5.S7I57 1979   574.5′2632   79-21632

Additional material to this book can be downloaded from http://extras.springer.com.

ISBN 978-1-4684-8615-5   ISBN 978-1-4684-8613-1 (eBook)
DOI 10.1007/978-1-4684-8613-1

Proceedings of the First International Symposium on
Regulated Streams held in Erie, Pennsylvania,
April 18–20, 1979.

Sponsored by
The North American Benthological Society
and
The National Science Foundation

This book is dedicated to Prof. Drs. Joachim Illies and Karl Müller, whose presence and valued participation in the First International Symposium on Regulated Streams was most appreciated.

Preface

The idea for an international symposium on regulated streams
was conceived over an open-faced sandwich at the Rådhus in Copenhagen
when we attended the Congress of the Societas Internationalis Lim-
nologiae in summer 1977. Although we were aware that various colleagues were working on ecological problems in reservoir tailwaters,
we did not fully comprehend the magnitude of worldwide stream regulation nor the extent of interest in the subject. Such revelations
are reflected in the 21 papers included in this book. The authors
have summarized current understanding of the ecology of regulated
streams and attempted to convey the importance and direction of
future scientific investigations in stream ecosystems altered by
upstream impoundments.

The First International Symposium on Regulated Streams was the
plenary event at the 27th annual meeting of the North American
Benthological Society, April 18-20, 1979, in Erie, Pennsylvania.
More than 500 colleagues attended. We gratefully acknowledge the
support granted by the National Science Foundation; these funds
permitted intellectual exchange between scientists from eight countries on four continents. We extend personal thanks to Dr. K. W.
Stewart, President of NABS, and the NABS Program Committee, including
Drs. E. C. Masteller, E. R. Brezina, and W. P. Kovalak. These
individuals and other officers and members of the Executive Committee
assisted us with the many details leading to organization and
staging of a scientific forum. Discussions with Dr. John Cairns,
Jr. and Dr. G. Richard Marzolf during the early planning stage were
most helpful.

We also sincerely thank all of the Symposium participants,
especially those who prepared manuscripts at our request. The
authors of papers included in this volume have summarized a vast
array of published and unpublished information. Some of the cita-
tions encompass literature that may be difficult for other investigators to locate; therefore, interested readers are encouraged to
contact the authors for assistance in tracking down such materials.
Manuscripts were improved by the thorough review of the many stream
limnologists who graciously refereed the articles. Because acknowl-

edgment of individual reviewers would, in some cases, negate the
anonymity of the review process, referees are not identified;
however, their careful attention to the accuracy and scientific
content of the manuscripts is most appreciated.

The North American Benthological Society provided publication
support for the volume and we acknowledge the help of the Society's
Publication Committee. We sincerely thank Mrs. Janice Ward for
valuable editorial assistance, and the Publications Section of the
Natural Resource Ecology Laboratory, Colorado State University,
for typing the final manuscript.

                                        James V. Ward
                                        Fort Collins

                                        Jack A. Stanford
June 1979   Flathead Lake

Contents

SECTION III: SPECIAL TOPICS

"I do not know much about gods; but I think that the river
Is a strong brown god--sullen, untamed and intractable,
Patient to some degree, at first recognized as a frontier;
Useful, untrustworthy, as a conveyor of commerce;
Then only a problem confronting the builder of bridges.
The problem once solved, the brown god is almost forgotten
By the dwellers in cities--ever, however, implacable,
Keeping his seasons and rages, destroyer, reminder
Of what men choose to forget. Unhonoured, unpropitiated
By worshippers of the machine, but waiting, watching and
waiting."

T. S. Eliot
"The Dry Salvages"

DAMMED RIVERS OF THE WORLD: SYMPOSIUM RATIONALE

Jack A. Stanford and James V. Ward

Department of Biological Sciences
North Texas State University
Denton, Texas   76203

Department of Zoology and Entomology
Colorado State University
Fort Collins, Colorado   80523

H. B. N. Hynes (1970, 1975) has eloquently established stream
ecology as a fundamental subdiscipline of limnology. In general
terms he describes a river as an expression of the valley through
which it flows; production of carbon in lotic habitats is greatly
influenced by input of allochthonous nutrients and detritus from the
drainage basin. The degree of heterotrophy and/or autotrophy in
stream communities is a function of this allochthonous input and the
extent of canopy development, which influences the amount of solar
energy reaching primary producer biomass in the water. A natural
river system may be conceptually perceived as an ecological continuum
in which the P/R ratio (i.e., the relationship between community
production and respiration) changes from much less than unity in
heavily canopied reaches of small headwater tributaries to slightly

greater than unity in the slow, meandering areas of the river near
its mouth. This idea is being developed in collaborative, holistic
studies in very different geographical areas for the purpose of
defining more satisfactory methods of quantifying ecological
processes and providing ultimate comparative data on lotic ecosystems
(Cummins, 1975; this volume). Streams have received a fair amount
of hypothesis-oriented study, although we are only now beginning to
use experimental techniques in stream limnology that have been
employed so successfully in the study of eutrophication in lentic
systems (Hynes, 1975).

The river continuum, however, is profoundly interrupted when
dams are employed by man to impound or divert river flow. It is in
this context that we use the term "regulated stream" to mean the
anthropogenic control of flow in lotic habitats (i.e., tailwaters)
by dams and/or diversions; the term "stream" is used broadly to
define any running-water segment within a particular river basin.
The World Register of Dams (Toran and Mermel, 1973) lists more than
12,000 structures greater than 15 m high that have been built on the
world's major rivers. Reservoirs totaling more than 300,000 km^2 in
surface area and impounding 4000 km^3 of water (i.e., greater
than one-third of the total water content of the earth's atmosphere)
have been created (Fels and Keller, 1973). In some countries (e.g.,
India) virtually every feasible dam site has been exploited. Most
of the major rivers of the world are now impounded at some point in
the basin. Organisms in lotic waters below reservoirs must adapt to
often-erratic discharges that embody the limnological characteristics of the impoundment, or be eliminated from tailwater areas.
Thermal and chemical characteristics of tailwaters depend in part on
the reservoir depth from which water is drawn. Downstream flows are
often derived from the hypolimnion of deep, stratified reservoirs;
the impoundment structures are referred to as hypolimnial- or
hypolimnetic-release dams.[1]

The Gunnison River in Colorado (Fig. 1) embodies many of the
characteristics of a stream continuum interrupted by impoundments
and altered by downstream flow regulation. *Pteronarcys californica*
(Insecta:Plecoptera), the lumbering willow fly, once characterized

---

[1]The terms hypolimnial and hypolimnetic are both used in current
 literature. Webster's Third New International Dictionary lists
 both as adjectives of hypolimnion. Those seeking linguistic
 purity may prefer to use the Greek suffix "etic" with a Greek root.

Fig. 1.   The Gunnison River, Colorado, in the Black Canyon,
           showing Curecanti Spire prior to impoundment (below)
           and after impoundment (above) by Morrow Point Dam.
           (Photos by L. S. Stanford and J. A. Stanford).

the River. Enough nymphs for a days fishing could be collected by
dislodging a few large rocks on the river bottom; nymphs were
abundant in leaf packs, which commonly accumulated on the upstream
side of obstructions or in crevices in the substrate.

The willow flies, leaf packs, and much of the river are gone
today (Fig. 1). The regulated stream in the Black Canyon of the
Gunnison is characterized by a stabilized shoreline and bottom;
caddisflies, blackflies, and crayfishes now dominate the benthos.
The changes are profound and transcend many limnological problems
that must be solved by the same hypothesis-testing techniques used
to study unregulated stream systems.

The recent emphasis on development of fossil fuel and nuclear
power generation facilities further justifies development of a
state-of-the-art synthesis on the ecology of regulated streams.
Present hydrogeneration facilities will be increasingly used as
"peaking units." Fossil fuel and nuclear power plants cannot be
operated precisely on a demand mode; therefore, a need exists (and
will likely become more profound) for mechanisms of "storing" excess
power in hydro-units. Withdrawals can be made from reservoirs to
meet additional power as needed above that produced by continuous
sources (i.e., fossil fuel and nuclear and possibly solar); or,
during periods below that produced by the continuous sources, excess
power may be used to pump water back into reservoirs from some point
downstream (i.e., "pumped storage"). This battery-like operation of
existing reservoirs will enhance the erratic nature of discharges, a
feature of tailwater ecology that seems to be most detrimental to
riverine biota.

The volumes by Hynes (1970) and Whitton (1975) grandly review
river ecology and propound the need and direction for further study,
especially holistic interpretation, in lotic ecosystems. Although
problems associated with impoundments have been addressed in a
variety of reviews (see Lowe-McConnell, 1966; Elder et al., 1968;
Obeng, 1969; Hall, 1971; Ackermann et al., 1973; Baxter, 1977;
Cairns et al., 1978), effects on the biota of downstream lotic
reaches are treated only cursorily, if at all.

The myriad of problems associated with stream regulation were
briefly summarized by Hagan and Roberts (1973) and Ridley and Steel
(1975). Holistic documentation of the effects of regulation on
specific riverine ecosystems is unavailable. A notably comprehensive study is that of Penáz et al. (1968), which involved detailed
documentation of the limnology of Vír Valley Reservoir tailwaters
(Svratka River, Czechoslovakia). No one has yet investigated
functional, as well as structural, aspects of tailwater ecosystems
from a hypothesis-oriented and experimentally defined approach.
Numerous reports and journal articles are, however, available that

describe various, site-specific effects of regulation on the lotic environment.

The reviews and contributed papers presented in these symposium proceedings represent a compilation of scientific knowledge on the ecology of regulated streams. No attempt was made to discuss the relevance of these data to societal statements on environmental impact (e.g., Goldman et al., 1973), water law (e.g., Nanda, 1977), or the economics and cost benefits of water development projects (e.g., James and Lee, 1971), because each of these is a treatise in itself. Authors, representing eight countries, were selected by the editors for their expertise in a subject area or geographical location, since nearly every major river in the world is now regulated to some extent. Russia and China, which contain large-scale and highly regulated stream systems, were not overlooked intentionally; rather, attempts to contact colleagues with knowledge on the subject in those regions proved futile. All manuscripts were sent to outside referees in addition to being subjected to editorial review.

We attempted to limit stated or implied advocacy or depreciation of specific water development projects and emphasize science. We acknowledge that science *must* at some critical point in the decision-making process influence logical management tactics in river basins. We agree with Wolman (1971) that stream regulation has exerted more profound and irrevocable effects on the character of the world's rivers than pollutants. Altered ecosystems below dams and diversions are now the most prevalent lotic environments on earth.

REFERENCES

Ackermann, W. C., White, G. F., Worthington, E. B., and Ivens, J. L., eds., 1973, "Man-Made Lakes: Their Problems and Environmental Effects," Am. Geophys. Union, Washington, D.C.
Baxter, R. M., 1977, Environmental effects of dams and impoundments, *Annu. Rev. Ecol. Syst.*, 8:255-283.
Cairns, J., Benfield, E. F., and Webster, J. R., eds., 1978, "Current Perspectives on River-Reservoir Ecosystems," N. Am. Benthol. Soc., and Virginia Polytechnic Inst. and State Univ., Blacksburg, Virginia.
Cummins, K. W., 1975, The ecology of running waters; theory and practice, p. 278-293, *in*: "Proc. Sandusky River Basin Symp.," D. B. Baker, W. B. Jackson, and B. L. Prater, eds., Int. Joint Commission.
Elder, R. A., Krenkel, P. A., Thackston, E. L., eds., 1968, "Current Research into the Effects of Reservoirs on Water Quality," Tech. Rep. No. 17, Dep. Environ. and Water Res. Eng., Vanderbilt Univ., Nashville, Tennessee.

Fels, E., and Keller, R., 1973, World register on man-made lakes,
    p. 43-49, *in*: "Man-Made Lakes: Their Problems and Environ-
    mental Effects," W. C. Ackermann, G. F. White, and D. B.
    Worthington, eds., Am. Geophys. Union, Washington, D.C.

Goldman, C. R., McEvoy, J., III, and Richerson, P. J., 1973, "Envi-
    ronmental Quality and Water Development," W. H. Freeman,
    San Francisco.

Hagan, R. M., and Roberts, E. B., 1973, Ecological impacts of water
    storage and diversion projects, p. 196-215, *in*: "Environmental
    Quality and Water Development," C. R. Goldman, et al., eds.,
    W. H. Freeman, San Francisco.

Hall, G. E., ed., 1971, "Reservoir Fisheries and Limnology," Spec.
    Publ. No. 8, Am. Fish. Soc., Washington, D.C.

Hynes, H. B. N., 1970, "The Ecology of Running Waters," Univ.
    Toronto Press, Toronto, 555 p.

Hynes, H. B. N., 1975, The stream and its valley, *Verh. Int. Verein.
    Limnol.*, 19:1-15.

James, D. L., and Lee, R. R., 1971, "Economics of Water Resources
    Planning," McGraw-Hill, New York.

Lowe-McConnell, R. H., ed., 1966, "Man-Made Lakes," Acad. Press,
    London.

Nanda, V. P., ed., 1977. "Water Needs for the Future. Political,
    Economic, Legal, and Technological Issues in a National and
    International Framework," Westview Press, Boulder, Colorado.

Obeng, L. E., ed., 1969, "Man-Made Lakes: The Accra Symposium,"
    Ghana Univ. Press, Accra.

Penáz, M., Kubícek, F., Marvan, P., and Zelinka, M., 1968, Influence
    of the Vír River Valley Reservoir on the hydrobiological and
    ichythological conditions in the River Svratka, *Acta Sci. Nat.
    Brno*, 2:1-60.

Ridley, J. E., and Steel, J. A., 1975, Ecological aspects of river
    impoundments, p. 565-587, *in*: "River Ecology," B. A. Whitton,
    ed., Blackwell Sci. Publ., Oxford.

Toran, J., and Mermel, T. W., 1973, "World Register of Dams," Int.
    Commission on Large Dams, Paris.

Whitton, B. A., ed., 1975, "River Ecology," Blackwell Sci. Publ.,
    Oxford.

Wolman, M. G., 1971, The nations rivers, *Science*, 174:905-918.

Section I
Topical Reviews

THE NATURAL STREAM ECOSYSTEM

Kenneth W. Cummins

Department of Fisheries and Wildlife
Oregon State University
Corvallis, Oregon 97331

GENERAL STREAM ECOSYSTEM STRUCTURE AND FUNCTION

In the past decade, theory of stream ecosystem structure and
function has come to emphasize the origins and fates of organic
resources and inorganic nutrients in running waters, which are
viewed as subsystems of their watersheds or drainage basins (Cummins,
1974, 1975, 1977; Hynes, 1970, 1975; Cummins and Spengler, 1978;
Minshall, 1978). Lotic research, conducted on isolated reaches
without a spatial-temporal perspective of general watershed processes, has often proven too narrow to be of extensive use either
for the development of theory or management application. Stream
order (Strahler, 1957), geomorphic and vegetational setting, and the
annual hydrograph pattern exert important controls on biological
function. Studies of detritus dynamics in headwater streams have
shown the importance of organic inputs from the riparian zone
(Fisher and Likens, 1973; Sedell et al., 1975; Cummins et al.,
1979). Management of the riparian zone exerts a major control on
headwater streams, and such management of headwaters is a primary
ingredient in the maintenance of community structure of larger
receiving streams and rivers in a drainage net.

The zone of riparian vegetation largely determines the balance
between autotrophy and heterotrophy in headwater streams, through
light alteration and supplies of allochthonous organic matter.

An expression of the autotrophy/heterotrophy balance is the
relationship in the rates of gross primary production to community
respiration (P/R) (Odum, 1956). Although the ratio does not indicate
the importance of organic inputs from upstream (e.g., Fisher and

Likens, 1973) or absolute values of P or R, it is a useful index
relating total community catabolic processes to the level of instream
photosynthesis. Because primary production in headwaters is largely
controlled by light levels (Minshall, 1978), streams with closed
canopies have annual P/R values < 1 (Cummins, 1974). The exception
to the generalization that headwaters (orders 1 to 2 or 3) are
characterized by P/R < 1 are those that are not light limited—
namely, high altitude and latitude and arid-land streams.

Even under light limitation, periphyton communities may be more
extensive without the scour of annual floods. For example, during
the severe 1977 drought in the northwestern U.S., unusually high
standing crops of attached algae were observed in headwater streams
during fall and winter, normally the period of peak flows. Floods
also excavate detritus buried in the sediments, release some lodged
in debris jams, and capture it from the flood plain. This redistribution of organics along with displacement of periphyton are key
stream ecosystem processes tied to the hydrographic regime. Although
the nature of the effects will be determined by such factors as
geomorphic characteristics and severity and timing of a given storm
event, floods undoubtedly represent important reset mechanisms in
all stream ecosystems.

Heterotrophic communities are characteristic of first to
second or third order headwaters with annual P/R < 1 (Fig. 1). The
zone of transition to P/R > 1 varies, depending upon light regime
and other factors (Cummins, 1974; Minshall, 1978). As shown by
Minshall (1978) for a second order arid-land stream, even if instream
primary production is the major energy input, the utilization of
this photosynthetically derived biomass, especially macrophytes, may
be largely in the detrital .compartment of stream ecosystems. As
streams widen to mid-sized rivers and the riparian vegetation exerts
less control over the light regime and contributes a reduced proportion of the total organic flux (Fisher, 1977), these systems,
approximately orders 4 through 6, receive significant quantities of
organic and inorganic material from the upstream drainage net and
have an annual P/R of > 1 (Fig. 1). Very large rivers, generally
orders 7 or greater, tend to be heterotrophic, primarily a feature
of reduced light penetration at increased depth and turbidity
(Fig. 1) (Cummins, 1975, 1977). The River Continuum Hypothesis,
proposed by R. L. Vannote, J. R. Sedell, G. W. Minshall, C. E.
Cushing, and K. W. Cummins, embodies the concept of ecological
changes, such as P/R, as a continuous drainage basin gradient from
headwaters to river mouths (Fig. 1). A basic element of the
hypothesis is the dependence of downstream communities on upstream
processes—communities in each successive stream order are dependent
upon the inefficiency or "leakage" from the preceding orders. This
storage-cycle-release nature of open flowing-water ecosystems is
embodied in the Nutrient Spiraling Concept (in contrast to nutrient

cycling) proposed by J. R. Webster, R. O'Neil, J. B. Wallace,
J. B. Wade, and coworkers. This cascading of nutrients, such as
particulate (POM) and dissolved (DOM) organic matter, in which some
is recycled and some released, is a fundamental feature of lotic
ecosystems and is tied to the flow regime and the physical and
chemical retention features of a given reach.

The balance between primary production and respiration (P/R)
and between storage-processing and export (S-P/E) constitute basic
features of lotic systems that change along a continuum with stream
order and would be significantly affected by altered flow regime.
The model of lotic ecosystem dynamics derived by McIntire and
Colby (1978) provides for simulation of ecological effects (e.g.,
P/R or S-P/E) resulting from changes in discharge, suspended load
and light regime. In addition, specific models deal with periphyton
primary production (McIntire, 1973) and particulate organic matter
processing (Boling et al., 1975a,b).

ORGANIC SUBSTRATES IN RUNNING WATER ECOSYSTEMS

Many recent conceptualizations of stream ecosystem structure
and function have emphasized the types, sources, and fates of the
spectrum of organic substances (Table 1) that drive the biological
system (e.g., Hynes, 1970, 1975; Cummins, 1974; Whitton, 1975;
Wallace et al., 1977; McIntire and Colby, 1978; Minshall, 1978;
Anderson and Sedell, 1979; Anderson and Cummins, 1979; Cummins and
Klug, 1979; Wallace and Merritt, 1980).

The four general categories of organic resources in stream
ecosystems (primary producers, detritus, dissolved organics, and
animal consumers) vary widely in their nutritive content (Table 1).
As discussed by Russell-Hunter (1970), herbivore and detrivore
metabolism often entails the ingestion of generally low-quality
substrate and the assimilation of a high-quality fraction of the
intake. In general, a protein content in excess of 16% (by dry
weight) and a carbon to nitrogen ratio (C/N) of less than 17 constitute the lower limit of high substrate quality for consumers
(Russell-Hunter, 1970). Although detritus may be outside this
quality range (C/N > 17), the microbial biomass associated with the
organic particles is not (C/N = 10-11). The C/N of fine detritus is
deceptively low, since a very significant portion of the nitrogen
present is in an extremely refractory form, not available to animals,
and is a poor substrate for microbial metabolism (Cummins and Klug,
1979).

Studies of organic budgets for small watersheds (e.g., Fisher
and Likens, 1973; Sedell et al., 1973) or reaches of larger rivers
(e.g., Fisher, 1977) have indicated that 40 to 75% of the annual

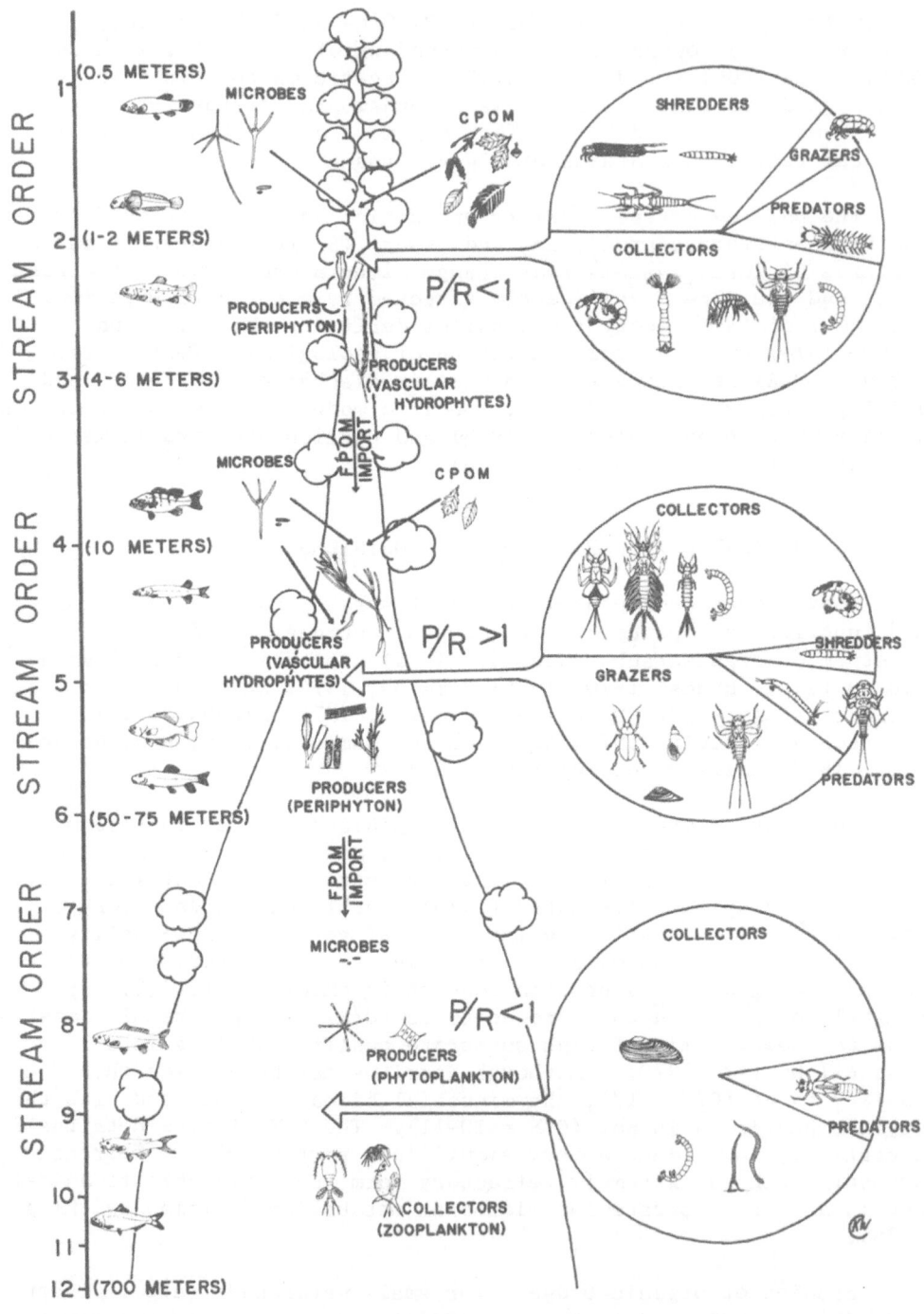

Fig. 1.   Diagrammatic representation of the river continuum shown
           as a single stem of increasing order (actually the number
           of branches increases with decreasing stream order).
           General range of stream widths (in meters) for each order
           is given, and orders have been roughly grouped into head-
           waters (orders 1-3), mid-sized rivers (4-6), and large
           rivers (7-12), considering the Mississippi River as order
           12 at its mouth. The headwaters and large rivers are
           shown as heterotrophic (P/R, or ratio of gross photosyn-
           thesis to community respiration, < 1) because of restricted
           light, a consequence of shading by riparian zone vegetation
           in the headwaters and attenuation from depth and turbidity
           in the large rivers. The mid-sized rivers are depicted as
           autotrophic, with a P/R > 1, through a combination of
           reduced riparian shading, relatively shallow and clear
           water. The importance of terrestrial inputs of coarse
           particulate organic matter (CPOM) decreases and the
           transport of fine POM increases down the continuum. The
           ratios of macroinvertebrate functional groups shift from
           shredder-collector headwaters to collector-grazer (scraper)
           mid-sized rivers to collector-dominated large rivers,
           which are lentic-like, with plankton communities. Fish
           populations are shown to shift from cold- to warm-water
           invertebrate feeders. Mid-sized rivers have piscivorous
           forms as well, and large rivers have both bottom feeders
           and planktivorous species.

Table 1. Categories of Organic Materials in Lotic Ecosystems (Modified from Cummins and Klug, 1979)

Organic Component	General Size Range	Primary Sources	Stream Order (In Which Most Prevalent)	C/N Range (Approx.)	Macroinvertebrate Functional Feeding Group Using the Resource
Periphyton (microproducers)	Approx. <500 µm >10 µm	Instream (light fixation)	4-6	5-10:1	Scrapers
Macrophytes (macroproducers)	Approx. >1 cm (some macroalgae <1 cm >1 mm)	Instream (light fixation)	4-6	13-70:1	Scrapers, piercers (herbivore shredders)
Detritus (particulate organic matter, POM)	>0.5 µm	Riparian zone; upstream tributaries	1-12	7-1300:1	Shredders, collectors
Coarse (CPOM)	>1 mm	Riparian zone	1-4	20:1300:1	Shredders
Woody		Riparian zone		200-1300:1[a]	Shredders (gougers)
Coarse	>10 cm		1-3		
Fine	<10 cm >1 mm		1-4		
Nonwoody (esp. leaves, needles)	>1 mm	Riparian zone	1-3	20-80:1[a]	Shredders (scrapers)
Fine (FPOM)	<1 mm >50 µm	Upstream; CPOM processing	2-8	7-40:1[b]	Collectors (esp. gathering)
Ultrafine (UPOM)	<50 µm >0.5 µm	Upstream; CPOM processing; DOM flocculation, adsorbtion	1-12	9-20:1[b]	Collectors (esp. filtering)
Dissolved organic matter (DOM)	<0.5 µm	Upstream: POM leaching; ground water; producer and consumer exudates and excretions	1-12	>17:1[c]	None
Animals		Instream (growth)	1-12	<17:1 (10)	Predators
Macro	Approx. >100 µm				
Micro	Approx. <100 µm				

[a] Microbial portion 10-11:1.

[b] Low values misleading, since a major portion of the nitrogen present may be highly biologically refractory. A better ratio might entail the use of nonfiber nitrogen only.

[c] Labile portion of DOM, which turns over rapidly, probably <17.

energy flux is in the form of dissolved organic matter. Of the 25 to 60% POM, at least half consists of CPOM and FPOM, the former being more abundant in the headwaters. POM inputs to headwater streams (orders 1-3) vary from 300 to 800 g AFDW m^{-2}, not including coarse wood debris (Anderson and Sedell, 1979). Measurements of wood standing crops range from 1-2 kg $\cdot$ m^{-2} in Michigan streams to 10-15 kg $\cdot$ m^{-2} in western Oregon streams (Anderson et al., 1978; Swanson and Lienkaemper, 1978).

Only the very general aspects of DOM processing, i.e., sources, utilization, and fates, have been elaborated. Input-output measurements for stream reaches show little change in total quantity (Fisher and Likens, 1973; Fisher, 1977), indicating that the DOM is physically, chemically, and biologically resistant. However, rapid physical and biological removal of labile fractions of DOM has been demonstrated (e.g., Cummins et al., 1972; Lock and Hynes, 1976; McDowell and Fisher, 1976; Lush and Hynes, 1978). The physical-chemical and biological mechanisms that convert DOM to UPOM and the availability of the resultant fine particles for microbial metabolism and ingestion by collectors are of major interest. Because the conversion of solutions to particles undoubtedly increases retention in a given stream reach, the process is of significance to the productivity of lotic communities.

The types of organic matter differ in quality, defined as the capacity for producing microbial and animal growth (Cummins and Klug, 1979). Changes in the sources of organic matter resulting in alterations of the relative abundance of various components can strongly influence lotic systems. For example, a shift from a detritus-based heterotrophic system, dependent upon terrestrial inputs, to an autotrophic community would enhance the production of grazing invertebrates (scrapers) if the producers were predominantly nonfilamentous periphytic algae. However, if macrophytes become dominant, the result would be an altered detrital system (Minshall, 1978), possibly less suitable for maintaining a diverse invertebrate community. As examples, modification of the riparian zone during timber harvest or alteration of the annual hydrographic pattern through impoundment would influence the quantity and quality of POM and DOM inputs and primary-producer distribution and abundance. In addition, such changes would directly and indirectly influence the structure of the macroinvertebrate community.

POM processing in running waters has been extensively studied with packs of leaf and needle litter CPOM. The method provides an analogue for the CPOM:fungal-bacterial:shredder association dominant in headwater streams (Fig. 1) (Cummins, 1977; Merritt et al., 1979). Processing, defined as conversion of leaves (or other CPOM) to DOM, FPOM-UPOM, CO_2, animal biomass, and microbial biomass lost from the substrate, is measured as weight loss. The conversion is primarily

biological, as indicated by lower processing rates in larger streams
and rivers, where the potential for mechanical breakage is greater.
The prediction, embodied in the River Continuum Hypothesis, of an
inverse relationship betweem CPOM processing and P/R is supported by
the data given in Fig. 2 (Cummins, 1974, 1975, 1977). As shown in
Fig. 3, there is a general inverse relationship between stream order
and processing efficiency per unit of temperature. The data show
that reaches below man-made impoundments were characterized by
extremely low processing rates--those that would be predicted for
much larger rivers by extending the CPOM processing-stream order
relationship.

Processing of leaf litter per unit temperature was higher at
lower temperatures (Fig. 4), largely reflecting the community structure related to CPOM utilization found in the cooler headwaters
(Cummins, 1974; Cummins and Spengler, 1978). The stream reaches
below impounded areas represented the highest temperatures but

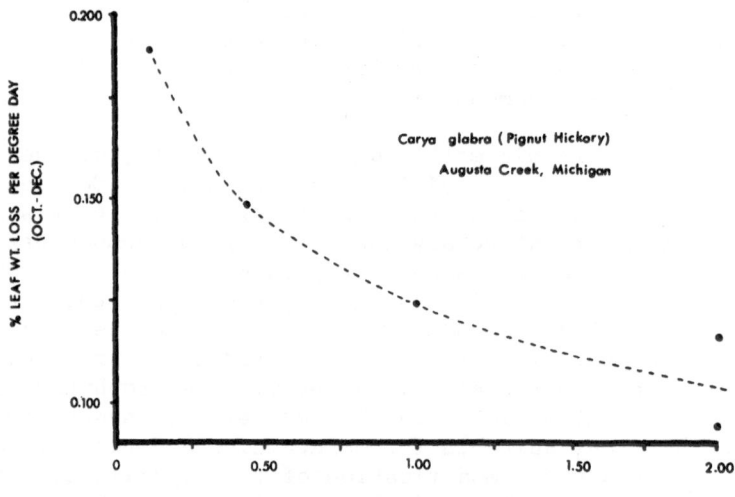

Fig. 2.   Comparison of P/R and leaf-litter temperature-specific
           processing at 5 sites (orders 1-3) in the Augusta Creek,
           Michigan, watershed (Mahan and Cummins, 1978) during the
           fall-winter. Leaf-pack data are from Petersen and Cummins
           (1974). P/R data from circulating chambers are from
           D. K. King, "Autotrophic-Heterotrophic Relationships in a
           Woodland Stream," Ph.D. Thesis, Michigan State Univ.,
           Kellogg Biological Station, in preparation. Line fit by
           introspection.

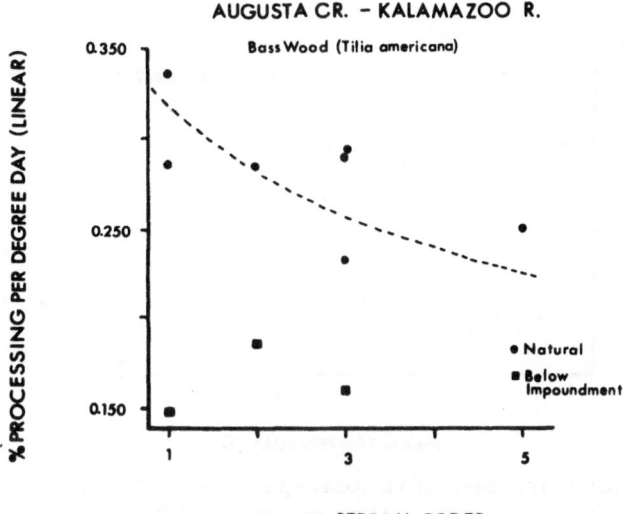

Fig. 3. Temperature-specific leaf-pack processing vs. stream order
         in the Augusta Creek watershed Kalamazoo River Basin,
         Michigan. Linear weight-loss model (see June-July in
         Fig. 5) rather than exponential (Petersen and Cummins,
         1974); line fit by introspection. Based on summer (June-
         July) dry-weight loss after leaching of 5-g packs. Total
         days varied from 8 to 16, degree days 147-398, number of
         packs per site = 3-4; coefficients of variation ranged
         from 3 to 41%; the rates for the three sites below man-
         made, low-head impoundments were significantly different
         ($P < 0.05$) from the non-impounded (natural) reaches.

lowest thermal-related processing. Again, the low rates were
attributed to lost biological function resulting from increased
temperature (Fig. 3) and other deviations from normal conditions
generally typical of a given stream order.

     The rate of leaf or needle litter processing is a function of
temperature in a given stream reach (Fig. 5), but differs between
reaches (orders) because of characteristics of community structure
and function. For example, the high processing rates observed
during the summer in the headwaters of Augusta Creek were related to
the feeding of *Lepidostoma costalis* (Trichoptera) larvae, which
were present only in the first-order sections of the watershed. The
reduced rate of biological processing of CPOM leaf litter below
headwater impoundments (Figs. 3 and 4) is similar to that measured
in lentic habitats (Fig. 6). It is apparent that the lotic-lentic
differences are not merely the result of temperature.

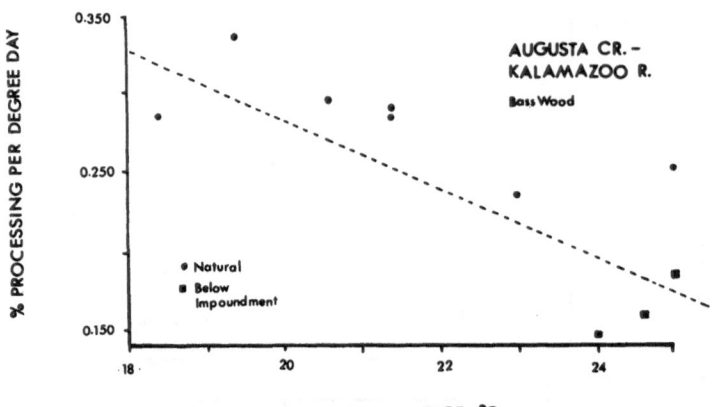

Fig. 4.   Temperature-specific leaf-pack processing as a function of
mean temperature during a summer (June-July) study period
showing an inverse relationship with temperature and the
lowest rates below man-made impoundments. Linear model,
data as in Fig. 3; line fit by introspection.

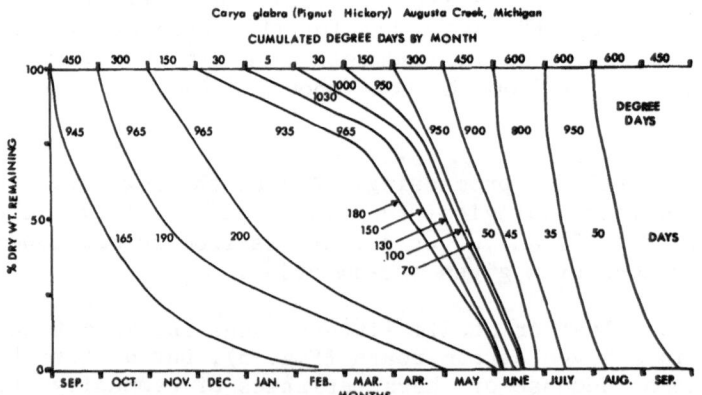

Fig. 5.   Seasonal processing of 5- and 10-g initial dry-weight leaf
packs showing the influence of temperature on the shape of
the weight-loss curves. Although processing time in days
varies 70% seasonally, the range in degree days is fairly
constant (13% variation). Data from Petersen and Cummins
(1974) and Suberkropp et al. (1975); lines fit by
introspection.

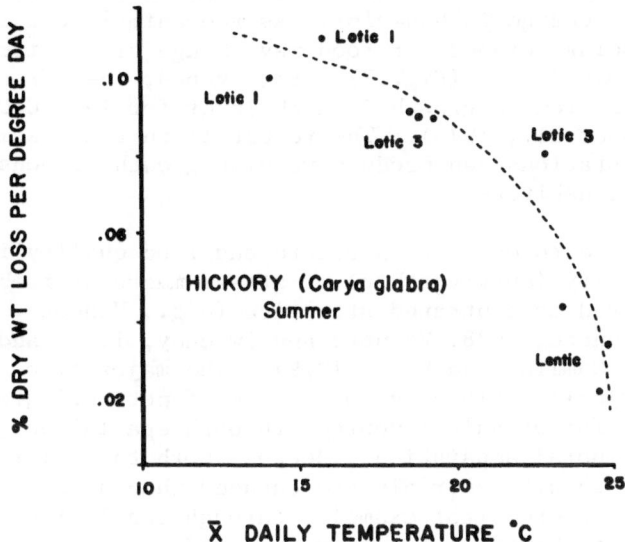

Fig. 6.   Summer (June-August), temperature-specific weight loss of
10-g initial dry-weight leaf packs in first- and third-
order lotic sites, and littoral lake (upper points) and
pond (lower point) lentic sites in the Augusta Creek and
Gull Lake watersheds, Kalamazoo and Barry counties,
Michigan. Based on 49-day processing period, 3 sampling
times (6- to 20-day intervals) of 2-4 leaf packs from each
site. Total degree days for the sites ranged from 769 to
1201; coefficients of variation for processing ranged from
20 to 60%; processing rates at lentic sites were significantly different from those at lotic sites ($P \leq 0.05$);
line fit by introspection.

INVERTEBRATE FUNCTIONAL FEEDING GROUPS

     Process-oriented interest in the sources and fates of organic
substrates in running waters has led to a classification of organisms
on the basis of function, a common practice in microbiology (e.g.,
aerobes vs. anaerobes, cellulose degraders vs. non-cellulose
degraders). Stream macroinvertebrates have been categorized on the
basis of morpho-behavioral adaptations for food acquisition (e.g.,
Cummins, 1973; Merritt and Cummins, 1978), which is linked to
various organic resources (Table 1). With regard to food ingested,
most stream invertebrates are omnivores (e.g., Coffman et al., 1971;
Berrie, 1976), but the mechanisms responsible for obtaining the food
are more restrictive. For example, a scraper such as *Stenonema*
(Ephemeroptera), adapted for shearing off food attached to surfaces,
especially diatoms on rocks, can ingest a variety of organic

substrates within the restriction of its food-harvesting morphology and without any change in behavior. As the animal gets larger, the microhabitat being scraped for food may change or the items scraped from surfaces may differ (Fig. 7). For nymphs, the efficiency of converting ingestion to growth is much lower for leaf CPOM than for algae (Cummins et al., 1973). The result is that the same morpho-behavioral adaptations can produce varying growth responses in different lotic habitats.

The relative roles of temperature and food quality in regulating growth and density (survivorship) of stream macroinvertebrates have recently received concentrated attention (e.g., Vannote, 1978; Sweeney and Vannote, 1978; Vannote and Sweeney, 1979; Anderson and Cummins, 1979; Cummins and Klug, 1979). The majority of macro-invertebrate species belonging to the same functional group appear not to compete for organic resources, through spatial and, most frequently, temporal separation. Because both thermal regime and the fluxes of organic materials are changed when natural stream conditions are altered (for example, through regulation of flow) the interplay of the role of temperature and food in controlling macroinvertebrate populations is of significant interest. Until the

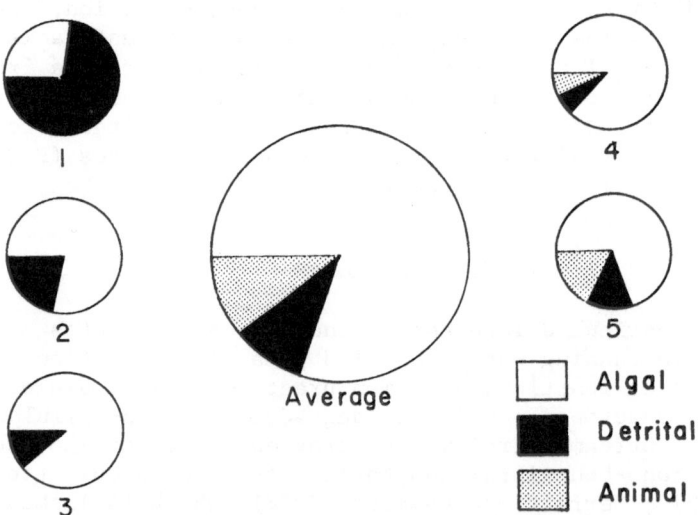

Fig. 7.   Relative algal, detrital, and animal ingestion by
           Stenonema fuscum averaged over the life cycle and for five
           age groups (each representing 3-5 instars). Food material
           converted to caloric equivalents for comparison (after
           Coffman et al., 1971).

interplay of these two control parameters is documented for natural, undisturbed running waters, prediction and interpretation relative to environmental changes are not possible.

Many groups of invertebrates have been tentatively classified into functional feeding groups, especially the aquatic insects (Merritt and Cummins, 1978). As shown in Table 2, the mayflies are predominantly gathering collectors (feeding on FPOM in the sediments) or scrapers; stoneflies are either CPOM shredders or predators; craneflies are shredders, gathering collectors, or predators; and blackflies and mosquitoes are filtering collectors (removing FPOM from transport). The caddisflies are well represented in all functional groups. In a generic analysis of the Trichoptera, Wiggins and Mackay (1979) demonstrated significant differences in representatives of the shredder, collector, and scraper functional feeding groups when small headwater and mid-sized rivers of the eastern deciduous and western coniferous montane biomes were compared. Dominance of the shredders in the headwaters and shifts to communities of collector-scraper genera in mid-sized rivers were indicated. Other analyses (e.g., Cummins and Klug, 1979; Cummins et al., 1979) have also demonstrated the general utility of community assessment based on functional groups.

As shown in Fig. 1, macroinvertebrate functional feeding groups change with differing stream characteristics, such as those associated with increasing stream order. The CPOM:fungal-bacterial:shredder association is typical of the riparian zone-dominated headwaters. FPOM-UPOM:bacterial:collector associations are well represented in all types of running waters. The general size range of fine particulates decreases downriver with increasing order (Sedell et al., 1978), and differences between filtering and gathering collectors reflect varying quantities and qualities of FPOM-UPOM in transport and in the sediments. Periphyton:scraper and macrophytic algal:piercer associations are most prevalent in mid-sized, autotrophic rivers. The predator component of the macro-invertebrate community appears to be relatively constant and ubiquitous (Fig. 1; Coffman et al., 1971).

MANAGEMENT IMPLICATIONS

Alterations that would change the patterns of the annual hydrograph, such as low-flow augmentation or flood control, or change light or thermal regimes and/or quantity and quality of POM-DOM inputs, can be predicted to produce major restructuring of running-water communities. In general, impoundment of smaller rivers and streams creates certain large-river characteristics. For example, a dam on a mid-sized river with hypolimnetic release might

Table 2. Estimated numbers of taxa of North American lotic insects categorized by functional feeding group. The numbers of families and genera containing species belonging to a given functional group are tabulated (i.e., a family or genus is listed in all categories for which species have been reported; species within each genus known to include more than one functional group have been assigned according to the expected dominance for that genus). (From data in Merritt and Cummins, 1978).

Functional Feeding Group	Ephemeroptera Fam	Gen	Sp	Plecoptera Fam	Gen	Sp	Trichoptera Fam	Gen	Sp	Coleoptera Fam	Gen	Sp
Shredders												
CPOM Shredders	1	1	1	5	14	260	7	41	267	1	1	1
Macrophyte Shredders	1	1	1	0	0	0	6	22	70	6	7	28
Collectors												
Filtering Collectors	3	3	32	0	0	0	5	27	208	0	0	0
Gathering Collectors	16	40	384	3	3	6	11	35	76	30	16	124
Scrapers	7	19	152	5	7	28	7	32	155	5	5	16
Piercers (of macrophytes)	0	0	0	0	0	0	1	9	74	3	4	14
Predators	5	10	16	2	14	187	8	15	183	6	39	273

	Diptera							
	Tipulidae		Simuliidae		Chironomidae[a]		Culicidae	
Functional Feeding Group	Gen	Sp	Gen	Sp	Gen	Sp	Gen	Sp
Shredders								
CPOM Shredders	10	45	0	0	1	10	0	0
Macrophyte Shredders	4	105	0	0	4	28	0	0
Collectors								
Filtering Collectors	0	0	8	143	8	58	2[b]	14[b]
Gathering Collectors	14	198	0	0	72	462+	0[b]	0[b]
Scrapers	1	5	2	4	13	24	0	0
Piercers (of macrophytes)	0	0	0	0	0	0	0	0
Predators	6	203	0	0	27	145	0	0

[a] Numbers of genera and species highly speculative because so few have been studied even superficially.

[b] Some genera and species may belong in this category, but not yet reported from streams (probably occur in backwater areas).

produce thermal conditions more typical of the headwaters, but in an autotrophic setting and with greatly altered POM fluxes.

Based on existing conceptual models of stream ecosystem structure and function and given the watershed or drainage basin setting, various alterations, such as those influencing the annual hydrograph pattern, should produce effects that are generally predictable or can be reasonably evaluated after the fact. Process-oriented ecological analyses are promising for both static (structural) and dynamic (functional) evaluations. Examples are seasonal and annual assessments of P/R; weight loss of CPOM, such as leaf and needle litter; quantities and qualities of POM and DOM fluxes; and the relative dominance of various invertebrate functional feeding groups.

Presently, the great majority of the major river systems have been dammed. We may never know how large rivers in the temperate zone functioned biologically as the result of hundreds of millions of years of evolution and at least 10 to 20 thousand years of acclimatization of resident populations. The next onslaught, already underway, is focused on smaller streams for low-flow augmentation, irrigation, and, most recently, proposed lowhead dams for power generation. The spread of impacts from large rivers toward headwaters will most certainly entail significant alterations in patterns and quantities of nutrient and sediment fluxes and attendant changes in community structure.

ACKNOWLEDGMENTS

Preparation of this manuscript, participation in the Symposium on Regulated Streams, and original research were supported by Contract 79 EV 10004 from the Ecological Research Division of the Department of Energy.

REFERENCES

Anderson, N. H., and Sedell, J. R., 1979, Detritus processing by macroinvertebrates in stream ecosystems, *Annu. Rev. Entomol.*, 24:351–377.

Anderson, N. H., and Cummins, K. W., 1979, The influences of diet on the life histories of aquatic insects, *J. Fish. Res. Board Can.*, 36:335–342.

Anderson, N. H., Sedell, J. R., Roberts, L. M., and Triska, F. J., 1978, The role of aquatic invertebrates in processing of wood debris in coniferous forest streams. *Am. Midl. Nat.*, 100:64–82.

Berrie, A. D., 1976, Detritus, microorganisms and animals in fresh-
     water, p. 323-338 *in*: "The Role of Terrestrial and Aquatic
     Organisms in Decomposition Processes," J. M. Anderson and
     A. Macfayden, eds., Blackwell Sci., Oxford.
Boling, R. H., Petersen, R. C., and Cummins, K. W., 1975, Ecosystem
     modeling for small woodland streams, p. 183-204 (Chapter 9),
     in: "Systems Analysis and Simulation in Ecology," Vol. 3,
     B. C. Patten, ed., Acad. Press, New York.
Boling, R. H., Goodman, E. D., Zimmer, J. O., Cummins, K. W.,
     Petersen, R. C., Van Sickle, J. A., and Reice, S. R., 1975,
     Toward a model of detritus processing in a woodland stream,
     Ecology, 56:141-151.
Coffman, W. P., Cummins, K. W., and Wuycheck, J. C., 1971, Energy
     flow in a woodland stream ecosystem: I. Tissue support
     trophic structure of the autumnal community, *Arch. Hydrobiol.*,
     68:232-276.
Cummins, K. W., 1973, Trophic relations of aquatic insects, *Annu.
     Rev. Entomol.*, 18:183-206.
Cummins, K. W., 1974, Structure and function of stream ecosystems,
     BioScience, 24:631-641.
Cummins, K. W., 1975, The ecology of running waters; theory and
     practice, p. 227-293, *in*: "Proc. Sundusky River Basin Sympo-
     sium," D. B. Baker, W. B. Jackson, and B. L. Prater, eds., Int.
     Joint Comm., Int. Ref. Gp. Great Lakes Pollution from Land Use
     Activities, 1976-653-346, U.S. Govt. Printing Office, Washington,
     D.C.
Cummins, K. W., 1977, From headwater streams to rivers, *Am. Biol.
     Teacher*, 39:305-312.
Cummins, K. W., and Klug, M. J., 1979, Feeding ecology of stream
     invertebrates, *Annu. Rev. Ecol. Syst.*, 10:in press.
Cummins, K. W., and Spengler, G. L., 1978, Stream ecosystems, *Water
     Spectrum*, 10:1-9.
Cummins, K. W., Petersen, R. C., Howard, F. O., Wuycheck, J. C., and
     Holt, V. I., 1973, The utilization of leaf litter by stream
     detritivores, *Ecology*, 54:336-345.
Cummins, K. W., Petersen, R. C., Spengler, G. L., Ward, G. M.,
     King, R. H., and King, D. L., 1979, Microbial-animal processing
     in a first order woodland stream, *Oikos*, submitted.
Cummins, K. W., Klug, M. J., Wetzel, R. G., Petersen, R. C.,
     Suberkropp, K. F., Manny, B. A., Wuycheck, J. C., and Howard,
     F. O., 1972, Organic enrichment with leaf leachate in experi-
     mental lotic ecosystems, *BioScience*, 22:719-722.
Fisher, S. G., 1977, Organic matter processing by a stream-segment
     ecosystem: Fort River, Massachusetts, U.S.A., *Int. Rev. Ges.
     Hydrobiol.*, 63:701-727.
Fisher, S. G., and Likens, G. B., 1973, Energy flow in Bear Brook,
     New Hampshire: An integrative approach to stream ecosystem
     metabolism, *Ecol. Monogr.*, 43:421-439.

Hynes, H. B. N., 1970, "The Ecology of Running Waters," Univ.
    Toronto Press, Toronto, 555 p.

Hynes, H. B. N., 1975, The stream and its valley, *Verh. Int. Verein.
    Limnol.*, 19:1-15.

Lock, M. A., and Hynes, H. B. N., 1976, The fate of "dissolved"
    organic carbon derived from autumn-shed maple leaves (*Acer
    saccharum*) in a temperate hard-water stream, *Limnol. Oceanogr.*,
    21:436-443.

Lush, D. L., and Hynes, H. B. N., 1978, The uptake of dissolved
    organic matter by a small spring stream, *Hydrobiologia*,
    60:271-275.

Mahan, C. D., and Cummins, K. W., 1978, "A Profile of Augusta Creek
    in Kalamazoo and Barry Counties, Michigan," Tech. Rep. 3, W. K.
    Kellogg Biol. Stn., Michigan State Univ., East Lansing,. p. 1-12.

McDowell, W. H., and Fisher, S. G., 1976, Autumnal processing of
    dissolved organic matter in a small woodland stream ecosystem,
    Ecology, 57:561-569.

McIntire, C. D., 1973, Periphyton dynamics in laboratory streams:
    A simulation model and its implications, *Ecol. Monogr.*,
    43:399-420.

McIntire, C. D., and Colby, J. A., 1978, A hierarchical model of
    lotic ecosystems, *Ecol. Monogr.*, 48:167-190.

Merritt, R. W., and Cummins, K. W., eds., 1978, "An Introduction to
    the Aquatic Insects of North America," Kendall/Hunt, Dubuque,
    Iowa, 441 p.

Merritt, R. W., Cummins, K. W., and Barnes, J. R., 1979, Demonstra-
    tion of stream watershed community processes with some simple
    bioassay techniques, *Tech. Rept. Ser. Can. Dept. Fish. Mar.
    Serv.* in press.

Minshall, G. W., 1978, Autotrophy in stream ecosystems, *BioScience*,
    28:767-771.

Naiman, R. J., and Sedell, J. R., 1979, Characterization of particu-
    late organic matter transported by some Cascade Mountain
    streams, *J. Fish. Res. Board Can.*, 36:17-31.

Odum, H. T., 1956, Primary production in flowing waters, *Limnol.
    Oceanogr.*, 1:102-117.

Petersen, R. C., and Cummins, K. W., 1974, Leaf processing in a
    woodland stream, *Freshwater Biol.*, 4:343-268.

Russell-Hunter, W. D., 1970, "Aquatic Productivity," MacMillan Co.,
    New York, 306 p.

Sedell, J. R., Triska, F. J., and Triska, N. S., 1975, The processing
    of conifer and hardwood leaves in two coniferous forest streams.
    I. Weight loss and associated invertebrates, *Verh. Int.
    Verein. Limnol.*, 19:1617-1627.

Sedell, J. R., Naiman, R. J., Cummins, K. W., Minshall, G. W., and
    Vannote, R. L., 1978, Transport of particulate organic material
    in streams as a function of physical processes, *Verh. Int.
    Verein. Limnol.*, 20:1366-1375.

Sedell, J. R., Triska, R. J., Hall, J. D., Anderson, N. H., and
    Lyford, J. H., 1973, "Sources and Fates of Organic Inputs in
    Coniferous Forest Streams," Contribution Coniferous Forest
    Biome No. 66, IBP, Oregon State Univ., Corvallis, p. 1-23.

Strahler, A. N., 1957, Quantitative analysis of watershed geomor-
    phology, *Trans. Am. Geophys. Union*, 38:913-920.

Suberkropp, K., Klug, M. J., and Cummins, K. W., 1975, Community
    processing of leaf litter in woodland streams, *Verh. Int.
    Verein. Limnol.*, 19:1653-1658.

Swanson, F. H., and Lienkaemper, G. W., 1978, "Physical Consequences
    of Large Organic Debris in Pacific Northwest Streams," USDA
    For. Serv. Gen. Tech. Rep. PNW-69, p. 1-12.

Sweeney, B. W., and Vannote, R. L., 1978, Size variation and the
    distribution of hemimetabolous aquatic insects: Two thermal
    equilibrium hypotheses, *Science*, 200:444-446.

Vannote, R. L., 1978, A geometric model describing a quasi-
    equilibrium of energy flow in populations of stream insects,
    Proc. Natl. Acad. Sci., 75:381-384.

Vannote, R. L., and Sweeney, B. W., 1979, Geographic analysis of
    thermal equilibria: A conceptual model for evaluating the
    effect of natural and modified thermal regimes on aquatic
    insect communities, *Am. Nat.*, in press.

Wallace, J. B., and Merritt, R. W., 1980, Filter feeding ecology of
    aquatic insects, *Annu. Rev. Entomol.*, 25:in press.

Wallace, J. B., Webster, J. R., and Woodall, W. R., 1977, The role
    of filter feeders in flowing waters, *Arch. Hydrobiol.*,
    79:506-532.

Whitton, B. A., ed., 1975, "River Ecology," Blackwell Sci. Publ.,
    Oxford.

Wiggins, G. B., and MacKay, R. J., 1979, Some relationships between
    systematics and trophic ecology in nearctic aquatic insects,
    with special reference to Trichoptera, *Ecology*, in press.

PHYTOBENTHIC ECOLOGY AND REGULATED STREAMS

Rex L. Lowe

Department of Biological Sciences
Bowling Green State University
Bowling Green, Ohio   43403

INTRODUCTION

Phytobenthos, the photosynthetic component of benthic ecosystems, plays a fundamental role in lotic food webs. Although the
importance of allochthonous energy sources has been established for
smaller streams of the first through third orders (Scott, 1958;
Hynes, 1963; Cummins, 1975), phytobenthos is of major importance in
medium-sized rivers (Blum, 1956, 1957; Cummins, 1975; Pryfogle and
Lowe, in press; Reid, 1961; Gale et al., in press). The most
important algal members of the phytobenthos are usually from one of
three divisions, Bacillariophyta (diatoms), Chlorophyta (green
algae), and Cyanophyta (blue-green algae). Mosses and liverworts
(Bryophyta) reach maximum importance in small, swift streams having
stony substrates (Haslam, 1978). Aquatic vascular plants, often
referred to as aquatic macrophytes, will be referred to as higher
plants in this paper. They belong to several plant families and
their occurrence is usually regulated by such chemical and physical
parameters as flow, turbulence, light, substrate, and dissolved
chemicals (Haslam, 1978).

Regulated streams exhibit many modifications of chemical and
physical parameters that profoundly influence phytobenthic communities. Physical controlling factors include temperature, substrate,
flow, and turbidity. Chemical factors can be divided into two large
categories, toxins and nutrients. The effect of impoundments on
each of these parameters and the ultimate effects on the phytobenthic
community will be systematically discussed.

PHYSICAL FACTORS

Temperature

Regulated streams below reservoirs having deep-release dams generally have lower summer temperatures and higher winter temperatures than unregulated streams in the same region (Spence and Hynes, 1971; Hannan and Young, 1974; Ward, 1974, 1976a,b; Neel, 1963). Ice formation may be postponed or entirely eliminated for varying distances downstream. This temperature modification, in combination with factors discussed later, usually leads to increased densities of benthic algae and changes the structure of the community. In a study below Cheesman Lake, a hypolimnial-release reservoir on the South Platte River, Ward (1976b) found epilithic standing crops 3-20 times greater in the regulated portion of the stream.

Relatively higher winter temperature is believed to be one of the factors responsible for enhanced algal standing crops. The absence of ice for several kilometers below deep-release dams has a major positive impact on algal density. Phinney and McIntire (1965) found that increased temperature within the normal temperature ranges for rivers (6-21°C) led to increases in respiration rates and O_2-evolution rates of lotic periphyton communities. One might then expect relatively lower summer densities of periphyton in the cool waters immediately below hypolimnial-release dams. Ward (1974), however, found temperature and periphyton to be negatively correlated or even unrelated. Regulated streams apparently favor the growth of cool-water stenotherms. Once established, the growth of these populations is stimulated by additional factors discussed later.

Diatoms normally dominate lotic phytobenthic microcommunities. In regulated streams the modified temperature regime seems to favor green algae. In Ward's (1976b) study of the South Platte River, green algae comprised 72% of the epilithon immediately below the reservoir but further downstream comprised only 14-23%. *Cladophora* and *Ulothrix* were the most common genera of green algae. *Microthamnion*, *Microspora*, and several desmids were common to both regulated and unregulated portions of the river. Diatoms were of similar relative abundance in both sections of the river. *Diatomella*, *Synedra*, *Frustulia*, and *Cocconeis* were the most common diatom genera in regulated sections, while *Gomphonema*, *Tabellaria*, *Frustulia*, *Cocconeis*, and *Hantzschia* were most abundant in unregulated sections. It is interesting to note that *Diatomella* is reputed to prefer cool water (Patrick and Reimer, 1966). *Hydrurus*, a cold-water stenotherm from the division Chrysophyta, was more abundant in regulated sections of the river. *Diatoma vulgare* and *Cocconeis pediculus* were the most abundant diatoms in epilithon samples collected from the regulated section of the South Platte

River in July 1978 during a low flow year with higher than normal summer temperatures (J. V. Ward, personal communication). Spence and Hynes (1971) considered lower summer temperatures partly responsible for abundant development of periphyton in a regulated portion of the Grand River, Ontario. They noted that the substrate was always covered with dense layers of diatoms, and from June to mid-September the communities became covered with growths of *Ulothrix zonata* (Weber and Mohr) Kütz. Blum (1960) considers this taxon a rheobiont of cool, well-oxygenated streams having population maxima in spring and late summer. Epiphytic diatoms from the genera *Diatoma, Navicula, Achnanthes, Gyrosigma, Cymbella*, and *Gomphonema* were associated with *U. zonata*. In unregulated portions of the Grand River, however, winter populations of diatoms were replaced in spring by encrusting marl-forming communities.

Environmental factors in regulated streams that favor periphyton would presumably also favor macrophytes. Indeed, increases in aquatic macrophytes below hypolimnial-release dams have been reported (Hilsenhoff, 1971; Ward, 1976b; Haslam, 1978). Such factors as flow and substrate seem to be most important in maintaining aquatic macrophytes; this community will be discussed later.

In addition to modifying temperatures on a seasonal basis, reservoirs also greatly dampen diurnal fluctuations in the temperature of regulated streams. The effect of this phenomenon on the phytobenthos is unknown.

Flow

Many large reservoirs are operated in such a way that flow downstream becomes more constant, eliminating large spates or severe droughts (Neel, 1963; Ward, 1976a,b). The relatively steady current in such regulated streams is one of the contributing factors that leads to increased densities of phytobenthos. The response of periphytic algae is particularly notable. Periphytic algae thriving in rivers evolved in lotic habitats and can be assumed to be tolerant of, if not dependent upon, considerable current. McIntire (1966), working with a laboratory stream, compared periphyton communities developed at current speeds of 9 and 38 cm/sec. He found that periphyton accumulates more rapidly, is more productive, and exports more biomass in faster currents. Other workers have reported similar findings in field observations (Butcher, 1940, 1946; Jones, 1951; Blum, 1960). Zimmerman (1961a,b, 1962) concluded, after studying artificial streams in Switzerland, that the effect of current was often more important than water quality. A moderately swift current without large fluctuations appears to favor luxuriant growth of periphyton. Hynes (1970) discusses the problems of scouring the periphyton community resulting from fluctuating current

speeds. The relatively steady flow from Cheesman Dam is certainly
partly responsible for large standing crops of periphyton in the
South Platte River, described by Ward (1976b). Part of the effect
of increased periphyton with steady current can be attributed to
increased bed stability. It was recently observed in several small,
turbulent streams in the Great Smoky Mountains National Park that
only large boulders (on the order of 0.5 m in diameter or larger)
develop substantial communities of periphyton. Smaller stones and
cobbles maintain only sparce communities. In a long-term study of
these stones and cobbles (E. Larson, personal communication), it was
observed that they are constantly changing position because of intermittent spates. The net movement is, of course, downstream. This
has a negative impact on would-be colonizers of these stones. A
community colonizing the lighted side of a large cobble in one month
may be turned upside down the next. The relatively constant flow of
a regulated stream lends stability to smaller stones and cobbles,
resulting in enduring periphyton communities.

An increase in bed stability is also advantageous to aquatic
vascular plants (Neel, 1963; Sculthorpe, 1967; Stober, 1964;
Hilsenhoff, 1971; Holmes and Whitton, 1977; Ward, 1974, 1976b;
Haslam, 1978). Broad-leafed species are usually absent in fast
current, which favors *Ranunculus* and *Myriophyllum*; and extreme
turbulence may eliminate aquatic vascular plants (Sculthorpe, 1967).
In addition to reducing the effects of scour on root systems, the
shoreline and bank become more stable in regulated streams. Regulated streams in upland areas often develop beds of *Callitriche*
spp., *Myriophyllum* spp., *Ranunculus* spp., and *Sparganium emersum*,
while regulated streams in lowlands often change from a *Ranunculus-
Zannichellia palustris*-dominated community to a *Enteromorpha* (green
alga)-*Sparganium emersum* community (Haslam, 1978).

In an extensive study on the River Tees, Holmes and Whitton
(1977) studied the effects of a new impoundment, Cow Green Reservoir,
on aquatic macrophytes. They noted that flow fluctuations in the
Tees have been reduced in amplitude. Before regulation, a particularly severe flood in 1968 removed most of the *Ranunculus fluitans*
from one section of river and also removed an island having relatively mature willow trees. Holmes and Whitton credit stream
regulation with a substantial upstream spread of four submerged
angiosperm species, *Potomogeton crispus, Zannichellia palustris,
Ranunculus penicillatus* var. *calcareus*, and *Myriophyllum spicatum*.

Peňáz et al. (1968) observed profound influences of the Vir
River Valley Reservoir on stream phytobenthos. Although not all the
changes noted were directly attributable to modifications of stream
flow, this appeared to be one of the major factors. Localities
below the dam had a relatively greater percentage of the stream bed
covered by phytobenthos than sites upstream from the dam. Total

streambed coverage increased from 32% immediately above the reservoir to 88% immediately below; 33 km below the dam, streambed coverage fell to 45%. The quality of the flora was also influenced by the dam. The bryophytes *Fontinalis antipyretica* and *Amblystegium riparium* reached maximum abundance immediately below the dam. The increase in phytobenthos in the regulated portion of the stream also involves species of the filamentous algae *Oedogonium*, *Microspora*, *Spirogyra*, *Vaucheria*, and *Tribonema*. Diatom species increasing in relative abundance below the reservoir include *Melosira varians*, *Fragilaria construens*, *F. bicapitata*, *Synedra rumpens*, *Diatoma elongatum*, and *Navicula gregaria*.

Some dams and reservoirs, particularly those used for irrigation or power, impart immense fluctuations in flow to regulated streams (Neel, 1973; Powell, 1958; Kroger, 1973). This often has disastrous effects on the phytobenthic community of the stream. Neel (1963) discusses the stranding and desiccation of algal communities in the North Platte River below an irrigation dam. Kroger (1973) presented similar data on the Snake River in Grand Teton National Park.

Turbidity

Suspended silt and other sestonic river particles are usually reduced to relatively low levels by settling in the lentic waters of reservoirs (Neel, 1973; Hannan and Young, 1974; Baxter, 1977). Suspended solids below Cheesman Reservoir on the South Platte River ranged from 2.2 to 7.4 mg/liter but increased (29.5-36 mg/liter) 40 km downstream (Ward, 1976b). The decreased turbidity leads to increased light penetration, which favors submerged phytobenthos. Much of the limited suspended matter exported from reservoirs to controlled streams is in the form of phytoplankton. The lentic conditions of reservoirs often lead to luxuriant growths of phytoplankton. However, the suspended material does not persist downstream in the regulated river (Maciolek and Tunzi, 1968). The selective release of highly turbid waters from reservoirs is often used to combat reservoir siltation. This would decrease light penetration, and silt would settle on the bed of the controlled stream. Both outcomes would negatively affect submerged phytobenthos. In one extreme example, Nisbet (1961) described the release of tons of sediment from behind the dam at Verbois as having disastrous effects on the upper Rhone River.

CHEMICAL FACTORS

Nutrients and Toxins

Reservoirs provide a relatively large surface area for evaporation and thus tend to accumulate and concentrate dissolved solids. The concentration of dissolved solids in reservoirs is a function of several factors, not the least of which are age and geographic location. If water is released from the hypolimnion of the reservoir, one might expect general increases in a number of nutrients important to phytobenthos in the regulated stream. Neel (1963) observed that phytoplankton or other algal growth is generally stimulated by the high nutrient content of waters released from the hypolimnia of reservoirs. The resulting increases may be responsible for increases in dissolved oxygen further downstream, although Spence and Hynes (1971) noted an oxygen sag at night below a large bed of phytobenthos in a regulated stream. In the system they studied, however, the reservoir provided the regulated stream with considerable amounts of organic matter. Increases in organic matter and/or inorganic nutrients would be expected to stimulate species of phytobenthos normally associated with nutrient-rich streams. Lawson and Rushforth (1975) found 16 diatom taxa to be confined to a site below an impoundment on the Provo River. The trophic affinities are known for only seven of these diatoms (Lowe, 1974), and they are all considered eutrophic indicators. These seven taxa are *Amphipleura pellucida* Kütz., *Cocconeis diminuta* Pant., *Cymbella prostrata* (Berk.) Cl., *Fragilaria crotonensis* Kitton, *Navicula oblonga* (Kütz.) Kütz., *N. reinhardtii* (Grun.) Grun., and *N. scutelloides* W. Sm. *ex* Greg.

Regulated streams below surface-release reservoirs would probably have decreased nutrients during periods of reservoir stratification. Lentic phytoplankton productivity can rapidly strip nutrients from the epilimnion. Such nutrient-poor water could limit periphyton productivity in the regulated stream.

Discharge from anoxic hypolimnia may occasionally release reduced substances into regulated streams. Hannan and Young (1974) observed small increases in ammonia and hydrogen sulfide in a regulated stream in Texas.

NEEDS FOR FUTURE RESEARCH

The limited data produced on the phytobenthos of regulated streams thus far has been, with a few exceptions, secondary information gathered during studies of benthos or fish. Some well-planned studies of phytobenthos are highly warranted in the following areas: (1) detailed description of changes occurring in the physical and

Table 1. A Summary of the Effects of Stream Regulation from a Deep-Release Reservoir on Lotic Phytobenthos

Parameter	Modification	Effect on Quantity	Effect on Quality
Temperature	Warmer winter, often ice free, cooler summer	Increased standing crop	Favors cool-water stenotherms, such as *Ulothrix zonata* and *Hydrurus* sp. Also, green filamentous algae in general is promoted
Flow	More constant under some management schemes	Increased standing crop	Green filamentous algae; *Callitriche, Myriophyllum, Ranunculus, Sparganium, Potomogeton,* and *Zannichellia*
Flow	Widely fluctuating, power and irrigation dams	Decreased standing crop	
Turbidity	Decreased by reservoirs, often relatively low in controlled stream	Increased standing crop	Favors submerged phytobenthos
Nutrients	Increased nutrients, some increase in ammonia and hydrogen sulfide	Increased standing crop	Favors species normally associated with nutrient-rich habitats, *Cladophora*

numerical structure of phytobenthic communities as a result of
stream regulation; (2) experiments documenting the energetics and
productivity of the community; and (3) experiments investigating the
nature and efficiency of energy transfer from the phytobenthic
community to heterotrophic stream inhabitants.

SUMMARY

The effect of stream regulation on the phytobenthos is summarized
in Table 1. Most of the chemical and physical parameters characteristic of regulated streams flowing from deep-release reservoirs
have a positive impact on the density of phytobenthos. Reduced
temperature fluctuation, reduced flow fluctuation, increased transparency, and increased nutrient concentrations all tend to stimulate
growth of periphyton and aquatic vascular plants. Stimulated species
are, for the most part, cool-water stenotherms characteristic of
nutrient-rich aquatic habitats.

LITERATURE CITED

Baxter, R. M., 1977, Environmental effects of dams and impoundments,
    Annu. Rev. Ecol. Syst., 8:255-283.
Blum, J. L., 1956, The ecology of river algae, *Bot. Rev.*, 22:291-341.
Blum, J. L., 1957, An ecological study of the algae of the Saline
    River, Michigan, *Hydrobiologia*, 9:361-408.
Blum, J. L., 1960, Algal populations in flowing waters, Spec. Publ.
    Pymatuning Lab. Field Biol., 2:11-21.
Butcher, R. W., 1940, Studies on the ecology of rivers, IV. Obser-
    vations on the growth and distribution of sessil algae in the
    River Hull, Yorkshire, *J. Ecol.*, 28:210-223.
Butcher, R. W., 1946, Studies on the ecology of rivers, VI. The
    algal growth in certain highly calcareous streams, *J. Ecol.*,
    33:268-283.
Cummins, K. W., 1975, The ecology of running waters; theory and
    practice, p. 278-293, *in*: "Proceedings Sandusky River Basin
    Symposium," Baker, Jackson, Prater, eds., Int. Joint Commission.
Gale, W. F., Gurzynski, A. J., and Lowe, R. L., in press, Coloniza-
    tion and standing crops of epilithic algae in the Susquehanna
    River (Pennsylvania), *J. Phycol.*
Hannan, H., and Young, W. J., 1974, The influence of a deep-storage
    reservoir on the physiochemical limnology of a central Texas
    river, *Hydrobiologia*, 44:177-207.
Haslam, S. M., 1978, "River Plants," Cambridge Univ. Press, London,
    England.
Hilsenhoff, W. L., 1971, Changes in the downstream insect and
    amphipod fauna caused by an impoundment with a hypolimnion
    drain, *Ann. Entomol. Soc. Am.*, 64:743-746.

Holmes, N. T. H., and Whitton, B. A., 1977, The macrophytic vegeta-
    tion of the River Tees in 1975: Observed and predicted changes,
    Freshwater Biol., 7:43-60.
Hynes, H. B. N., 1963, Imported organic matter and secondary produc-
    tivity in streams, *Int. Congr. Zool.*, 16(4):324-329.
Hynes, H. B. N., 1970, "The Ecology of Running Waters," Univ.
    Toronto Press, Toronto, Canada, 555 p.
Jones, J. R. E., 1951, An ecological study of the River Towy, *J.
    Anim. Ecol.*, 20:68-86.
Kroger, R. L., 1973, Biological effects of fluctuating water levels
    in the Snake River, Grand Teton National Park, Wyoming, *Am.
    Midl. Nat.*, 89:478-481.
Lawson, L. L., and Rushforth, S. R., 1975, The diatom flora of the
    Provo River, Utah, U.S.A., *Bibl. Phycol.*, 17:149 p.
Lowe, R. L., 1974, "Environmental Requirements and Pollution Toler-
    ance of Freshwater Diatoms," EPA-670/4-74-005, Off. Res. Dev.,
    U.S. Environmental Protection Agency, Cincinnati, 333 p.
Maciolek, J. A., and Tunzi, M. G., 1968, Microseston dynamics in a
    simple Sierra Nevada lake-stream system, *Ecology*, 49:60-75.
McIntire, C. D., 1966, Some effects of current velocity on periphyton
    communities in laboratory streams, *Hydrobiologia*, 27:559-570.
Neel, J. K., 1963, Impact of reservoirs, p. 575-593, *in*: "Limnology
    in North America," D. G. Frey, ed., Univ. Wisconsin Press,
    Madison.
Nisbet, M., 1961, Un example de pollution de riviére par vidage
    d'une retenue hydroélectrique, *Verh. Int. Verein. Theor. Angew.
    Limnol.*, 14:678-680.
Patrick, R., and Reimer, C. W., 1966, "The Diatoms of the United
    States, Volume 1," Acad. Nat. Sci., Philadelphia, Monograph 13,
    688 p.
Peňáz, M., Kubicek, F., Marvan, F., and Zelinka, M., 1968, Influence
    of the Vir River Valley reservoir on the hydrobiological and
    ichthyological conditions in the River Svratka, *Acta Sci. Nat.
    Brno*, 2(1):1-60.
Phinney, H. K., and McIntire, C. D., 1965, Effects of temperature on
    metabolism of periphyton communities developed in laboratory
    streams, *Limnol. Oceanogr.*, 10(3):341-344.
Powell, G. C., 1958, "Evaluation of the Effects of a Power Dam Water
    Release Pattern Upon the Downstream Fishery," M.S. Thesis,
    Colorado State Univ., Fort Collins, 149 p.
Pryfogle, P. A., and Lowe, R. L., in press, Sampling and interpre-
    tation of epilithic lotic diatom communities, *in*: "Methods and
    Measurements of Attached Microcommunities: A Review," Am. Soc.
    Testing and Materials, Philadelphia, Pennsylvania.
Reid, G. K., 1961, "Ecology of Inland Waters and Estuaries," Reinhold
    Publ. Corp., New York, 375 p.
Scott, D. C., 1958, Biological balance in streams, *Sewage Ind.
    Wastes*, 1958, 1169-1173.

Sculthorpe, C. D., 1967, "The Biology of Aquatic Vascular Plants,"
    St. Martins Press, New York, 610 p.

Spence, J. A., and Hynes, H. B. N., 1971, Differences in benthos
    upstream and downstream of an impoundment, *J. Fish. Res. Board
    Can.*, 28:35-43.

Stober, Q. J., 1964, Some limnological effects of Tiber Reservoir
    on the Marias River, Montana, *Proc. Mont. Acad. Sci.*,
    23:111-137.

Ward, J. V., 1974, A temperature-stressed stream ecosystem below a
    hypolimnial release mountain reservoir, *Arch. Hydrobiol.*,
    74:247-275.

Ward, J. V., 1976a, Effects of thermal constancy and seasonal temper-
    ature displacement on community structure of stream macroinverte-
    brates, p. 302-307, *in*: "Thermal Ecology II," G. W. Esch and
    R. W. McFarlane, eds., ERDA Symp. Ser. (CONF-750425).

Ward, J. V., 1976b, Comparative limnology of differentially regulated
    sections of a Colorado mountain river, *Arch. Hydrobiol.*,
    78(3):319-342.

Zimmerman, P., 1961a, Experimentelle Untersuchungen über die
    ökologische Wirkung der Strömgeschwindigkeit auf die
    Lebensgemeinschaften des fliessenden Wassers, *Schweiz. Z.
    Hydrol.*, 23:1-81 and 63:200.

Zimmerman, P., 1961b, Experimentelle Untersuchungen über den Einfluss
    der Strömungsgeschwindigkeit auf die Fliesswasserbiozönose,
    Verh. Int. Verein. Theor. Angew. Limnol., 14:396-399 and
    63:200.

Zimmerman, P., 1962, Der Einfluss der Strömung auf die Zusammensetzung
    der Lebensgemeinschaften im experiment, *Schweiz. Z. Hydrol.*,
    24:408-411, 63.

ECOLOGICAL FACTORS CONTROLLING STREAM ZOOBENTHOS WITH EMPHASIS ON

THERMAL MODIFICATION OF REGULATED STREAMS

James V. Ward and Jack A. Stanford

Dept. Zoology & Entomology, Colorado State University,
Fort Collins, Colorado 80523; Dept. Biological Sciences,
North Texas State University, Denton, Texas 76203

INTRODUCTION

A myriad of factors, including temperature, flow, substrate, aquatic and riparian vegetation, dissolved substances, food and biotic interactions, determine the composition and abundance of stream zoobenthos (Macan, 1961, 1974; Hynes, 1970a,b). The influence of the watershed on many of these factors has only recently been fully appreciated (e.g., Hynes,, 1975; Cummins, this volume).

A much more exhaustive list of controlling factors could be developed, and each of those listed may be further subdivided. However, for the vast majority of unpolluted streams, temperature, flow, and substrate, and their ramifications, may be considered the major factors controlling the macroinvertebrates. These major controlling factors are, of course, often interrelated. For example, current influences substrate and substrate influences the composition and abundance of aquatic plants, all of which directly or indirectly affect zoobenthic organisms. Little is known regarding biotic interactions in streams, but even these are influenced by temperature, current and substrate (Pattee and Bournaud, 1970, Edington and Hildrew, 1973).

Because the substrate of streams in a given region is largely a function of the flow regime, temperature and flow (and their ramifications) remain perhaps the two most important controlling factors for zoobenthos of unpolluted streams. Therefore, following a brief overview of the general effects of reservoirs on the receiving stream, the major thrust of this review paper deals with effects of thermal and flow phenomena on benthic macroinvertebrates in lotic reaches downstream from impoundments, with special emphasis on the influence

of temperature regime alterations. Discussion concentrates on results of comprehensive investigations of the zoobenthos of regulated
streams in temperate latitudes, although only occasional reference
is made to the numerous investigations conducted in the Tennessee
Valley, USA (reviewed by Krenkel, et al., this volume).

GENERAL ENVIRONMENTAL MODIFICATIONS

    Although the literature on reservoirs is voluminous, the few
reviews dealing broadly with the ecological effects of damming rivers
(Baxter, 1977; Ridley and Steel, 1975; Neel, 1963; Ackermann et al.,
1973) primarily treat conditions in the reservoir, giving only brief
general consideration to downstream effects.

    Impoundments may result in a variety of downstream modifications
of significance to stream zoobenthos (Table 1). The specific changes
depend on a complex series of interactions resulting from operation
and construction variables (e.g., position of outlet ports, purpose
of the reservoir) and limnological variables (e.g., trophic state,
extent of stratification), and are a function of geographical, climatic, geochemical and topographic factors.

    Some effects, such as clarification (sedimentation of particles
in the reservoir), generally occur irrespective of the mode of operation or additional variables. Other modifications of the receiving
stream depend, at least in part, on the release depth. The influence
of surface-release dams, for example, may be similar to that of
natural lakes, which rarely release water low in dissolved oxygen;
whereas anaerobic waters may be discharged from the hypolimnion of
stratified reservoirs. However, not all reservoirs develop anoxic
hypolimnia; and, even if they do, air drafts in the release ports or
turbulent conditions below the dam function to raise dissolved oxygen
levels in the receiving stream (Young et al., 1976). There are
exceptions to all of the generalizations presented in Table 1. Turbidity currents in the reservoir, for example, may maintain high
levels of turbidity in the receiving stream (Neel, 1963). Under some
conditions most dissolved constituents may decrease in concentration
during passage through a deep-release reservoir, despite the higher
values in bottom waters (Soltero et al., 1973). Additional perturbation, such as thermal and organic pollution, and acid mine drainage,
have confounded the interpretation of effects of regulation on some
stream systems (e.g., Simmons and Voshell, 1978).

ZOOBENTHOS OF REGULATED STREAMS

    Examination of the results of studies of zoobenthos in regu-
lated streams reveal some common features despite vast differences
in geographical locale, limnological conditions in the reservoir,
release depth, flow modification, and a variety of other factors

Table 1. Some General Effects of Deep-Release Storage
        Reservoirs on Conditions in the Receiving
        Stream, Exclusive of Temperature, Flow, and Substrate.
        D = Decrease, I = Increase, V = Variable Effects

	Modification[a]	Selected References
Turbidity	D	Wright, 1967; Soltero et al. 1973; Ward, 1974
TDS, hardness	I	Neel, 1963; Wright, 1967
Nutrients	I	Neel, 1963; Hilsenhoff, 1971; Hall et al., 1976
Dissolved oxygen	V	Neel, 1963; Isom, 1971; Crisp, 1977
Hydrogen sulfide	V	Wright, 1967; Hannan and Young, 1974
Submerged angiosperms	I	Hall and Pople, 1968; Ward, 1976a, Holmes and Whitton, 1977
Periphyton	I	Spence and Hynes, 1971; Armitage, 1976; Ward, 1976a
Plankton	I	Müller, 1962; Ward, 1975; Armitage and Capper, 1976
Transport detritus	D	Ward, 1976a; Armitage, 1977

[a]Relative to unregulated lotic reaches or surface-release.

(Table 2). Macrobenthic diversity is generally reduced compared
with that of the stream above the reservoir, unregulated tributaries,
or locations farther downstream. The Brazos River in Texas is the
only regulated stream known to the authors in which zoobenthic diversity is increased downstream (McClure and Stewart, 1976; Coulter
and Stanford, unpublished). Deep releases from Possom Kingdom reservoir have sluiced the stream bottom, changing a warm sandy-bottom
river into a cooler river with a rubble substrate. The former
conditions and associated impoverished zoobenthic community prevail
above the reservoir and downstream after ca. 80 km.

Table 2.　Some effects[a] of stream regulation on lotic zoobenthos and controlling factors. I = Increased, D = Decreased, S = Similar, L = Lower level release depth, U = Upper level release depth.

Study and Location	Standing Crop	Diversity	Amphipoda	Isopoda	Mollusca	Turbellaria	Oligochaeta	Plecoptera	Ephemeroptera	Trichoptera	Coleoptera	Diptera	Release Depth	Angiosperms	Algae	Plankton	Biotic Interaction	Substrate	Transport Detritus	Sediment Detritus	Chemistry	Dissolved Oxygen	Constant Flow	Fluctuating Flow	Diel Constancy	Seasonal Constancy	Summer Cool	Summer Warm	Winter Warm	Pattern
Armitage 1976, 1978 (England)	I	S	I		I		I	D	D	S	I	S	U/L	X	X	X		X	X				X		X		X	X		X
Briggs 1948 (California)	I						D	D	I				L					X					X				X			X
Stewart & Stanford, unpubl. (Texas)	I	I			I			I	I	I	I	I	L		X	X		X			X	X	X	X				X		X
Chutter 1963 (S. Africa)	I	D			I		D	D	I				L	X	X	X	X						X							
Fraley, this volume (Montana)	I	D	I				D	S	I				U		X								X	X	X			X		
Gore 1977 (Montana)	D	D			I	I	D	D	S	I			L	X			X								X		X			
Henricson & Müller, this volume (Sweden)	S	D	S		S	S	S	S	S	S			L										X	X	X	X			X	X
Hilsenhoff 1971 (Wisconsin)		D	I				D	D	D	I			L			X			X	X	X		X					X		X
Hoffman & Kilambi 1971 (Arkansas)	I	D	I	I	D		I	D	D	D	I		L	X	X		X						X	X	X	X				
Isom 1971 (Tennessee Valley)	D	D	I	*			D	D	D		I		L	X		X		X	X				X		X					
Lehmkuhl 1972 (Saskatchewan)	D	D	D			D	D	D	D	I			L															X		X X
Merkley 1978 (Iowa)	D	D					D	I		D			L							X			X							
Müller 1962[d] (Sweden)	D	D	S	D	D		D	D	D	D	D	D	L			X							X	X	X	X			X	X
Pearson et al. 1968 (Utah)	I	D	I				I	D	D	D	D	I	L	X		X		X					X	X			X			
Penáz et al. 1968 (Czechoslovakia)	I	D	I		I		I	S	I	I	D	I	L	X	X	X	X	X	X		X		X				X	X	X	X
Radford & Hartland-Rowe 1971 (Alberta)	D	D					S	I	D				L	X	X		X		X				X							
Simmons & Voshell 1978 (Virginia)	I	D			I		D	S	I		I	U		X	X					X	X									
Spence & Hynes 1971 (Ontario)	I	D	I		D		D	S	I	I	I	I	L	X	X	X	X		X		X	X	X					X	'	X
Trotsky & Gregory 1974 (Maine)	D	D					S	S	D	D	I		L										X							
Ward 1976a (Colorado)	I	D	I		I	I	I	D	S	D		I	L	X	X		X		X	X	X		X		X		X	X	X	X	X
Ward & Short 1978 (Colorado) (JWC)	D	D				D	I	S	S	D	D	I	L		X			X	X				X							
Ward & Short 1978 (Colorado) (TC)	I	D	I		I		D	D	I		I	U		X	X	X	X		X	X			X					X		
Young et al. 1976 (Texas)	I	D			I						I		L							X			X		X	X				

[a]Relative to unregulated streams or stations farther downstream. Missing taxonomic data indicate either that the taxon is not present or abundant, or that the author(s) made no mention of the effect.

[b]Factors indicated by the author(s) as significantly modified by regulation.

[c]Increased seasonal constancy was associated with increased diurnal fluctuation in some cases.

[d]Dam construction phase.

*Gastropods increased; pelecypods decreased.

The standing crop in the receiving stream may be enhanced or reduced compared to unregulated locations, reductions being characteristic of tailwaters with rapid flow fluctuations. However, standing crop may be enhanced in tailwaters with rapid daily flow fluctuations if regulation reduces the high flows associated with spring runoff (Pearson et al., 1968, Hoffman and Kilambi, 1971). Merkley (1978) reported decreased zoobenthic numbers, but increased biomass and production, downstream from a dam on an organically polluted prairie river.

All investigators report major alterations in macroinvertebrate composition at regulated sites. Table 2 shows the changes reported for major taxa; Ward and Short (1978) address the more specific taxonomic and functional group modifications induced by upstream impoundments. The relative abundance of some major groups (e.g., Ephemeroptera) may not change appreciably, although composition within the group will be greatly modified in the receiving stream. For example, heptageniid mayflies are generally reduced or absent in regulated sections. Some major taxa vary in their response to regulation (e.g., Trichoptera), whereas others exhibit a fairly consistent response irrespective of operational schemes or other variables. Amphipods, for example, are invariably enhanced by regulation, often appearing in streams lacking amphipods in unregulated sections (Ward, 1974, 1976a). Plecopterans are normally reduced or eliminated for some distance below dams, although certain hyporheic forms (e.g., chloroperlids) may increase in relative abundance under conditions of rapid flow fluctuations (Ward and Short, 1978; Trotsky and Gregory, 1974; Radford and Hartland-Rowe, 1971). The type of life cycle may largely determine whether a species can withstand the modified conditions (see Henricson and Müller, this volume).

Changes in the composition and abundance of zoobenthos have been attributed to a variety of factors resulting from regulation (Table 2). The enhancement of Trichoptera in lake outfalls (e.g., Müller, 1962) has been attributed to the contribution of lake plankton as a food source for filter-feeding species. Trichoptera and other filter feeders, such as Simuliidae, are enhanced in the receiving stream if surface water is released. Streams below dams that release deep water throughout the year may have reduced populations of filter feeders if reduction of transport detritus by clarification and mineralization in the reservoir is not rectified by the release of plankton (Ward, 1975). Many more data are needed on the dynamics of organic carbon in regulated streams.

The factors important to stream zoobenthos that are influenced by flow modification are shown in Figure 1. Submerged angiosperms and benthic algae are enhanced by decreased turbidity, increased nutrients, and increased bank and bed stability resulting from

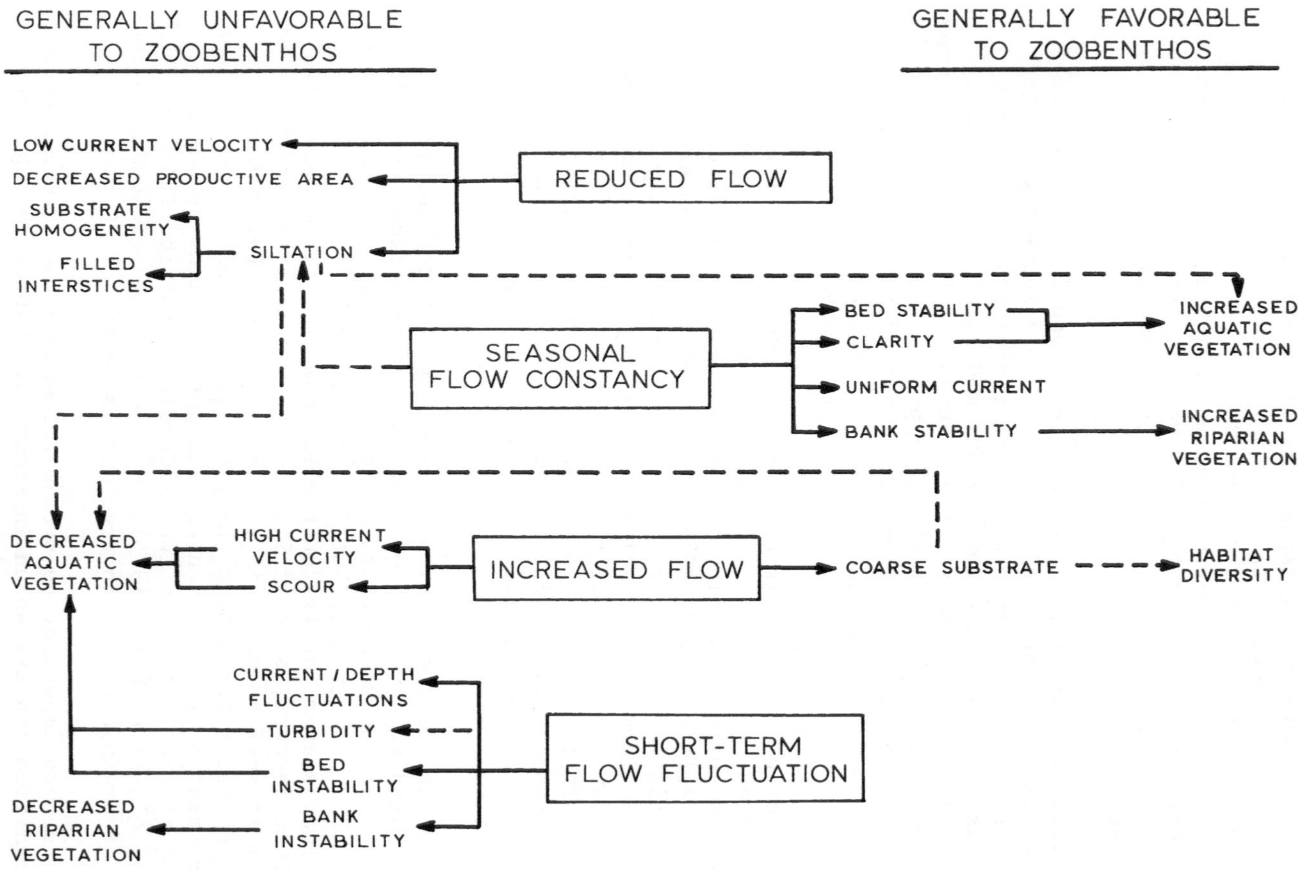

Figure 1. Potential effects of flow regime modifications below dams on ecological factors important to erosional zone zoobenthos. Dashed lines indicate less definite relationships (modified from Ward, 1976b).

increased flow constancy. Angiosperms and algae provide current
refugia for species not otherwise able to maintain populations in
rocky streams. Dense mats of epilithic algae may increase the availability of food to detritus feeders (Spence and Hynes, 1971), although forms utilizing holdfasts may be reduced or eliminated by the
absence of smooth rock surfaces (Ward, 1976a). Flow constancy may
also enhance riparian vegetation, with thermal and trophic implications for the stream system. Flow constancy combined with reduced
flow may result in stream habitat alteration by vegetation encroachment and siltation. Riparian species that are adapted to fluctuating
stream levels could be replaced by other vegetation under conditions
of long-term flow constancy; the ecological implications of this
are unclear.

Rapid short-term fluctuations reduce algae and macrophytes and
deplete sedimentary detritus. Species restricted to pools, as well
as those requiring rapid water, may be eliminated. Large areas may
be alternately dewatered and flooded. Species requiring relatively
constant flow or a narrow range of current velocity for efficient
feeding are eliminated directly or placed at a competitive disadvantage.

The chemical effects in Table 2 primarily refer to the release
of hydrogen sulfide and to changes in nutrient and oxygen levels.
Simmons and Voshell (1978) studied a stream influenced by thermal
pollution, acid mine drainage and heavy-metal pollution and concluded
that an impoundment significantly improved conditions in the receiving stream. Other authors (e.g., Isom, 1971; Young et al., 1976)
have reported inducement of deleterious chemical conditions. Many
of the effects are site-specific. For example, higher levels of
nutrients resulting from deep release may be beneficial in an oligotrophic stream, whereas eutrophication, depleted oxygen and toxic
compounds may result in the stream below a highly productive reservoir. Henricson and Müller (this volume) report that, for Swedish
rivers, downstream chemical changes are important mainly during the
first 5-10 years after impoundment, but eventually return to near-
normal values

UNREGULATED STREAM TEMPERATURE

Stream temperatures under natural conditions have been reviewed
by Smith (1972), who includes important Japanese investigations
rarely cited by European or North American workers. The following
account only briefly summarizes thermal conditions of unregulated
streams (see also Hynes, 1970a; Cummins, this volume).

Thermal changes downstream from the source are determined by
meteorological conditions, topography, riparian vegetation, and
hydrological conditions. The seasonal thermal range is positively

correlated with stream order. Small streams are subject to considerable influence from local factors, such as shading by canyon walls
and riparian vegetation, or from short-term changes in meteorlogical
conditions, such as cloud cover (Kamler, 1965). Therefore, in upper
reaches, the diel thermal range increases downstream as the small
volume of water responds to atmospheric vagaries and local conditions. However, as discharge increases, a point is reached where
the greater heat capacity of the river increases thermal stability,
resulting in increased thermal constancy of lower reaches. During
higher discharge, the thermal conditions of the headwaters extend
farther downstream.

Small streams in temperate latitudes often exhibit annual ranges
of more than 20°C. Diurnal ranges of around 6°C are fairly common
during summer (Hynes, 1970a); changes as great as 14°C have been
reported (Mackichan, 1967). Temperature extremes are reduced by
shading from riparian vegetation and by groundwater inputs. Reduced
flows result in more extreme temperatures, enhancing the formation
of anchor ice in winter.

THERMAL MODIFICATION BY DAMS

The extent to which impoundments modify the temperature regime
of the receiving stream depends primarily upon the release depth
(which changes relative to the water surface as a function of reservoir level), the thermal stratification pattern of the reservoir,
the retention time, and dam operation.

Thermal modification of the receiving stream may be considered
under six categories (Table 2): (1) increased diurnal constancy,
(2) increased seasonal constancy, (3) summer depression, (4) summer
elevation, (5) winter elevation, and (6) thermal pattern changes.
Despite great modification of the thermal regime, mean annual stream
temperatures may not be greatly modified by impoundment (Jaske and
Goebel, 1967; Lavis and Smith, 1971; Ward, 1976a).

The influence of surface-release dams on the temperature of
receiving streams is similar to that of natural lakes, although
reservoir retention times are often less. Despite great differences
in stream size, zoobenthic communities below dams with upper-level
release in Montana (Fraley, this volume) and Colorado (Ward and
Short, 1978) were similarly modified by regulation. At regulated
locations in both streams, increased density was associated with
reduced zoobenthic diversity, plecopterans were severely reduced,
and hydropsychid caddisflies comprised the majority of the macroinvertebrates. Elevated summer temperatures of these previously
cold-water streams were partly responsible for the altered zoobenthic communities.

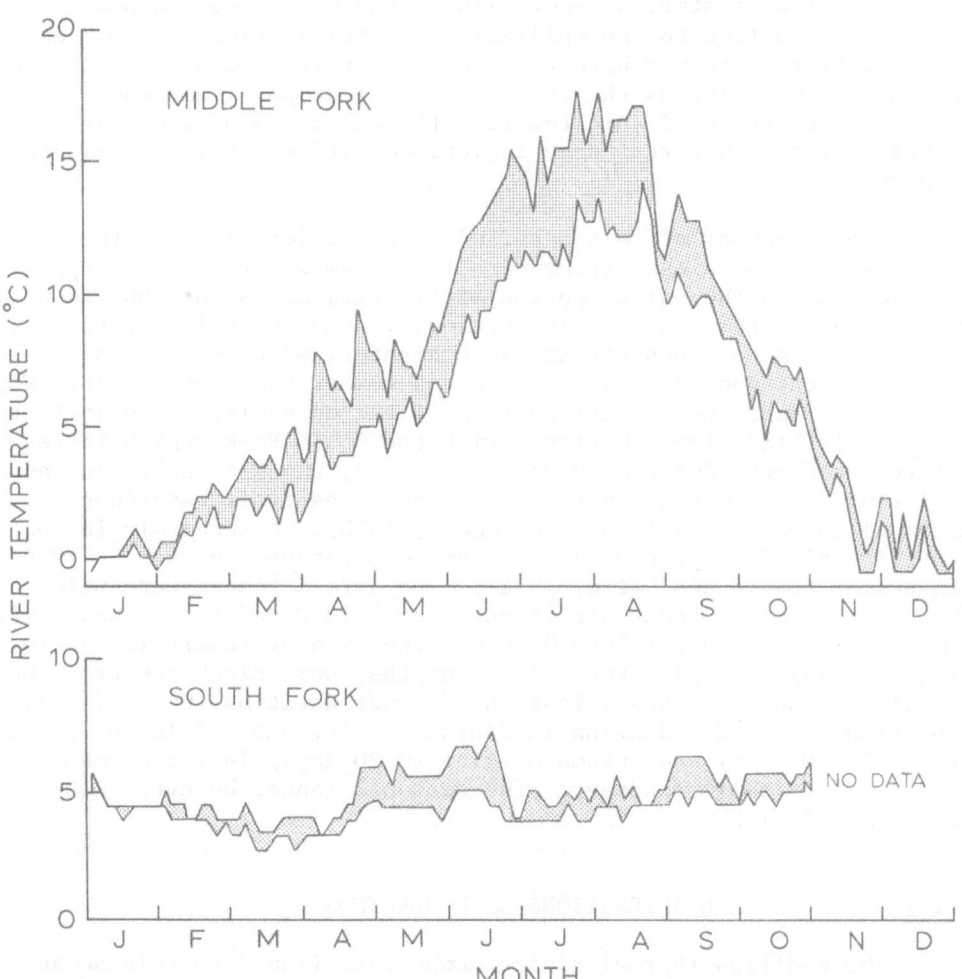

Figure 2.   Thermal regimes of the unregulated Middle Fork and the
regulated South Fork of the Flathead River during 1977
(Stanford, unpublished).

Severe thermal fluctuations over short periods may occur in streams below dams. Because of rapid heating during low flow periods, summer tailwater temperatures below deep-release dams may fluctuate 6-8°C as power releases peak and wane, which may occur two or three times each day (Pfitzer, 1967). Thermal shock may be induced below stratified reservoirs as the water level drops and release is shifted to the epilimnion. A combination of lower and upper outlet ports may have the same effect if discharge is shifted to the lower outlet as the reservoir level drops. Comprehensive data are not yet available relating the operation of dams with multi-level outlet structures to ecological conditions in the receiving stream.

The remainder of this paper will deal primarily with thermal effects of deep-release reservoirs that draw water from the hypo-limnion during the entire period of stratification and thus do not produce rapid temperature fluctuations in the receiving stream. The thermal regime of such streams is characterized by diurnal and seasonal constancy and winter warm and summer cool conditions. Figure 2 contrasts the thermal pattern of the unregulated Middle Fork of the Flathead River, Montana, with the South Fork, which is regulated by Hungry Horse Reservoir (Stanford, Unpublished). The annual range 5 km downstream from a Japanese reservoir was reduced from 21 to 12°C (Nishizawa and Yamabe, 1970). A reservoir in the British uplands depressed downstream temperatures as much as 12°C in summer (Lavis and Smith, 1971). The stream temperature below a Colorado mountain reservoir ranged from 3-13°C and the seasonal maximum was delayed until late October, the time of reservoir overturn (Ward, 1974). Despite two release depths, some discharge from the spillway, and weak stratification of short duration, Crisp (1977) reported a marked reduction in diurnal range, and a delayed spring rise (20-50 days) and autumn decline (0-20 days) in the stream below a British impoundment. The seasonal range, however, was reduced only 1-2°C.

EFFECTS OF THERMAL ALTERATIONS ON ZOOBENTHOS

The modified thermal regime downstream from deep-release impoundments may be a major factor contributing to modifications of the zoobenthos community, especially in streams with otherwise favorable environmental conditions (Briggs, 1948; Pearson et al., 1968; Hoffman and Kilambi, 1971; Spence and Hynes, 1971; Lehmkuhl, 1972; Ward, 1974, 1976c; Gore, 1977).

Figure 3 shows the interrelationships, resulting from thermal regime modifications, hypothesized as responsible, in part, for the selective elimination of macroinvertebrate species in streams below deep-release dams.

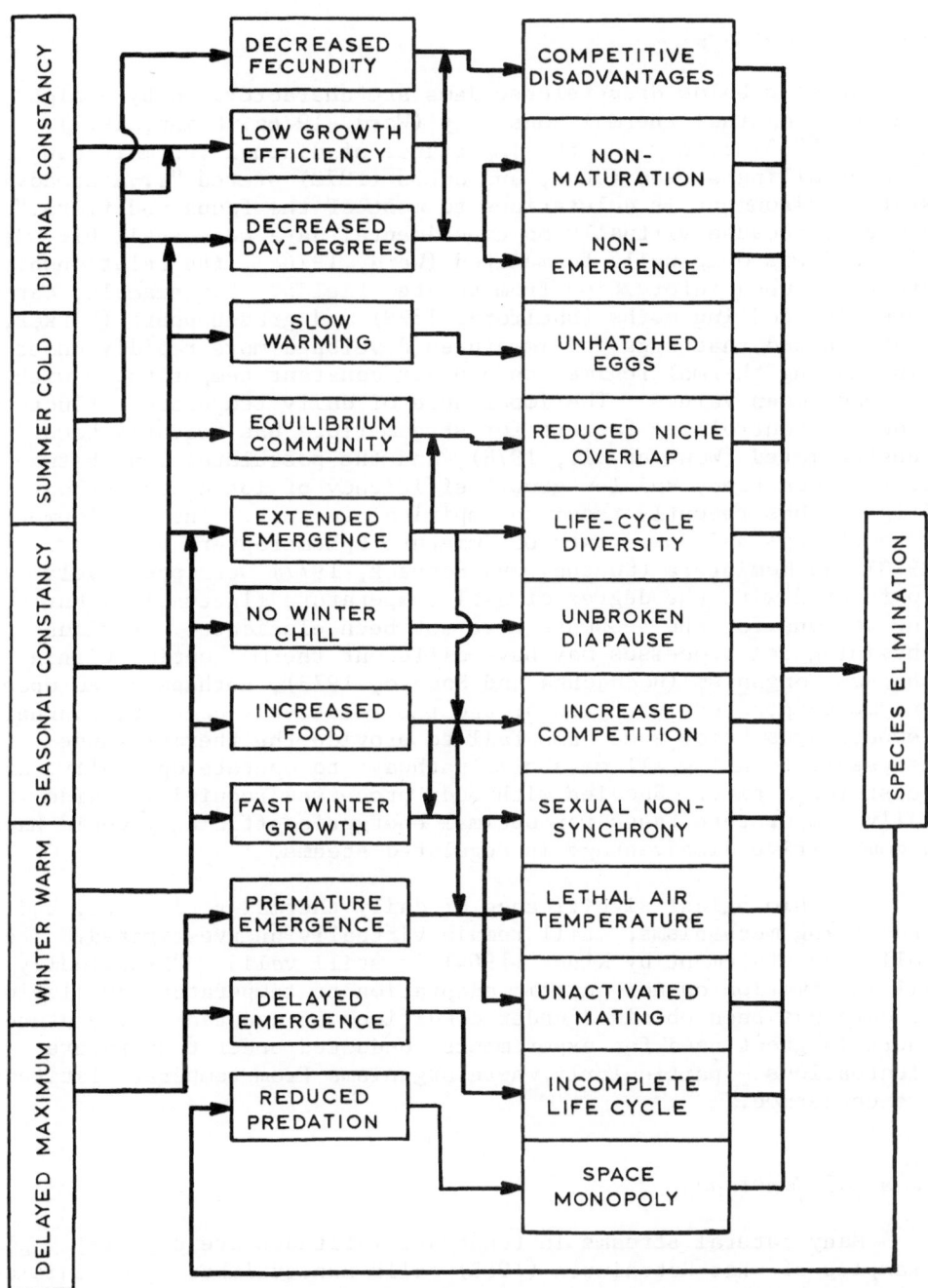

Figure 3. Thermal modifications below deep-release dams and resulting interrelationships hypothesized as partly responsible for selective elimination of zoobenthic species (modified from Ward, 1976c).

Diurnal Constancy

Streams below deep-release dams are characterized by a high degree of diurnal thermal constancy (Ward, 1976c, Crisp, 1977). Kamler (1965) emphasized the importance of diurnal thermal changes in controlling stream fauna, and Hubbs (1972) stated "...a steady-state environment is deleterious to most of the fauna and flora." However, because virtually no experimental data were available when Figure 3 was originally formulated (Ward, 1976c), the relationships are based upon information from related fields. For example, early work with codling moths (Shelford, 1929) and grasshoppers (Parker, 1930) showed that eggs and immatures developed more rapidly under fluctuating thermal regimes than under constant temperatures with the same mean values. The importance of daily temperature fluctuations as controlling factors for stream flora has recently been substantiated (Wong et al., 1978), and the postulated link between diurnal constancy and low growth efficiency of lotic zoobenthos (Fig. 3) has recently received empirical support. The development rate of eggs and the growth of immature Ephemeroptera (Sweeney, 1978) and Hemiptera (Sweeney and Schnack, 1977) were positively correlated with the degree of diel temperature fluctuation, but the reasons for the response have not been elucidated. Various physiological processes may have different thermal optima within the same organism (Hochachka and Somero, 1973), perhaps based upon enzyme temperature preferenda; and the relatively constant diurnal temperatures below some dams fail to provide the thermal range necessary to allow all metabolic pathways to operate optimally, at least for a time. Species with body processes requiring a wide daily temperature range for optimal energetic efficiency would have a competitive disadvantage in regulated steams.

The biological significance of daily thermal variations, and underlying mechanisms, still remain virtually uninvestigated. The following statement by Kinne (1964) is still valid: "Practically all information on non-genetic adaptation to temperature available to date has been obtained under conditions of constant temperature. There is great need for experiments conducted under temperature fluctuations - particularly where organisms from temperate latitudes are concerned."

Seasonal Constancy

Many natural streams in temperate latitudes are especially good examples of what Hutchinson (1953) calls nonequilibrium communities, since temporal fluctuations in environmental conditions are great and often unpredictable. Different species are favored as environmental conditions change, which allows considerable niche overlap,

since competitive exclusion is not given sufficient time to eliminate species. Ide (1935) reported that stream mayfly diversity was
positively correlated with the temperature range, and Paine (1966)
found that the most diverse benthic intertidal community was associated with the greatest annual temperature fluctuation.

The relatively constant environmental conditions of streams
below deep-release dams may cause a shift toward an equilibrium
community, especially if factors in addition to temperature (e.g.,
chemical parameters, flow, turbidity, and the concentration and
particle size of transport detritus) exhibit increased constancy
and predictability. Thermal constancy also has indirect effects on
zoobenthos. For example, epilithic algae may be abundant year-round
in regulated streams, in contrast to the great temporal variability
exhibited in unregulated streams resulting from ice and sediment
abrasion and other controlling factors.

Whereas species such as *Gammarus lacustris*, which are able to
complete their life cycle under constant thermal conditions (Smith,
1973), are favored below dams, species that depend upon seasonal
cycles of temperature to cue various life cycle phenomena will be
eliminated. The eggs of some organisms require a rapid rise of water
temperature in spring to initiate hatching (Britt, 1962), and the
spring rise in temperature is important in synchronizing emergence
in some species (Lutz, 1968). The very gradual rise in temperature
throughout the spring and summer below a dam on a Colorado mountain
river contrasted greatly with the rapid spring increase at a location farther downstream, which was characterized by a much more
diverse faunal assemblage (Ward, 1976a). Except for the unusual
situation in the Brazos River referred to earlier, the only regulated
stream known to the present authors in which zoobenthic diversity was
not lowered was located below a dam that simultaneously draws water
from upper and lower levels, thus reducing the seasonal range by
only 1-2°C (Table 2).

Summer Cold Conditions

Water released from the hypolimnion of a deep stratified reservoir may result in summer temperatures below dams that are up to 20°C
lower than those of unregulated streams of the region. According
to Illies' (1952) "Entwicklungsnullpunkt" theory, growth is not only
temperature dependent, but it ceases altogether if the temperature
falls below the minimum critical temperature (the Entwicklungsnullpunkt) for a given species. Total degree days may not be adequate
for the completion of the life cycle of some species, or the temperature may not attain the level necessary for nymphal maturation
or emergence. For example, growth and maturation of the stonefly
Pteronarcella badia is based on thermal summation, although emergence in dependent upon the attainment of an absolute water temperature (Stanford and Gaufin, unpublished). The failure of most

stoneflies in the tailwaters below Hungry Horse Reservoir can be
explained by lack of appropriate thermal criteria. The time between
oviposition and hatching and the length of the hatching period may
be greatly extended by low summer temperatures (e.g., Elliott, 1972).
Reduced growth efficiency at low temperatures may eliminate species
even though the temperature is within the tolerance range of the
organism (Edington and Hildrew, 1973), presumably by causing a
competitive disadvantage. Delayed emergence caused by slower growth
could eliminate species, although this would depend on the particular
life-cycle pattern. The emergence period may be lengthened under
summer-cold conditions (Macan, 1957), or adult longevity may be
increased (Nebeker, 1971a). Sweeney (1978) found that reduced temperatures during spring and summer lowered fecundity of winter and
summer subimagos, apparently because of a reduction in the size of
adults (Sweeney and Vannote, 1978). Species specific differences
in the effects of summer cold conditions on fecundity would place
some species at a competitive disadvantage. Species capable of
metabolic adaptation to a wide range of thermal conditions (see,
e.g., Fahy, 1973) would be favored in streams below deep-release dams.

Winter Warm Conditions

     Increases in winter stream temperatures below deep-release dams
may be of considerable ecological significance, especially in regions
where streams normally develop an ice cover.

     Species requiring winter chill (temperatures at or near 0°C) to
break egg or larval diapause will be eliminated if winter temperatures are elevated (Lehmkuhl, 1972). Enhanced growth and resulting
premature emergence (as much as five months early) from increased
winter temperatures has been well documented (Coutant, 1968; Nebeker,
1971b; Lillehammer, 1975; but see Langford, 1975). Fey (1977) found
that warming of a river by a power station eliminated the quiescent
stage of *Hydropsyche pellucidula* larvae so that accelerated growth
in February advanced emergence by 3-4 months.

     Premature emergence may eliminate species if air temperatures
are lethal to the adults (Nebeker, 1971b). Surviving adults may not
reproduce, because mating mechanisms are not activated at low air
temperatures (e.g., drumming in stoneflies; Rupprecht, 1975) or
because of nonsynchronous emergence of males and females (Nebeker,
1971b). It may be significant that groups without aerial adults,
such as amphipods, isopods, gastropods, oligochaetes, and turbella-
rians, often increase in relative abundance in streams below dams
(Ward and Short, 1978) and in springbooks (Ward and Dufford, in
press).

     Winter warm conditions and seasonal constancy may cause extended
emergence and thus increased life cycle diversity and increased and

more constant productivity in regions with warm winter air temperatures (Ward, 1976c). Because the niche of a species may change throughout the life cycle, the temporal spacing of overlapping generations may occupy niches that would have been available for other species. Newell and Minshall (1978) found that a normally bivoltine mayfly exhibited multivoltinism in a stream section that was 18°C year-round, due to warm springs. Coactive patterns may increase under conditions of increased population density and more constant resource availability (Hutchinson, 1953; Yount, 1956; Patrick, 1970), resulting in reduced niche overlap and elimination of rare species.

Effects of increased winter temperatures will likely be of less ecological import in regions with more equable climates, such as Great Britain, where 0°C temperatures do not normally occur in streams and thus the fauna has not evolved a requirement for winter chill to break egg or larval diapause. This may explain, at least in part, some of the geographical differences in faunal response to stream regulation.

Delayed Thermal Maximum

In addition to the thermal modifications discussed, the seasonal maximum temperature may be greatly delayed (Jaske and Goebel, 1967; Ward, 1976a,c). The delayed seasonal maximum may conceivably delay or cause precocious emergence, depending on specific life history requirements. However, because empirical data are lacking, the effects of the delayed thermal maximum must rest on speculation. Research is badly needed in this area.

Reduced Predation

Predation often exerts a major role in determining the structure of stream communities (Patrick, 1970). A reduction of predation pressure may reduce the number of species in a community; even those not directly preyed upon are affected (Paine, 1966).

If one of the species eliminated by regulation is a "keystone species" (sensu Paine, 1969), that is, a top carnivore that prevents the monopolization of resources by a few species, zoobenthos diversity may be further reduced.

CONCLUSIONS

Although our understanding of the effects of reservoirs on biota of downstream lotic reaches has increased substantially in recent years, it has not kept pace with the proliferation of dams. Only a handful of investigators have published comprehensive ecological data on a relatively few regulated streams. Dams with multilevel outlet structures offer nearly unlimited potential for experimental manipulations as a means to more fully understand the structure and function of stream ecosystems. In addition to the great need for experimental research, much descriptive work remains to be done. The differential response to regulation of streams along latitudinal or altitudinal gradients could provide consider- able insight into ecological phenomena, but too few sites have been intensively studied to offer valid comparative data.

Ward and Short (1978) stressed the value of benthic organisms as integrators of ecological conditions and developed a preliminary classification system based upon species response to regulation. An understanding of why certain taxa are affected in a given way would contribute not only to knowledge of ecological requirements of species but ultimately would provide data of predictive value in assessing impacts and beneficial modification of future stream regu- lation projects.

ACKNOWLEDGMENTS

The authors wish to thank Dr. H. B. N. Hynes (University of Waterloo) and Dr. R. W. Pennak (University of Colorado) for review- ing the manuscript. Dr. K. W. Stewart (North Texas State University) provided useful comments on the manuscript and kindly supplied unpublished data on the Brazos River.

REFERENCES

Ackermann, W. C., White, G. F., and Worthington, E. B., eds., 1973, "Man-Made Lakes: Their Problems and Environmental Effects," Geophys. Monogr. 17, Am. Geophys. Union, Washington D.C., 847 p.

Armitage, P. D., 1976, a quantitative study of the invertebrate fauna of the River Tees below Cow Green Reservoir, *Freshwater Biol.*, 6:229-240.

Armitage, P. D., 1977, Invertebrate drift in the regulated River Tees, and an unregulated tributary Maize Beck, below Cow Green dam, *Freshwater Biol.*, 7:167-183.

Armitage, P. D., 1978, Downstream changes in the composition, num- bers and biomass of bottom fauna in the Tees below Cow Green Reservoir and an unregulated tributary Maize Beck, in the first five years after impoundment, *Hydrobiologia*, 58:145-156.

Armitage, P. D., and Capper, M. H., 1976, The numbers, biomass and transport downstream of microcrustaceans and *Hydra* from Cow Green Reservoir (upper Teesdale), *Freshwater Biol.*, 6:425-432.

Baxter, R. M., 1977, Environmental effects of dams and impoundments, *Annu. Rev. Ecol. Syst.*, 8:255-283.

Briggs, J. C., 1948, The quantitative effects of a dam upon the bottom fauna of a small California stream, *Trans. Am. Fish. Soc.* 78:70-81.

Britt, N. W., 1962, Biology of two species of Lake Erie mayflies, *Ephoron album* (Say) and *Ephemera simulans* Walker, *Bull. Ohio Biol. Surv.*, 5:1-70.

Chutter, F. M., 1963, Hydrobiological studies on the Vaal River in the Vereenigung area. Part I. Introduction, water chemistry and biological studies on the fauna of habitats other than muddy bottom sediments, *Hydrobiologia*, 21:1-65.

Coutant, C. C., 1968, Effect of temperature on the development rate of bottom organisms, p. 9.13-9.14, *in*: "Annual Report for 1967," USAEC Div. Biology and Medicine, Batelle-Northwest, Richland, Washington.

Crisp, D. T., 1977, Some physical and chemical effects of the Cow Green (upper Teesdale) impoundment, *Freshwater Biol.* 7:109-120.

Edington, J. M., and Hildrew, A. H., 1973, Experimental observations relating to the distribution of net-spinning Trichoptera in streams, *Verh. Int. Verein. Limnol.*, 18:1549-1558.

Elliott, J. M., 1972, Effects of temperature on the time of hatching in *Baetis rhodani* (Ephemeroptera:Baetidae), *Oecologia*, 9:47-51.

Fahy, E., 1973, Observations on the growth of Ephemeroptera in fluctuating and constant temperature conditions, *Proc. R. Irish Acad.*, 73:133-149.

Fey, J.M., 1977, Die Aufheizung eines Mittelgebirgsflusses und ihre Auswirkungen auf die Zoozönose-dargestellt an de Lenne (Sauerland), *Arch. Hydrobiol. Suppl.*, 53:307-363.

Gore, J. A., 1977, Reservoir manipulations and benthic macroinvertebrates in a prairie river, *Hydrobiologia*, 55:113-123.

Hall, A., Davies, B. R., and Valente, I., 1976, Caboro Bassa: Some preliminary physico-chemical and zooplankton pre-impoundment survey results, *Hydrobiologia*, 50:17-25.

Hall, J. B., and Pople, W., 1968, Recent vegetational changes in the lower Volta River, *Ghana J. Sci.*, 8:24-29.

Hannan, H. H., and Young, W. J., 1974, The influence of a deep-storage reservoir on the physicochemical limnology of a central Texas river, *Hydrobiologia*, 44:177-207.

Hilsenhoff, W. L., 1971, Changes in the downstream insect and amphipod fauna caused by an impoundment with a hypolimnion drain, *Ann. Entomol. Soc. Am.*, 64:743-746.

Hochachka, P. W., and Somero, G. N., 1973, "Strategies of Biochemical Adaptation," W. B. Saunders Co., Philadelphia.

Holmes N. T. H., and Whitton B. A., 1977, The macrophytic vegetation of the River Tees in 1975: Observed and predicted changes *Freshwater Biol.*, 7:43-60.

Hoffman, C. E., and Kilambi, R. V., 1971, "Environmental Changes Produced by Cold-Water Outlets from Three Arkansas Reservoirs, Water Resources Res. Center Publ. No. 5, Univ. Arkansas, Fayetteville.

Hubbs, C., 1972, Some thermal consequences of environmental manipu-
    lations of water, *Biol. Conserv.*, 4:185–188.
Hutchinson, G. E., 1953, The concept of pattern in ecology, *Proc.
    Acad. Nat. Sci.* Phila. 105:1–12.
Hynes, H. B. N., 1970a, "The Ecology of Running Waters," Univ.
    Toronto Press, Toronto, 555 p.
Hynes, H. B. N., 1970b, The ecology of stream insects, *Annu. Rev.
    Entomol.*, 15:25–42.
Hynes, H. B. N., 1975, The stream and its valley, *Verh. Int. Verein.
    Limnol.*, 19:1–15.
Ide, F. P., 1935, The effect of temperature on the distribution of
    the mayfly fauna of a stream, *Publ. Ontario Fish. Res. Lab.*,
    50:1–76.
Illies, J., 1952, Die molle Faunistischokologische Untersuchungen
    an einem Forellenbach in lipper Bergland, *Arch. Hydrobiol.*,
    46:424–612.
Isom, B. G., 1971, Effects of storage and mainstream reservoirs on
    benthic macroinvertebrates in the Tennessee Valley, p. 179–191,
    in: "Reservoir Fisheries and Limnology," G. E. Hall, ed.,
    Spec. Publ. No. 8, Am. Fish. Soc., Washington, D.C.
Jaske, J. T., and Goebel, J. B., 1967, Effects of dam construction
    on temperature of Columbia River, *J. Am. Water Works Assoc.*,
    59:935–942.
Kamler, E., 1965, Thermal conditions in mountain waters and their
    influence on the distribution of Plecoptera and Ephemeroptera
    larvae, *Ekol. Pol. Ser. A.*, 13:377–414.
Kinne, O., 1964, Non-genetic adaptation to temperature and salinity,
    Helgol Wiss. Meeresunters., 9:433–458.
Langford, T. E., 1975, The emergence of insects from a British river,
    warmed by power station cooling-water. Part II. The emergence
    patterns of some species of Ephemeroptera, Trichoptera, and
    Megaloptera in relation to water temperature and river flow,
    upstream and downstream of the cooling-water outfalls, *Hydro-
    biologia*, 47:91–133.
Lavis, M. E., and Smith K., 1971, Reservoir storage and the thermal
    regime of rivers, with special reference to the River Lune,
    Yorkshire, *Sci. Total Environ.*, 1:81–90.
Lehmkuhl, D. M., 1972, Change in thermal regime as a cause of
    reduction of benthic fauna downstream of a reservoir, *J. Fish.
    Res. Board Can.*, 29:1329–1332.
Lillehammer, A., 1975, Norwegian stoneflies. IV. Laboratory studies
    on ecological factors influencing distribution, *Norw. J.
    Entomol.*, 22:99–108.
Lutz, P. E., 1968, Effects of temperature and photoperiod on larval
    development in *Lestes eurinus* (Odonata:Lestidae), *Ecology*,
    49:637–644.
Macan, T. T., 1957, The life histories and migrations of the
    Ephemeroptera in a stony stream, *Trans. Soc. Br. Entomol.*,
    12:129–154.

Macan, T. T., 1961, Factors that limit the range of freshwater animals, *Biol. Rev.*, 36:151-198.

Macan, T. T., 1974, Running water, *Mitt. Int. Verein. Limnol.*, 20:301-321.

Mackichan, K. A., 1967, Diurnal temperature variations of three Nebraska streams, *U.S. Geol. Surv. Pap.*, 575B:233-234.

McClure, R. G., and Stewart, K. W., 1976, Life cycle and production of the mayfly *Choroterpes (Neochoroterpes) mexicanus* Allen (Ephemeroptera:Leptophlebiidae), *Ann. Entomol. Soc. Am.*, 69:134-144.

Merkley, W. B., 1978, Impact of Red Rock Reservoir on the Des Moines River, p. 62-67, *in*: "Current Perspective on River-Reservoir Ecosystems," J. Cairns, Jr., E. F. Benefield and J. R. Webster eds., N. Am. Benthol. Soc.

Müller, K., 1962, Limnologisch-Fischereibiologische Untersuchungen in regulierten Gewässern Schwedisch-Laplands, *Oikos*, 13:125-154.

Nebeker, A. V., 1971a, Effect of water temperature on nymphal feeding rate, emergence, and adult longevity of the stonefly *Pteronarcys dorsata, Kansas Entomol. Soc. J.*, 44:21-26.

Nebeker, A. V., 1971b, Effect of high winter water temperatures on adult emergence of aquatic insects, *Water Res.*, 5:777-783.

Neel, J. K., 1963, Impact of reservoirs, p. 575-593, *in*: "Limnology in North America," D. G. Frey, ed., Univ. Wisconsin Press, Madison.

Newell, R. L., and Minshall, G. W., 1978, Life history of a multi-voltine mayfly, *Tricorythodes minutus*: An example of the effect of temperature on the life cycle, *Ann. Entomol. Soc. Am.*, 71:876-881.

Nishizawa, T., and Yamabe, K., 1970, Change in downstream temperature caused by the construction of reservoirs. Part I . *Tokyo Univ. Sci. Rep. Sect. C.*, 10:27-42.

Paine, R. T., 1966, Food web complexity and species diversity, *Am. Nat.*, 100:60-75.

Paine, R. T., 1969, The *Pisaster-Teguia* interaction: Prey patches, predator food preferences, and intertidal community structure, *Ecology*, 50:950-961.

Parker, J. R., 1930, Some effects of temperature and moisture upon *Melanoplus mexicanus* and *Camnula pellucida* Scudder (Orthoptera), *Bull. Univ. Mont. Agric. Exp. Sta.*, 223:1-132.

Patrick, R., 1970, Benthic stream communities, *Am. Sci.*, 58:546-549.

Pattee, E., and Bournaud, M., 1970, Etude experimentale de la rhéophilie chez les planaires triclades d'eau courante, *Schweiz. Z. Hydrol.*, 32:181-191.

Pearson, W. D., Kramer, R. H., and Franklin, D. R., 1968, Macro-invertebrates in the Green River below Flaming Gorge Dam, 1964-1965 and 1967, *Proc. Utah Acad. Sci. Arts, Lett.*, 45:148-167.

Penáz, M., Kubíček, F., Marvan, P., and Zelinka, M., 1968, Influence
    of the Vír River Valley Reservoir on the Hydrobiological and
    ichthyological conditions in the River Svratka, *Acta Sci. Nat.
    Brno*, 2:1-60.
Pfitzer, D. W., 1967, Evaluation of tailwater fishery resources
    resulting from high dams, p. 477-488, *in*: "Reservior Fishery
    Resources Symposium," Am. Fish. Soc., Washington D.C.
Radford, D. S., and Hartland-Rowe, R., 1971, A preliminary invest-
    igation of bottom fauna and invertebrate drift in an unregu-
    lated and a regulated stream in Alberta, *J. Appl. Ecol.*,
    8:883-903.
Ridley, J. E., and Steel, J. A., 1975, Ecological aspects of river
    impoundments, p. 565-587, *in*: "River Ecology," B. A. Whitton,
    ed., Blackwell Sci. Publ., Oxford.
Rupprecht, R., 1975, The dependence of emergence-period in insect
    larvae on water temperature, *Verh. Int. Verein. Limnol.*,
    19:3057-3063.
Shelford, V. E., 1929, "Laboratory and Field Ecology," Williams and
    Wilkins Co., Baltimore, 608p.
Simmons, G. M., Jr., and Voshell, J. R., Jr., 1978, Pre- and post-
    impoundment benthic macroinvertebrate communities of the North
    Anna River, p. 45-61, *in*: "Current Perspectives on River-
    Reservoir Ecosystems," J. Cairns, Jr., E. F. Benefield, and
    J. R. Webster, eds., N. Am. Benthol. Soc.
Smith, K., 1972, River water temperatures--an environmental review
    Scott. Geogr. Mag., 88:211-220.
Smith, W. E., 1973, Thermal tolerance of two species of *Gammarus*,
    Trans. Am. Fish. Soc., 102:431-433.
Soltero, R. A., Wright, J. C., and Horpestad, A. A., 1973, Effects
    of impoundment on the water quality of the Bighorn River,
    Water Res. Pergamon Press, 7:343-354.
Spence, J. A., and Hynes, H. B. N., 1971, Differences in benthos
    upstream and downstream of an impoundment, *J. Fish. Res.
    Board Can.*, 28:35-43.
Sweeney, B. W., 1978, Bioenergetic and developmental response of a
    mayfly to thermal variation, *Limnol. Oceanogr.*, 23:461-477.
Sweeney, B. W., and Schnack, J. A., 1977, Egg development, growth,
    and metabolism, of *Sigara alternata* (Say) (Hemiptera:Corixadae)
    in fluctuating thermal environments, *Ecology*, 58:265-277.
Sweeney, B. W., and Vannote, R. L., 1978, Size variation and distri-
    bution of hemimetabolous aquatic insects: Two thermal equilib-
    rium hypotheses, *Science*, 200:444-446.
Trotsky, H. M., and Gregory, R. W., 1974, The effects of water flow
    manipulation below a hydroelectric power dam on the bottom
    fauna of the upper Kennebec River, Maine, *Trans. Am. Fish. Soc.*,
    103:318-324.
Ward, J. V., 1974, A temperature-stressed stream ecosystem below a
    hypolimnial release mountain reservoir, *Arch. Hydrobiol.*,
    74:247-275.

Ward, J. V., 1975, Downstream fate of zooplankton from a hypolimnial release mountain reservoir, *Verh. Int. Verein, Limnol.*, 19:1798-1804.

Ward, J. V., 1976a, Comparative limnology of differentially regulated sections of a Colorado Mountain river, *Arch. Hydrobiol.*, 78:319-342.

Ward, J. V., 1976b, Effects of flow patterns below large dams on stream benthos: A review, p. 235-253, *in:* "Instream Flow Needs Symposium," Vol. II, J. F. Orsborn and C. H. Allman, eds., *Am. Fish. Soc.*, Bethesda, Maryland.

Ward, J. V., 1976c, Effects of thermal constancy and seasonal temperature displacement on community structure of stream macroinvertebrates, p. 302-307, *in:* "Thermal Ecology, II," G. W. Esch and R. W. McFarlane, eds., ERDA Symp. Ser. (CONF.-750425).

Ward, J. V., and Dufford, R. G., In press, Longitudinal and seasonal distribution of macroinvertebrates and epilithic algae in a Colorado springbrook-pond system, *Arch. Hydrobiol.*

Ward, J. V., and Short, R. A., 1978, Macroinvertebrate community structure of four special lotic habitats in Colorado, U.S.A., *Verh. Int. Verein. Limnol.*, 20:1382-1387.

Wong, S. L., Clark, B., Kirby, M., and Kosciuw, R. F., 1978, Water temperature fluctuations and seasonal periodicity of *Cladophora* and *Potamogeton* in shallow rivers, *J. Fish. Res. Board Can.*, 35:866-870.

Wright, J. C., 1967, Effects of impoundments on productivity, water chemistry and heat budgets of rivers, p. 188-199, *in:* "Reservoir Fishery Resources Symposium," Am. Fish. Soc., Washington, D.C.

Young, W. C., Kent, D. H., and Whiteside, B. G., 1976, The influence of a deep storage reservoir on the species diversity of benthic macroinvertebrate communities of the Guadalupe River, Texas, *Texas J. Sci.*, 27:213-224.

Yount, J. L., 1956, Factors that control species numbers in Silver Springs, Florida, *Limnol. Oceanogr.*, 1:286-295.

ECOLOGY OF RIVERINE FISHES IN REGULATED STREAM SYSTEMS WITH

EMPHASIS ON THE COLORADO RIVER

Paul B. Holden

BIO/WEST, Inc.
P.O. Box 3226
Logan, UT 84321

INTRODUCTION

Riverine fishes can be defined rather simply as species that inhabit rivers. This simplistic definition may suffice for general informational needs, but is insufficient when describing impacts. Some fishes that live in rivers require the river's environment for sustenance of their populations, e.g., Pacific salmon (*Oncorhynchus* spp.), paddlefish (*Polyodon spathula*), and blue sucker (*Cycleptus elongatus*). Other riverine species can do just as well, or better, in lakes, ponds, or other lentic environments, e.g., carp (*Cyprinus carpio*), fathead minnow (*Pimephales promelas*), and bass (*Micropterus* sp.). Therefore, changes in a natural river's flow, temperature, water quality, or physical morphometry may affect certain species more than others; more likely it will affect those species that require lotic conditions to a greater extent than those which also tolerate lentic conditions.

Riverine fishes include obligate and faculative riverine species. Obligate riverine fishes may be further subdivided into those requiring rivers for all their ecological needs, such as most darters (*Etheostoma* sp., *Percina* sp.) and the blue sucker, versus those that require rivers for only a portion of their life history, such as salmon, paddlefish, and other migrating species. Further subdivisions could be made but are not required for this discussion.

The initial intent of this paper was to concentrate on the effects of dams on obligate riverine species, especially those that require natural rivers for all their ecological requirements. It soon became evident that this was not practical, because most of these species are not game fish but, rather, are minnows, suckers, darters, and other non-game fish. Therefore, most of the information concerning the ecology of fishes in regulated streams comes from sport or commercial fishes, such as salmon and trout, which require rivers for only part of their ecological requirements.

EFFECTS OF DAMS

The impact of damming, or regulation, of rivers on obligate riverine fishes is generally negative. Approximately 60% of the fishes presently listed as endangered or threatened in the U.S. are obligate riverine species. Although not all were decimated by regulation of their habitat, dams are one of the major causes in the decline of obligate riverine species.

The impacts of regulation can be classified as Immediate—those impacts that become immediately apparent when a dam becomes operational; and Delayed—those impacts that do not become evident for several years after dam completion.

Immediate Impacts of Regulation

One of the most obvious immediate impacts of dams is the blockage of upstream and/or downstream migration. This has been identified as a major problem to Pacific salmon (Raymond, 1968; Trefethen, 1972; Collins, 1976; Robinson, 1978), Atlantic salmon (Dominy, 1973), paddlefish (Branson, 1974; Pflieger, 1975), and several Atlantic species of herring (*Alosa* sp.) and bass (*Morone* sp.) (Nichols, 1968). These species all need to spawn in flowing water, and dams have blocked their natural routes to upstream spawning areas. Shikhshabekov (1971) indicated that several species of fish in Russia resorbed eggs when their migrations were slowed or stalled due to regulation. Dams present a special problem to downstream migrating juvenile salmon and steelhead trout. Since many dams these fish encounter are for hydroelectric generation, much of the water passing the dam flows through turbines. Turbine-associated mortality is high. Also, passage through a dam often ends in stilling basins where predators, such as northern squawfish (*Ptychocheilus oregonensis*) and sea gulls, await the disoriented fish (Long and Krcma, 1969). The construction of fishways, to aid upstream migrating adults, and the use of screens or the actual trucking of juveniles around dams have decreased the negative impact of some dams on Pacific salmon and steelhead trout (Long and Krcma, 1969; Park and Farr, 1972; Collins, 1976; Raymond, 1976).

The other major immediate effect of dams is habitat alteration. The alteration of a free-flowing river to an impoundment is a

considerable change. Impounded waters preclude obligate riverine
fishes, since they require flowing water for all their ecological
requirements (Fraser, 1972). There is some evidence that white
sturgeon prefer free-flowing areas and would not maintain viable
populations in reservoirs (Haynes et al., 1978). Such is also the
case for blue suckers (Pflieger, 1975; Smith, 1979) and for the
widely publicized snail darter (*Percina tanasi*) (Williams and
Finnley, 1977). Other species that require only rivers for spawning
may be found in the reservoir for a few years following impoundment
but will disappear due to a lack of recruitment if appropriate
spawning areas are inundated. For species such as the paddlefish in
the Missouri River, the adults often live in reservoirs but migrate
upstream to spawn in tributary rivers (Pflieger, 1975). These populations will be adversely affected if reservoirs cover their
spawning areas. Lewis (1974) noted a decrease in fishes of the
family Mormyridae, which are primarily riverine species, when the
River Niger was dammed to form Lake Kainji and attributed the change
primarily to food availability. Bottom feeding insectivores no
longer had a plentiful food supply in the reservoir. Blake (1977)
indicated that the loss of riverine species in the lake was primarily
due to a reduction in reproductive success. Balon (1978) and Petr
(1978) review similar situations in other tropical reservoirs.

        Therefore, it appears that reservoirs impact adult riverine
fishes primarily by not providing basic living conditions, such as
food supply, and by not providing adequate habitat for successful
reproduction. Other factors that may be important include changes
in water chemistry and interactions with other species that become
abundant in the reservoir.

        Reservoirs also create adverse situations for downstream
migrating juvenile fish. Passage of young Pacific and Atlantic salmon through large reservoirs is only one-third as fast as that
through free-flowing sections (Raymond, 1969; Dominy, 1973; Raymond,
1976). This delay causes increased mortality due to increased
exposure to disease and predation (Collins, 1976). The thermal
stratification of large reservoirs can also cause passage problems
to young salmon (Collins, 1976). The upper layers may be too warm
for the fish, and the lower layers too low in oxygen. Furthermore,
reservoirs can accumulate pollutants to levels that become toxic to
the juvenile fishes (Dominy, 1973).

        Dams result in habitat alteration in tailwater reaches, which
may have major impacts on riverine fishes. Downstream changes are
not nearly as obvious as the change seen in a reservoir but are
often just as effective in eliminating riverine fishes. A major
effect of dams on downstream habitat is a change in temperature.
The temperature below a dam usually depends on the level of withdrawal from the reservoir. A dam with a high intake will generally

produce warm summer temperatures; deep intakes generally produce
cold summer temperatures. Summer temperatures have generally been
lower than natural in most instances where native fishes have been
adversely affected. Winter temperatures may be elevated below
deep-release dams.

Most obligate riverine fishes are adapted to a certain daily,
seasonal, and yearly temperature regime, as are most fishes in general (Allen, 1969; Hubbs, 1972). Therefore, changes in the preferred
thermal regime will affect them, and the greater the change, the
greater the impact or effect. Species that are relatively specialized
in terms of temperature requirements may disappear altogether because
they are unable to tolerate the "new" regime (Pfitzer, 1963; Spence
and Hynes, 1971; Trautman and Gartman, 1974; Edwards, 1978). They
are often found further downstream, where preferred temperatures are
still available. Other species that are preadapted to the regulated
temperatures often become more abundant. In many situations, the
altered thermal regime is sufficient for adults, but temperatures
are not adequate for reproduction (Pfitzer, 1963; Hubbs, 1972;
Zakharyan, 1972). Therefore, the tailwater population is either
replenished from downstream recruitment or, in the case of species
with short life spans and little movement capabilities, the species
disappear immediately below the dam.

The regulation of rivers also tends to alter the turbidity and
general water chemistry in the tailwaters. Turbidity is generally
lowered and, therefore, could affect fishes that require turbidity.
No information in the literature indicates that this has actually
been the cause of a species or population decline. Smith (1976)
indicated that increased turbidity below dams in the Trinity River,
California, caused siltation of salmonid spawning areas.

Water chemistry changes can also be great. Generally, rivers
below dams reflect the water chemistry from the intake level of the
upstream reservoir. Therefore, the types of chemical changes vary
between dams, and seasonally at any one dam, as the reservoir
stratifies and overturns. Again, there is little information to
show that these kinds of chemical changes have influenced fishes.

Of great concern in most large dam situations is gas supersaturation in tailwaters. The waters become supersaturated when
spilled over high dams, trapping air and plunging it to depths
where high pressures enhance solubility. This has been a major
problem in the Pacific Northwest, where gas-bubble disease has been
one of the major mortality factors on young salmon and trout (Ebel,
1969; Collins, 1976; Raymond, 1976; Robinson, 1978). It has been
solved in some cases by the use of deflectors in the spillways
(Collins, 1976).

Gas supersaturation in the St. Johns River of New Brunswick occurred when air was injected into the water as it passed through the turbines of a dam during low flows (MacDonald and Hyatt, 1973). Effects on young Atlantic salmon were similar to those found in Pacific salmon.

A major habitat alteration that occurs in regulated streams in arid regions is dewatering, or reduced flows. Dams in the Southwest are often built for irrigation storage, and much of the water may be diverted into canals. The impact of reduced flows on fishes has received much recent attention. The primary impact of reduced streamflows is a reduction in usable habitat, especially critical types of habitat such as spawning areas (Smith, 1976). A considerable amount of work has been conducted in the last few years delineating the effects of reduced streamflows on species of the family Salmonidae (Banks et. al, 1974; Hazel, 1976; Mullan et al., 1976; Nickelson, 1976; Bovee, 1978). Also, substantial effort has been expended in trying to formulate methodologies to adequately predict the impact of reduced flows (Orsborn and Allman, 1976; Stalnaker and Arnette, 1976; Bovee and Cochnauer, 1977). Zakharyan (1972), studying sturgeon in the Kura River system of Russia, indicated that reduced flows below irrigation dams dewatered spawning areas, hence impacting reproduction. The low flows also negatively impacted downstream migrating juveniles.

Another type of habitat alteration below hydroelectric dams that may impact riverine fishes is daily fluctuation due to power peaking. Daily fluctuations in water level as great as 2 m are found in some areas. These changes in water surface elevation cause tremendous changes in available habitat. There is little information in the literature regarding the effect of this phenomenon, perhaps because altered temperature or some other factor is also present. One major concern is that daily fluctuations preclude the establishment of warm, rich, slow-moving areas usually favored by young riverine fishes. Kroger (1973) noted that abrupt reductions in flow stranded fish in the Snake River below Jackson Lake (Wyoming).

While cold tailwaters may be detrimental to native warm-water species, the artificial environment often provides conditions suitable for trout fisheries. The primary reason is that trout are maintained by stocking rather than by natural reproduction. Mullan et al. (1976) summarized the problems with tailwater trout fisheries in the upper Colorado River Basin. Major problems revolved around temperatures that were too cold, flows that were less than optimum, and excessive water velocities in relation to available shelter. The cold temperatures had a secondary effect of reducing the availability of invertebrate food organisms.

The effects of regulation discussed above have generally considered negative impacts. A few situations exist where conditions
for fish have been improved by regulation. The constant summer
irrigation flows and warmer winter temperatures of some Pacific
Northwest (U.S.A.) streams have actually increased native salmonid
production.

The perturbations discussed above are the major documented
factors affecting riverine fish ecology below dams. These effects
are fairly well known and are also generally quite predictable if
sufficient life history information is known about the riverine
fishes present.

Delayed Impacts of Regulation

Information from several areas suggests that some of the effects
of dams may noc appear immediately after closure, but tend to slowly
appear several years later. The general effect of these impacts on
riverine fishes is a rather slow reduction in populations rather
than the immediate loss usually associated with dams. These so-
called delayed impacts are poorly understood and have not been well
documented. Therefore, the following discussion should be viewed
as a series of hypotheses that need testing before they can be
considered valid. Also, many of these factors operate many miles
below the dam, usually outside the boundaries of studies intended
to delineate downstream effects of a dam.

One delayed impact is the habitat change that occurs after
several years of low or reduced flows in areas far downstream from
the dam itself. Reservoirs generally tend to store water during
high spring flows, which means that yearly high flows are generally
missing. The loss of these flows, which are needed for scouring and
channel maintenance, may be especially important in maintaining fish
habitat. Reduced flows in some areas may reduce the braidedness of
a channel, but perhaps increase it in other situations. Channels,
therefore, may become deeper or, perhaps, shallower, changing the
type and quantity of fish habitat. I suspect these changes do occur,
and that most are detrimental to native riverine fishes, but there
are few data presently available to support this hypothesis.

It is fairly well understood that the amount of fine sediments
below dams is generally reduced because the reservoir acts as a
silt trap. The effect of reduced silt deposits, and the consequent
armoring, may affect the variety of habitat available to fishes. In
general, it can probably be correctly stated that these processes
tend to reduce the habitat variability below dams, especially the
slow, silt substrate areas usually preferred by young fishes. The
distance below a dam such phenomena may take place depends on the

number and size of inflowing streams. The channel of the Colorado
River was lowered for 148 km below Hoover Dam by the removal of
over 115 million cubic meters of bed material from 1935 to 1951
(Gottschalk, 1964).

Another factor that influences riverine fishes in regulated
streams is predation or competition from introduced species (Kimsey,
1957; Long and Krcma, 1969). At times, exotic fish are introduced
for their sporting potential, and, therefore, the impact is immediate.
In situations where there is no mass injection of fish, that natural
movement and establishment of piscivorous or competitive species may
take several years.

Buildup of toxic gases (H_2S) (Ford, 1963) or the depletion of
a required gas (O_2) (Koryak, 1976) are delayed impacts in tailwaters
due to maturation in the upstream reservoir. These problems generally affect tailwaters for only a short distance, as aeration tends
to normalize the situation.

The synergistic effect of two or more altered factors in downstream areas appears to be a potential problem worthy of future
study. A rather slow change in downstream habitat could very gradually shift conditions so that an exotic or natural competitor or
predator gradually becomes more favored and therefore displaces a
riverine species. As the habitat becomes marginal for the obligate
riverine species, it becomes more preferred for the exotic competitor. At that point, the obligate species becomes more detrimentally
impacted by the exotic species than by the habitat change.

SUMMARY

Whether a factor is immediate or delayed, the ultimate impact
of regulation on obligate riverine fishes depends on the degree of
change and the tolerance level of the fish to that change. In
general, the least tolerant (most highly adapted) species tend to
sustain the greatest impact of regulation (Pfitzer, 1963; Lewis,
1974; Trautman and Gartman, 1974; Blake, 1977; Edwards, 1978). In
warm-water areas in the eastern and southern U.S., where diverse
fish faunas are common, tailwater species are usually replaced by
other native species where changes are not drastic (Pfitzer, 1963;
Spence and Hynes, 1971). Drastic changes, such as a reduction in
summer temperature from warm to cold, usually displace all native
fishes, and exotic species such as planted rainbow trout usually
predominate. In the more arid Southwest, with its relatively
depauperate but highly specialized fish fauna, regulation tends
to replace native fishes with exotic species (Minckley and Deacon,
1968).

The Colorado River System--A Regulated Stream

The Colorado River system is a classic example of a regulated stream, not only because of the numerous mainstream dams, but also because most of its native fish fauna are obligate riverine species. The Colorado system starts as cold, clear mountain streams, but the mainstem and larger tributaries are very warm and turbid. Discussion will concentrate on four species, all "large, warm river" forms that have undergone drastic population reductions in the 20th Century.

The Colorado squawfish (*Ptychocheilus lucius*) is North America's largest cyprinid, reportedly reaching 36 kg and 150 cm in length. It is piscivorous and was originally found in the mainstream Colorado River and all large tributaries. It reportedly migrates into tributaries to spawn (Miller, 1961), but recent evidence suggests that this may not be the case (Holden, 1977). It is presently considered endangered by the U.S. Department of the Interior (1977).

The humpback chub (*Gila cypha*) is a rather bizarre minnow with a large nuchal hump, thin caudal peduncle, and large falcate fins, adaptions for life in swift water (Miller, 1946). It reaches lengths of 45 cm and was one of the last large species of North American freshwater fish to be discovered (Miller, 1946). The humpback chub is found only in several canyon areas in the upper Colorado Basin and Grand Canyon. It apparently prefers swift, deep water and seldom ventures out of these areas. It also is listed as an endangered species by the Department of the Interior (1977).

The bonytail chub (*Gila elegans*) is closely related to the humpback chub, but lacks the abrupt nuchal hump. It was more widespread than the humpback chub, living in most of the mainstem and larger tributaries. It attains lengths of about 60 cm and is generally considered a swift-water form. It is presently protected by most states in the Colorado system and has been recommended for federal endangered listing. It is nearly extinct, with no known reproducing populations.

The razorback sucker (*Xyrauchen texanus*) is a large sucker with a distinct bony keel along its back, hence its common name. It had an original range similar to that of the bonytail chub. Earlier reports suggested that it migrated, apparently to spawn, but no recent evidence supports this hypothesis (Minckley, 1973).

Regulation of the Colorado River began in the late 1800s, when water was diverted for irrigation in both the upper and lower basins. Regulation through mainstem dams began with the completion of Hoover Dam in 1935. At present, the lower Colorado Basin, from Lake Mead downstream, is a succession of reservoirs and dams

(Fig. 1). The Gila River system of Arizona, a major tributary, is
dammed and diverted, so no flow enters the Colorado much of the year.

The upper Colorado Basin remained free flowing until the 1960s,
when the Colorado River Storage Project dams were completed. The
dams of greatest significance to the native riverine fishes were
Flaming Gorge Dam on the upper Green River; the Curecanti Dams on
the upper Gunnison River, the major tributary to the upper Colorado
mainstem; Navajo Dam on the San Juan River; and Glen Canyon Dam on
the Colorado River (Fig. 1). All are high dams, creating large
reservoirs and releasing cold downstream flows in summer. All but
Navajo Dam are hydroelectric, with daily fluctuating flows.

The dams in both the upper and lower basins have adversely
affected the four native species and are a major reason for their

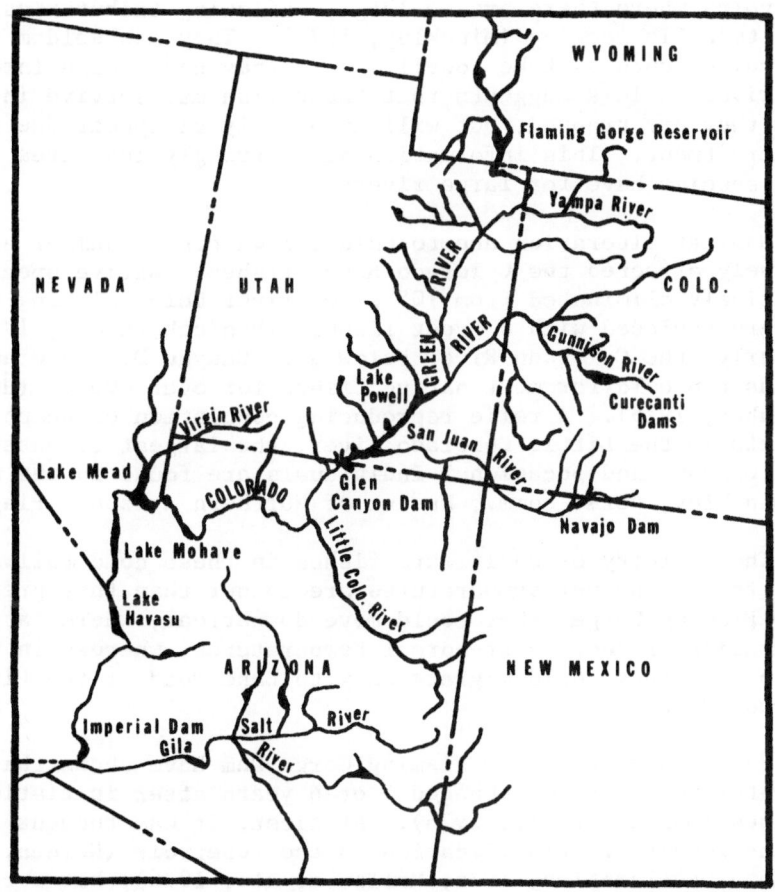

Fig. 1. Map of the Colorado River Basin showing major impoundments.

decline. The dams blocked movement of the riverine fish. Colorado
squawfish and razorback suckers both reportedly migrated up tributaries to spawn (Minckley, 1973), but recent evidence suggests that
this may not be the case (Holden, 1977). Therefore, whether the
dams actually impeded necessary migration is questioned. Short-term
movements were definitely impeded.

Habitat alteration was the major detrimental factor wrought by
the dams. The large number of reservoirs changed hundreds of kilometers of river to still water (Minckley and Deacon, 1968). Most of
this change occurred in areas inhabited by the large river fishes.
Colorado squawfish and humpback chub seldom are found in reservoirs,
except immediately after impoundment or sporadically in later years.
Colorado squawfish are extinct in the lower Colorado Basin (Miller,
1961; Minckley and Deacon, 1968; Minckley, 1973). Adult bonytail
chubs and razorback suckers are found in some of the lower basin
reservoirs where there are no large, inflowing tributaries, but
recruitment is lacking (Minckley, 1973). They are seldom found in
reservoirs, such as Lake Powell, where they can escape into large,
warm rivers. This suggests that these fish may survive in reservoirs
where they are trapped, but will eventually disappear due to a loss
of recruitment. This information also strongly indicates the need
these species have for large rivers.

Habitat alteration due to cold tailwaters in summer also has
adversely affected the Colorado River fishes. Native species were
effectively eliminated from 105 km of river below Flaming Gorge Dam
and were replaced with a trout fishery (Vanicek et al., 1970).
Similarly, the Colorado River below Glen Canyon Dam in Grand Canyon
remains too cold for most native fishes for over 400 km (Holden and
Stalnaker, 1975). A relic reproducing population of humpback chubs
persists in the Little Colorado River, the largest tributary in the
Grand Canyon; and occasional individuals are found in the mainstream
(C. Minckley, pers. comm., Museum of Northern Arizona, Flagstaff).

The scarcity of adult rare fishes in these cold tailwaters
suggests that summer temperatures are colder than they prefer.
Below Flaming Gorge, they could move downstream, where large
tributaries produced near-normal temperatures, whereas in the Grand
Canyon area they could migrate only to Lake Mead or the Little
Colorado River.

Recent studies below Flaming Gorge Dam have shown that the
tailwater temperatures changed 5 or 6 years after initiation of dam
releases (Mullan et al., 1976). At first, it was thought that this
was the result of stratification in the reservoir (Holden, 1973;
Bureau of Reclamation, 1976; Mullan et al., 1976), but a recent
analysis indicates that the temperature change is due mainly to an
increase in flow (Holden and Crist, 1979). Releases from Flaming

Gorge Dam between 1962 and 1966 were relatively low to allow the reservoir to fill. These low flows were warmed by ambient conditions, so summer temperatures were increased from about 5°C to near 13°C shortly below the dam. After 1966, the releases were doubled and ambient warming was reduced to only a few degrees. This change in flow, and the resultant temperature change, reduced the tailwater trout fishery substantially and also eliminated successful reproduction of Colorado squawfish for 40 additional kilometers below the dam (Holden and Stalnaker, 1975; Holden, 1977; Holden and Crist, 1979).

The Bureau of Reclamation installed inlet modifications on Flaming Gorge Dam in 1978 to draw water from higher, warmer levels. This project was initiated to help the failing tailwater trout fishery. It is expected that the warmer summer flows will also bring back successful reproduction of Colorado squawfish to the 40 km of river mentioned. Studies are being conducted to monitor the effect of the inlet modification on both the trout fishery and the native warm-water species. If this project is successful, it may provide a methodology for alleviating negative temperature problems below other dams.

Changes in native fish populations in the Colorado River system caused by changes in turbidity-water chemistry or gas supersaturation have not been documented. The Colorado River historically fluctuated tremendously in turbidity and water chemistry, therefore, it is probable that the native species are fairly well adapted to extremes in these parameters. Gas supersaturation, caused by spillage over the dam, has occurred at Navajo Dam, with resulting kills of planted trout (W. McNall, pers. comm., New Mexico Game and Fish Department, Santa Fe).

The effect of reduced flows on the Colorado River fauna has not been adequately studied. The Colorado River Storage Project dams reduced flows significantly in all of the upper basin. Flows were reduced most drastically in the San Juan River and the Gunnison-Colorado rivers (Joseph et al., 1977), where mean monthly May flows were reduced to the lowest levels on record, even lower than the drought years of the 1930s. Flows in the Green River were not as severely reduced, primarily because of the natural inflow of the Yampa and White rivers, two of the three non-regulated large tributaries in the entire Colorado River system. Colorado squawfish still reproduce successfully in much of the Green River (Holden and Stalnaker, 1975; Holden, 1977; Holden and Crist, 1979), but reproduction in the Colorado River has not been documented since 1964 (Taba et al., 1965; Holden and Stalnaker, 1975; Seethaler, 1978). No studies have been conducted to determine reasons for this loss of reproduction. Temperatures in the Colorado River remain normal.

It appears possible that the major cause of the loss of reproduction is reduced flows and resulting ramifications.

The detrimental effect of low flows on reproduction of Colorado squawfish was recently noted in a portion of the Green River in Utah. Young-of-the-year squawfish had been found in the Green River between Jensen, Utah, and Desolation Canyon, a distance of about 160 km, in the early 1970s (Holden and Stalnaker, 1975). Studies in 1977, a drought year with low natural flows, located no young-of-the-year squawfish in that area, but numerous juveniles of the 1975 and 1976 year classes were found (Holden, 1977). A study in 1978, a normal water year, found young-of-the-year near Jensen, but the remaining area was not sampled (Holden and Crist, 1979). These data indicate that in this portion of the Green River, especially at Jensen, squawfish reproductive success is correlated with flow and that during a low-flow year (1977), reproduction was not successful, whereas it was successful during more normal- or high-flow years.

Glen Canyon Dam has altered the Colorado River in Grand Canyon from a warm, turbid river with tremendous seasonal flow variations to a cold, clear river that fluctuates 2-3 m daily due to power demands. The native fishes below Glen Canyon Dam have largely been replaced by stocked rainbow trout. Much of the change in the fish fauna can be attributed to the cold summer tailwater temperatures (Holden and Stalnaker, 1975), although the tremendous daily fluctuations must also be partly responsible. Spawning of most of the native suckers and minnows is presently restricted to tributaries. Juveniles of these species are seldom found in the mainstem. It appears probable that the ephemeral nature of the habitat, caused by the rapid, daily fluctuations, is the major factor preventing juvenile fishes from utilizing the Colorado mainstem.

The importance of delayed impacts of regulation on native Colorado basin fishes is difficult to assess. The major area where such factors may be important is the Colorado River below the inflow of the Gunnison River at Grand Junction, Colorado, over 160 km below the Curecanti Dams. As mentioned, Colorado squawfish no longer reproduce in this area, although adults are present. Several factors other than decreased flows are present that may be affecting the squawfish.

Exotic species, especially green sunfish (*Lepomis cyanellus*) and largemouth bass (*Micropterus salmoides*), are relatively common in this area (Kidd, 1977), much more so than in other areas of the upper Colorado Basin (Holden and Stalnaker, 1975). The primary reason for their abundance is the presence of flooded gravel pits that are joined to the river near Grand Junction, Colorado. These permanent backwaters afford the centrarchids with spawning and

rearing areas that are not found in undisturbed areas. The red
shiner (*Notropis lutrensis*), also an exotic, is very abundant in
the Colorado River, but is also very abundant in the Green River.
The channel morphology of the Colorado River in this area appears
to be somewhat entrenched compared to the lower Green River. This
observation is not based on any quantitative data and, therefore,
is the opinion of the author. Because of the great decrease in May
flows in this system, it appears probable that the channel morphology
may have changed over the last 10-15 years.

Therefore, it appears possible that a combination of factors,
such as low flows, abundant exotic competitors, predators, and
long-term habitat changes, may be working in combination to the
detriment of squawfish. Because our ability to determine successful
reproduction is limited to finding young-of-the-year, it is feasible
that sufficient habitat for fry is no longer available, therefore
leaving them very susceptible to exotic centrarchid predators.
Also, present information indicates that red shiners probably compete primarily with the young of native fishes and that in areas of
marginal habitat for the native juveniles, red shiners predominate
(Holden and Irvine, 1975; Holden, 1977). Therefore, it is also
possible that, in the Colorado River, altered flows have reduced
the amount of preferred habitat for native species to the point
where red shiners are favored.

The major reason for making these speculations is to suggest
that the regulation of rivers by dams may well cause more detrimental problems to riverine fishes than is presently known. It is
hoped that future investigations will include more in-depth studies
in sections of regulated rivers far below the dams and at time
periods fairly well removed from the initiation of the regulations.

One other factor that often affects fishes when dams are built
on rivers is the rather widespread practice of preimpoundment fish-
eradication programs. This has little to do with the ecology of
riverine fishes in regulated streams, except that these programs
may well cause the extirpation of a riverine species before impoundment. Eradication programs in the Colorado River Basin were
conducted at the Flaming Gorge Reservoir site and the Navajo Dam
site. Both were conducted to eradicate "rough fish," which were
primarily native species. No preimpoundment eradication program
was conducted in the Glen Canyon (Lake Powell) area. After closure
of Flaming Gorge and continuing until the present, large numbers of
exotic Utah chub (*Gila atraria*) have filled the reservoir and have
been a major management problem. In Navajo Reservoir, large numbers
of native roundtail chubs (*Gila robusta*) were abundant for several
years after formation, but have since declined in numbers drastically. At Glen Canyon, no increase in rough fish was noted. This
points out a fairly understandable ecological axiom: Obligatory

riverine fishes do not live in reservoirs. The eradication programs did little more than give exotic species a head start on game fishes. Where native riverine fish were not eradicated, they probably acted as a buffer against the buildup of unwanted species, allowing planted game fish populations a chance to become established. The downstream effects of eradication programs may create similar conditions in tailwaters, essentially leaving the habitat wide open for opportunistic exotic species.

CONCLUSIONS

As a concluding remark, it must be said that regulation of streams has had an enormous detrimental impact on riverine fishes, especially those species that require flowing water. The greatest impact is usually in those species that are fairly specialized, especially if they have a narrow temperature tolerance. The need in future research primarily revolves around predicting the effects of stream regulation on target species. This involves two types of studies. First, the habitat requirements of target species must be determined in detail. Second,, methods must be devised to predict physical, chemical, and biological changes in rivers due to damming. These kinds of data are necessary to predict impacts of regulation and will be valuable in attempts to correct existing problems.

LITERATURE CITED

Allen, K. R., 1969, Distinctive aspects of the ecology of stream fishes: A review, *J. Fish. Res. Board Can.*, 26:1429-1438.

Balon, E. K., 1978, Kariba: The dubious benefits of large dams, *Ambio*, 7(2):40-48.

Banks, R. L., Mullan, J. W., Wiley, R. W., and Dufek, D. J., 1974, "The Fontenelle Green River Trout fisheries--Considerations in its Enhancement and Perpetuation, Including Test Flow Studies of 1973," Fish Wildl. Serv., Salt Lake City, Utah.

Blake, R. F., 1977, The effect of the impoundment of Lake Kainji, Nigeria, on the indigenous species of mormyrid fishes. *Freshwater Biol.*, 7(1):37-42.

Bovee, K. D., 1978, "Probability-of-Use Criteria for the Family Salmonidae," Fish Wildl. Serv., Coop. Inst. Flow Serv. Group, Info. Pap. 4.

Bovee, K. D., and Cochnauer, T., 1977, "Development and Evaluation of Weighted Criteria, Probability-of-Use Curves for Instream Flow Assessments: Fisheries," Fish Wildl. Serv., Coop. Inst. Flow Serv. Group, Info. Pap. 3.

Branson, B. A., 1974, The American paddlefish: Signs of distress, *Natl. Parks Conserv. Mag.*, 48(1):21-23.

Collins, G. B., 1976, Effects of dams on Pacific salmon and steelhead trout, *Mar. Fish. Rev.*, 38(11):39–46.

Dominy, C. L., 1973, Recent changes in Atlantic salmon (*Salmo salar*)
runs in the light of environmental changes in the Saint John
River, New Brunswick, Canada, *Biol. Conserv.*, 5(2):105–113.

Ebel, W. J., 1969, Supersaturation of nitrogen in the Columbia
River and its effect on salmon and steelhead trout, *Fish. Bull.*,
68(1):1–11.

Edwards, R. J., 1978, The effect of hypolimnion reservoir releases
on fish distribution and species diversity, *Trans. Am. Fish.
Soc.*, 107(1):71–77.

Ford, M. E., 1963, Air injection for control of reservoir limnology,
J. Am. Water Works Assoc., 55:267–274.

Fraser, J. C., 1972, Regulated discharge and the stream environment,
p. 263–285, *in*: "River Ecology and Man," R. T. Oglesby,
C. A. Carlson, and J. A. McCann, eds., Acad. Press, New York.

Gottschalk, L. C., 1964, Reservoir sedimentation, p. 17-5 and 17-6,
in: "Handbook of Applied Hydrology," V. T. Chow, ed.,
McGraw-Hill, New York.

Hamilton, M. P., and Carothers, S. W., Co-directors, 1978, "Fish,
Wildlife and Habitat Assessment, San Juan River, New Mexico
and Utah," VTN Consolidated, Inc., and the Museum of Northern
Arizona.

Hatch, M. D., 1978, Fishes, p. E-1 to E-58, *in*: "Handbook of Species
Endangered in New Mexico," New Mexico Dep. Game Fish, Santa Fe.

Haynes, J. M., Gray, R. H., and Montgomery, J. C., 1978, Seasonal
movements of white sturgeon (*Acipenser transmontanus*) in the
mid-Columbia River, *Trans. Am. Fish. Soc.*, 107(2):275–280.

Hazel, C. R., 1976, The reservation of instream flow for fish in
California--a case study, *in*: "Proceedings, Instream Flow
Needs, Vol. II," J. F. Orsborn and C. H. Allman, eds., Am.
Fish. Soc., Bethesda, Maryland.

Holden, P. B., 1973, "Distribution, Abundance and Life History of
the Fishes of the Upper Colorado River Basin," Ph.D. Diss.,
Utah State Univ., Logan.

Holden, P. B., 1977, "A Study of the Habitat Use and Movement of
the Rare Fishes in the Green River from Jensen to Green River,
Utah, August and September, 1977," BIO/WEST, Logan, Utah,
PR-13-1.

Holden, P. R., and Crist, L. W., 1979, "Documentation of Changes in
the Macroinvertebrate and Fish Populations in the Green River
Due to Inlet Modification of Flaming Gorge Dam," BIO/WEST,
Logan, Utah, PR-16-2.

Holden, P. B., and Irvine, J., 1975, "A Study of the Aquatic Fauna
and Flora of Escalante Canyon, Utah," Final Rep. to Natl. Park
Service, Denver, Colorado.

Holden, P. B., and Stalnaker, C. B., 1975, Distribution and abundance of mainstream fishes of the middle and upper Colorado River Basins, 1967-1973, *Trans. Am. Fish. Soc.*, 104(2):217-231.

Hubbs, C., 1972, Some thermal consequences of environmental manipulation of water, *Biol. Conserv.*, 4(3):185-188.

Jordan, D. S., 1891, Report of explorations in Colorado and Utah during the summer of 1889, with an account of the fishes found in each of the river basins examined, *U.S. Fish. Comm. Bull.*, 9:1-40.

Joseph, T., Sinning, J., Behnke, R., and Holden, P., 1977, "An Evaluation of the Status, Life History, and Habitat Requirements of Endangered and Threatened Fishes of the Upper Colorado River System," FWS/OBS Rep. No. 24.

Kidd, G., 1977, "An Investigation of Endangered and Threatened Fish Species in the Upper Colorado River as Related to Bureau of Reclamation Projects," Final Rep. to Bureau Reclam., Grand Junction, Colorado.

Kimsey, J. B., 1957, Fisheries problems in impounded waters of California and the lower Colorado River, *Trans. Am. Fish. Soc.*, 87:39-57.

Koryak, M., 1976, The influence of mainstem navigation dams on water quality and fisheries in the upper Ohio River Basin, p. 158-173, *in*: "Proceedings, Instream Flow Needs, Vol. II," J. F. Orsborn and C. H. Allman, eds., Am. Fish. Soc., Bethesda, Maryland.

Kroger, R. L., 1973, Biological effects of fluctuating water levels in the Snake River, Grand Teton National Park, Wyoming, *Am. Midl. Nat.*, 89(2):478-481.

Lewis, D. S. C., 1974, The effects of the formation of Lake Kainji (Nigeria) upon the indigenous fish population, *Hydrobiologia*, 45(2-3):281-301.

Long, C. S., and R. F. Krcma, 1969, Research on a system for bypassing juvenile salmon and trout around low-head dams, *Comm. Fish. Rev.*, 31(6):27-29.

McDonald, J. R., and Hyatt, R. A., 1973, Supersaturation of nitrogen in water during passage through hydroelectric turbines at Mactaguac Dam, *J. Fish. Res. Board Can.*, 30:1392-1394.

Miller, R. R., 1946, *Gila cypha*, a remarkable new species of cyprinid fish from the Colorado River in Grand Canyon, Arizona, *J. Wash. Acad. Sci.*, 36:409-415.

Miller, R. R., 1961, Men and the changing fish fauna of the American Southwest, *Pap. Mich. Acad. Sci., Arts, Lett.*, 46:365-404.

Minckley, W. L., 1973, "Fishes of Arizona," Ariz. Game Fish Dep., Phoenix.

Minckley, W. L., and Deacon, J. E., 1968, Southwestern fishes and the enigma of "endangered species," *Science*, 159:1424-1432.

Mullan, J. W., Starostka, V. J., Stone, J. L., Wiley, R. W., and Wiltzius, W. J., 1976, Factors affecting upper Colorado River reservoir tailwater trout fisheries, p. 405-428, *in*: "Proceedings, Instream Flow Needs, Vol. II," J. F. Orsborn and C. H. Allman, eds., Am. Fish. Soc., Bethesda, Maryland.

Nichols, P. R., 1968, "Passage Conditions and Counts of Fish at the Snake Island Fishway, Little Falls Dam, Potomac River, Maryland, 1960-63," Fish. Wildl. Serv. Spec. Rep. 565.

Nickelson, T., 1976, Development of methodologies for evaluating instream flow needs for salmonid rearing, p. 588-596, *in*: "Proceedings, Instream Flow Needs, Vol. II," J. F. Orsborn and C. H. Allman, eds., Am. Fish. Soc., Bethesda, Maryland.

Orsborn, J. F., and Allman, C. H., eds., 1976, "Proceedings, Instream Flow Needs, Vols. I and II," Am. Fish. Soc., Bethesda, Maryland.

Park, D. L., and Farr, W. E., 1972, Collection of juvenile salmon and steelhead trout passing through orifices in gatewalls of turbine intakes at Ice Harbor Dam, *Trans. Am. Fish. Soc.*, 101(2):381-384.

Petr, T., 1978, Tropical man-made lakes--their ecological impacts, *Arch. Hydrobiol.*, 8(3):368-385.

Pfitzer, D. W., 1963, "Investigations of Waters Below Large Storage Reservoirs in Tennessee," Dingel-Johnson Rep., Proj. F-1-R, Tenn. Game Fish Comm.

Pflieger, W. L., 1975, "The Fishes of Missouri," Miss. Dep. Conserv., Jefferson City.

Raymond, H. L., 1968, Migration rates of yearling chinook salmon in relation to flows and impoundments in the Columbia and Snake rivers, *Trans. Am. Fish. Soc.*, 97(4):356-359.

Raymond, H. L., 1969, Effect of John Day Reservoir on the migration rate of juvenile chinook salmon in the Columbia River, *Trans. Am. Fish. Soc.*, 98(3):513-514.

Raymond, H. L., 1976, Effect of dams and river regulation on runs of anadromous fish to the mid-Columbia and Snake rivers, p. 444-465, *in*: "Proceedings, Instream Flow Needs, Vol. II," J. F. Orsborn and C. H. Allman, eds., Am. Fish. Soc., Bethesda, Maryland.

Robinson, W. L., 1978, The Columbia: A river system under siege, *Oreg. Wildl.*, 33(6):3-7.

Seethaler, K., 1978, "Life History and Ecology of the Colorado Squawfish (*Ptychocheilus lucius*) in the Upper Colorado River Basin," Unpubl. M.S. Thesis, Utah State Univ., Logan.

Shikhshabekov, M. M., 1971, Resorption of the gonads in some semi-diadromous fishes of the Arakum Lakes (Dagestan ASSR) as a result of the regulation of discharge, *J. Ichthyol.*, 11(3):427-431.

Smith, F. E., 1976, Water development impact on fish resources and
    associated values of the Trinity River, California, p. 98-111,
    in: "Proceedings, Instream Flow Needs, Vol. II," J. F. Orsborn
    and C. H. Allman, eds., Am. Fish. Soc., Bethesda, Maryland.
Smith, P. W., 1979, "The Fishes of Illinois," Univ. Illinois Press,
    Illinois St. Nat. Hist. Surv., Chicago.
Spence, J. A., and Hynes, H. B. N., 1971, Differences in fish
    populations upstream and downstream of a mainstream impound-
    ment, *J. Fish. Res. Board Can.*, 28:45-46.
Stalnaker, C. B., and Arnette, J. L., eds., 1976, "Methodologies
    for the Determination of Stream Resource Flow Requirements:
    An Assessment," U.S. Fish Wildl. Serv., Off. Biol. Serv.,
    West. Water Allocation.
Taba, S. S., Murphy, J. R., and Frost, H. H., 1965, Notes on the
    fishes of the Colorado River near Moab, Utah, *Proc. Utah Acad.
    Sci., Arts, Lett.*, 42(2):280-283.
Trautman, M. B., and Gartman, D. K., 1974, Re-evaluation of the
    effects of man-made modifications on Gordon Creek between 1887
    and 1973 and especially as regards its fish fauna, *Ohio J.
    Sci.*, 74(3):162-173.
Trefethen, P., 1972, Man's impact on the Columbia River, p. 77-98,
    in: "River Ecology and Man," R. T. Oglesby, C. A. Carlson,
    and J. A. McCann, eds., Acad. Press, New York.
U.S. Bureau of Reclamation, 1976, "Negative Determination of
    Environmental Impact, Penstock Modification, Flaming Gorge
    Dam, Utah," Salt Lake City, Utah.
U.S. Department of the Interior, 1977, Endangered and threatened
    wildlife and plants, republication of list of species, *Fed.
    Reg.*, 42(135):36420-36431.
Vanicek, C. D., Kramer, R. H., and Franklin, D. R., 1970, Distri-
    bution of Green River fishes in Utah and Colorado following
    closure of Flaming Gorge Dam, *Southwest. Nat.*, 14:297-315.
White, R. J., Hansen, E. A., and Alexander, G. A., 1976, Relation-
    ship of trout abundance to stream flow in midwestern streams,
    p. 597-615, *in*: "Proceedings, Instream Flow Needs, Vol. II,"
    J. F. Orsborn and C. H. Allman, eds., Am. Fish. Soc., Bethesda,
    Maryland.
Williams, J. D., and Finnley, D. K., 1977, Our vanishing fishes:
    Can they be saved, Acad. Nat. Sci., Philadelphia, *Frontiers*,
    Summer 1972.
Zakharyan, G. B., 1972, The natural reproduction of sturgeons in
    the Kura River following its regulation, *J. Ichthyol.*,
    12(2):249-259.

CHEMICAL MODIFICATIONS IN RESERVOIR-REGULATED STREAMS

Herbert H. Hannan

Aquatic Station
Southwest Texas State University
San Marcos, Texas   78666

INTRODUCTION

It has long been recognized that as stream water is impounded
in reservoirs, the water chemistry often changes before it becomes
outflow at the dam (Wiebe, 1938; Love, 1961; Symons, 1969). This
modification, in both the quality and quantity of chemical conditions, varies widely in the same reservoir over time and from one
reservoir to another. The economic and ecological complexity and
significance of these and other changes have resulted in several
symposia and literature reviews on reservoirs (Symons et al., 1964;
U.S. Public Health Service, 1965; Lowe-McConnel, 1966; Lane, 1967;
Elder et al., 1968; Symons, 1969; Obeng, 1969; Hall, 1971; Ridley
and Symons, 1972; Ackermann et al., 1973; Hill and Summerfelt,
1974; Cairns et al., 1978; Driver and Wunderlich, 1978). These
symposia along with several comprehensive studies on specific
reservoirs over time, symposia on eutrophication (Rohlich, 1969;
Likens, 1972), and current books on limnology (Wetzel, 1975;
Golterman, 1975) have provided for a better understanding of the
hydrological and limnological concepts controlling reservoir
dynamics.

It is evident from these works that a reservoir must be considered as a dynamic trophic system in which natural and regulated
flow dynamics interact with thermal stratification and community
metabolism to control the chemical dynamics within the impoundment
and chemical composition of the outflow. The factors controlling
the chemical dynamics are often as important to the chemical conditions at the outflow as the quantities and qualities of the material
in the sources supplying the reservoir. Chemical changes at the

outflow are generally accentuated if the water in the reservoir becomes thermally stratified (Ridley and Symons, 1972).

This review deals with deep-storage reservoirs that thermally stratify during the summer. An attempt is made to provide a generalized understanding of the limnological and hydrological factors associated with impoundment behavior that results in changes in the chemical conditions downstream from the reservoir.

CHEMICAL COMPOSITION OF RESERVOIR WATERS

Most research concerning the chemical composition of reservoirs has involved dissolved inorganic ions and compounds. The major cations (Ca^{++}, Mg^{++}, Na^+, and K^+) and anions (HCO_3^-, $CO_3^=$, $SO_4^=$, and Cl^-) constitute most of the salinity of natural fresh waters. The minor and essential trace ions or elements are those that limit plant growth. Many of the dissolved inorganic ions and elements have been separated into nonconservative (Ca^{++}, Mn^{++}, Fe^{++}, Si, NO_3^-, H^+, NO_2^-, NH_4^+, $SO_4^=$, HCO_3^-, and $PO_4^=$) and conservative (Na^+, K^+, Mg^{++}, and Cl^-) categories based on whether their concentrations are changed or not changed, respectively, by biological metabolism. The essential trace elements with other metals such as Ag, Cd, Hg, Al, and Pb constitute the heavy metals. .

Recently, the importance of dissolved organic matter in controlling the dynamics of lake metabolism (Wetzel, 1975) and of particulate organic matter and dissolved organic matter to the ecology of lotic waters (Cummins, 1975) has been theorized. These limnological concepts and theories, though formulated from lake and river studies, are applicable to reservoirs. The chemistry of natural waters has been discussed in recent books on limnology by Golterman (1975) and Wetzel (1975).

TYPES OF RESERVOIRS

There are many reservoirs throughout the world that differ greatly in flow dynamics, morphometry, physiocochemical, and biological characteristics. There are, however, basic differences that should be considered when studies of reservoirs are undertaken. Two general types of reservoirs exist. The off-stream, or closed, reservoirs have no natural inflow or outflow. They usually receive pumped water from a nearby lotic system and are typically more saline, since they are more modified by evaporation than open systems. The on-stream, or open, reservoirs are located on streams so that the discharge of the stream must pass through the impoundment. Open reservoirs are often conveniently classified as main-stream (run-of-the-river), transitional, and deep-storage impoundments. Most of the impounded water in main-stream reservoirs is restricted to the old river channel, whereas in deep-storage reservoirs it is extended well beyond the original river bed, thereby forming numerous coves and inlets. The transitional reservoirs are intermediate between main-stream and deep-storage reservoirs. All three types may or may not exhibit direct thermal stratification (Lockett, 1977). Generally, permanent summertime thermal stratification is exhibited only in deep-storage reservoirs.

FACTORS INFLUENCING CHEMICAL CONDITIONS IN RESERVOIR-REGULATED STREAMS

The chemical composition of stream water entering a reservoir usually undergoes significant chemical change during impoundment, and these changes can be observed in the outflow at the dam. These chemical changes have been attributed to a variety of factors that are typically related in some way to the action or an interaction of natural or regulated inflow and/or outflow, thermal stratification, density currents, overturn, and biological activities (Table 1). Little attention has been given to the latter factor. Most of the published and available field studies on the influence of impoundment on the physicochemical conditions in reservoirs and at outflows have emphasized the effect on temperature, dissolved oxygen (DO), and specific conductance.

Detailed reservoir studies on impoundment behavior are beyond the scope of this review. Three factors that have been shown by field studies to influence the chemical conditions at the outflow of deep-storage reservoirs will be discussed. Selected field studies of several reservoirs are then cited to illustrate these controlling conditions. It is hoped that some understanding of what occurs within a deep-storage reservoir can be developed and applied to the study of ecological conditions downstream from a reservoir.

General Flow Dynamics and Chemical Conditions

Several different physical events that cause variation in the chemistry of stream water entering a reservoir occur in a drainage basin. A watershed event, such as rainfall, may cause hourly (Westerdahl et al., 1975) and monthly (Hannan and Young, 1974; Hannan and Broz, 1976) changes in the chemistry of the inflowing water. Other watershed events that cause temporal variation in water chemistry include agricultural runoff (Prochazkova et al., 1973), effluent from domestic sewage and industrial pollution (Young et al., 1972; Hannan et al., 1973), and seepage from groundwater (Segura, 1978).

Table 1.   Principal Factors Influencing Chemical Conditions of
           Reservoir Outflow

Factors	Reference
Density flow; reduction in velocity of inflow; evaporation; wind movements; dissolution or precipitation of mineral species; biological activity	Love, 1961
Age of impoundment; extent and duration of thermal stratification; frequency of density currents; depth of water release through or over dams; operational objectives; extent of drawdown and refill; types of release structure	Neel, 1963
Water depth; extent of shallows; storage volume in relation to quantity of inflow; length to width ratio; depth of water withdrawal; orientation of axis with the prevailing wind direction; characteristics of the soils in the reservoir flow area; quality of incoming water; climate conditions of rainfall and temperature; depth of reservoir drawdown; whether or not stratification exists	Sylvester, 1968
Geometry and size of reservoir; storage management of the yearly cycle; outlet geometry and location; amount, distribution, and quality of the inflow; internal mixing processes; optical properties of the water; climate of the environment; heat and mass transfer process across the water surface and the ground; biological and chemical processes in the water	Wunderlich and Elder, 1973

The inflowing river water, with its varied chemical composition, will either mix and become homogeneous with the reservoir water or, in the presence of density gradients, will maintain its identity in a separate layer within the reservoir. The retention time, the route taken by the density flow through the reservoir, and the location of the outlet through which the water is discharged determine to a considerable degree the spatial and temporal distribution of the chemical constituents of water within the impoundment and its quality and quantity in the dam outflow (Wunderlich, 1971). The density flow may move through the reservoir or be stored at the surface as an overflow, at an intermediate depth as an interflow, or at the bottom as an underflow, as determined by the density of the layers in relation to that of the reservoir water (Ackermann et al., 1973). Vertical mixing or movement of the stratified water is often suppressed, while horizontal movement is enhanced (Wunderlich, 1971).

The location of the outlet at the dam and the level of the density current within the reservoir determines which water becomes outflow. For example, if the overflow, interflow, or underflow is at the same elevation as the outlet, relatively little change, except that caused by sedimentation, will occur in the chemical conditions of the inflow by physical processes before the water is discharged from the reservoir. If, however, overflow occurs in a reservoir with a deep-water outlet, the hypolimnetic zone is drawn off and the overflow becomes part of a deepening epilimnion throughout thermal stratification during the summer. If the rate of withdrawal is high, then the water impounded during the prior winter or spring is rapidly drawn off and the warmer epilimnetic spring or summer water could reach the bottom and be withdrawn. If, however, the rate of withdrawal is low, a portion of the spring or winter water may remain in the hypolimnion throughout most of the summer months or even until fall overturn (Hannan, Cole, and Wiedenfeld, unpublished data).

The combination of inflow and outlet types and discharges result in a variety of flow patterns (Wunderlich, 1971). Additional variations in flow and chemical conditions are the result of natural or man-made inner dams (inundated dam within a reservoir) and water outlets (Fiala, 1966; Vick, 1976). Such structures and outlets result in layers with restricted water movement (akinetic spaces) in the hypolimnion. For example, an outlet some distance upriver from the main dam within the body of the reservoir will often develop an akinetic space between the outlet and the dam, whereas an inner dam will develop an akinetic space on its upriver side.

It must be realized that currents do not always move down-reservoir (Ebel and Koski, 1968). Sometimes they vary with seasons (Anderson and Pritchard, 1951). This situation appears more common

when two different streams of different density flow into the same
reservoir. These streams may be naturally flowing streams (Eley
et al., 1968) or have a source from an upstream reservoir (Churchill,
1947). Under certain conditions of relative water temperature,
three separate and distinct levels of water movement can occur in a
cross-section. Mid-depth density currents have been found to flow
upreservoir, whereas surface and bottom currents moved downreservoir
(Churchill, 1947).

The history of the inflowing water and the path of the flow in
the reservoir can directly determine the quality and quantity of
conservative ions at the outlet if surface evaporation, groundwater
exchange, and sedimentation are low. The longitudinal and vertical
conditions of the nonconservative ions in the reservoir are, however, altered by community metabolism. The biological influence on
chemical conditions within the reservoir is often greater than the
influence of other factors combined.

The reader interested in a more complete analysis of reservoir
hydrodynamics is referred to Love (1961), Wunderlich and Elder
(1967), Ebel and Koski (1968), Wunderlich (1971), and Wunderlich
and Elder (1973).

Flow and Biologically Induced Chemical Conditions

Biologically induced chemical changes occur within the three
different vertical layers of a reservoir that result from thermal
stratification. Photosynthetically induced decalcification and
nutrient depletion often occur within the epilimnion. Both supersaturation and depletion of DO and associated chemical changes
occur in the metalimnion. These changes are caused by community
metabolism associated with organisms trapped within the mid-depth
zone or by respiration associated with an interflow from upreservoir
(Gordon, 1978; Segura, 1978). Hypolimnetic chemical changes are
associated with community respiration.

The spatial and temporal occurrence of these biologically
induced chemical conditions within the reservoir and their appearance at the outlet are controlled by natural inflow, regulated
outflow, thermal stratification, density currents, and overturn.
An interaction of these physicochemical and biological factors
result in longitudinal and vertical profiles of chemical conditions,
which are at times vastly and often progressively different from
one end of the reservoir to the other.

Often the trend throughout the year is for the greatest amount
of metabolic activity and sedimentation to occur upreservoir,
resulting in a progressive downreservoir decrease in concentration

of the different chemicals. These changes in longitudinal and vertical conditions are not always associated with the often-used temperature isotherms. For example, an upreservoir site may have a vertical temperature profile of the same shape as a downreservoir site but have considerably different chemical conditions. This is of special importance in predicting chemical conditions at the outflow.

Pattern of Hypolimnetic Dissolved-Oxygen Depletion

In reservoirs, the pattern of DO depletion is highly variable horizontally, vertically, and seasonally (Straskraba et al., 1973). There is, however, a pattern of DO depletion that is common to many reservoirs (Table 2) and which has been found to have a profound influence on chemical conditions at the outlet (Hannan et al., in press, a; Segura, 1978). This pattern shows that once a hypolimnion is established it will soon become anoxic in the upreservoir reach. As the summer continues, hypolimnetic anoxia develops in a down-reservoir progression until it reaches the dam or until overturn occurs. By the time the lower end of the reservoir becomes anoxic, the upreservoir region can become oxygenated. This oxygenation of the upreservoir region is caused by inflow and by the deepening of the epilimnion as a result of hypolimnetic withdrawal.

The downreservoir development of the anoxic hypolimnion is the result of an interaction of progressively larger amounts of hypolim-netic DO per unit of surface area downreservoir, autochthonous plant production throughout the reservoir in the photogenic zone, benthic oxygen demand (especially in the sediment zone), and down-reservoir movement of anoxic water from upstream as a result of drawdown at the outlet. Downward vertical migration of water also occurs, which transports photosynthetically produced organic matter to the anoxic zone as summer progresses.

As the DO approaches zero and as anoxic conditions appear in the hypolimnion of impoundments, changes occur in the chemical conditions (HCO_3^-, O_2, NO_3^-, NO_2^-, NH_4^+, Fe^{++}, Mn^{++}, $PO_4^\equiv$, $SO_4^=$, H_2S, pH, CH_4) associated with redox processes. The rate of withdrawal for a specific reservoir and relative redox and associated water chemistry of the different ions and compounds would determine the time of year the anoxic water and various differing chemical conditions would reach the dam. This is obvious in the differences in the time of year H_2S odor is prevalent at the outlet of many reservoirs. The longer it takes the anoxic conditions to get to the dam, the more time will have elapsed for the chemical reactions to have occurred, and the greater the concentration of most of the noncon-servative ions. During a drought year, the anoxic water may never

Table 2.   Reservoirs in Which Hypolimnetic Anoxic Conditions Have
           Been Observed to Develop in the Riverine End before the
           Dam End

Reservoir	Reference
Norris Reservoir, Tennessee	Wiebe, 1938; Dendy, 1945
Williamette River, Oregon	Fish and Wagner, 1950
Klicava Reservoir, Czechoslovakia	Fiala, 1966
Brownlee Reservoir, Oregon-Idaho	Ebel and Koski, 1968
Long Lake, Washington	Soltero et al., 1974a
Bighorn Lake, Montana-Wyoming	Soltero et al., 1974b
DeGray Reservoir, Arkansas	Nix, 1974
Center Hill Reservoir, Tennessee	Gnilka, 1975
Lake Kariba, Zambia-Rhodesia	Bowmaker, 1976
Cherokee Reservoir, Tennessee	Gordon and Nicholas, 1977; Iwanski et al., 1978
Canyon Reservoir, Texas	Hannan et al., in press, b

become part of the outflow. If the anoxic water moves quickly to
the dam, some chemical changes might not have time to take place.

This pattern of DO depletion is probably common to many reservoirs fed by unregulated streams. The time of development would
range from months for long oligotrophic reservoirs with small
allochthonous input and a great downreservoir hypolimnetic volume
to a few hours for shallow eutrophic reservoirs with large
allochthonous input and a small downreservoir hypolimnetic volume.

Case Studies

Norris Reservoir, Tennessee (Wiebe, 1938). Norris Reservoir
is a deep-storage reservoir with a deep outflow. Wiebe's work on
Norris Reservoir is a classic study that contributes greatly to our
understanding of impoundment behavior and chemical conditions at
the outlet of a dam. Hydrodynamics of the reservoir were correlated
with DO and temperature throughout the length of the reservoir.
Wiebe observed the interaction of natural inflow to the reservoir
and drawdown at the dam with the deepening of the epilimnion and
thermocline and noted that these events were related to the development of the anoxic zone from upreservoir to the dam. Although his
data included only observations on temperature and DO, he suggested
that those factors would influence free CO_2, pH, and H_2S. Wiebe's
findings were confirmed in a subsequent study by Dendy (1945) on

the effect of flow, temperature, and DO on fish distribution in
Norris Reservoir.

Fontana Reservoir, North Carolina (Dendy and Stroud, 1949).
Large, deep reservoirs exhibiting density currents not only change
the water chemistry downstream of rivers but often dominate the
water chemistry at the outlets of a series of reservoirs. Fontana
Reservoir, a large, deep impoundment with a deep-water outlet had
such a great influence on downstream Cheoah and Calderwood Reservoirs that outlet water at times passed through both downstream
reservoirs with little change in temperature and DO. Although
reservoirs in series in different locales are expected to have
different water chemistry at their outlets, the underlying principles are the same and should be considered in determining chemical
conditions.

Both the time of year the anoxic zone reaches the dam at the
most upstream reservoir and the lag time for the water to reach the
downstream reservoirs are key factors in determining chemical
condition at these reservoir outlets.

Douglas Reservoir, South Carolina (Ingols, 1959). Changes in
the chemical composition at the outlet of a dam may be related to
the events of a previous year. Managanese was extremely high in
the river below Douglas Reservoir during a year of normal flow but
was not found the previous year or the two subsequent years. It
was postulated that lush vegetation produced on the exposed flats
of the reservoir bottom during the previous dry year caused an
excessive organic-matter load in the hypolimnion when the vegetation was submerged by water during the spring following a dry
season. The increased organic matter caused a depletion of hypolimnetic DO, resulting in the solution of Mn from the sediment.
When the mud flats were not exposed with subsequent flooding, Mn
occurred in the outflow later in low concentrations or not at all.

Brownlee Reservoir, Oregon - Idaho (Ebel and Koski, 1968).
Brownlee Reservoir, a deep-storage impoundment with an outlet
approximately one-third of the depth below the surface, was studied
in a two-year project, including a detailed field study on factors
affecting current throughout the reservoir.

Isopleths of physicochemical conditions measured were restricted
to DO, temperature, and conductivity. These figures showed that
during both years DO depletion occurred upreservoir first and
developed toward the dam throughout the summer months. It appeared
that at least part of the zone of DO depletion left the bottom and
moved upward toward the outlet as it moved downreservoir. Later in
the summer an anoxic akinetic space developed below the outlet.

Planktonic productivity was high during the summertime as indicated by a high pH, high phenolphthalein alkalinity, and low HCO_3^- alkalinity. These conditions, together with anoxia in the hypolimnion, caused typical vertical stratification characteristic of eutrophic conditions. Physicochemical conditions changed at the outlet, depending upon surface and subsurface currents being significantly affected by changes in reservoir level and volumes of inflow and outflow.

Bighorn Lake, Montana-Wyoming (Soltero et al., 1973, 1974b). Bighorn Lake is a deep-storage impoundment with a mid-depth outlet. Inflow water drains from sedimentary rocks. The major cations in the impoundment were $Ca^{++} > Na^+ > Mg^{++} > K^+$ and the major anions were $SO_4^= > HCO_3^- > Cl^-$.

Dissolved solids at the dam decreased slightly from levels found in the inflow. The downstream decrease was attributed to rainfall and runoff from snow-fed tributaries (Soltero et al., 1973). The reservoir served as a sink for Fe, Mn, Cu, and Zn. The major cause of the decrease was attributed to the sedimentation of suspended load in the inflow (Soltero et al., 1973).

Bighorn Lake acted as a sink for some of the plant nutrients (Soltero et al., 1973). Twenty-five percent of the total nitrogen in the inflow was retained in the reservoir. Concentrations of NO_3^- increased and NH_4^+ decreased at the outlet compared to concentrations found in the inflow. This change in the form of nitrogen was attributed to nitrification and nitrogen fixation being in excess of nitrogen assimilation and denitrification. Eighty-six percent of the total phosphorus in the inflow was retained in the reservoir. Although the reservoir was a sink for both nitrogen and phosphorus, the concentration of the essential nutrients in the outflow were high enough to be considered non-limiting for plant production.

The study of Soltero et al. (1974b) was conducted over a three-year period. One year an interflow was established that was not evident the other two years of the study. A turbidity current was associated with the internal density current. Nitrate and orthophosphate maxima and DO and pH minima occurred in the interflow. This nutrient-rich intermediate-level withdrawal apparently resulted in algal growths and increased fish production downstream.

The low DO concentrations and associated chemical conditions of the internal density layer were attributed to respiratory processes *in situ* or the downreservoir pattern of anoxic development. The values of pH and DO in the density layer were similar to those of the stagnant hypolimnetic water at the upreservoir station, which suggested that the DO decrease and CO_2 increase occurred while on the bottom of the reservoir at the upreservoir station and subsequently moved to the dam with the density current (Soltero et al., 1974b). This demonstrates the importance of the upreservoir-downreservoir progresɛion of hypolimnetic DO change to chemical conditions of an intermediate-level withdrawal reservoir. Had DO depletion occurred at the upreservoir station and the pH lowered, other chemical changes, such as an increase in dissolved Fe and Mn, elevated H_2S, and a change in the nitrogen species and $PO_4^=$, could have occurred in the interflow and outflow.

Long-Lake, Washington (Soltero et al., 1974a). Long Lake is a deep-storage reservoir with an intermediate-level outflow. A density current modified the longitudinal and vertical distribution of physicochemical conditions. The Spokane River spring runoff of low salinity flowed throughout the entire depth of the reservoir to the outflow. Inflow decreased and salinity increased during the summer, causing a more dense downreservoir high salinity interflow to be discharged at the intermediate-level penstock.

The summer stratification isolated an akinetic space below the penstock, which then became anoxic. During the period of anoxia, phosphate-phosphorus and nitrogen maxima were present in the hypo-limnion. Total inorganic nitrogen concentrations were lower, while organic nitrogen was higher, in the discharge as a result of assimilation by the biota.

During and after fall overturn, DO concentrations and pH values of the surface waters were significantly lowered because of mixing with stagnant bottom waters. Orthophosphate increased during the fall overturn as a result of the higher concentration present in the anoxic zone mixing with the epilimnetic water. There was an orthophosphate depletion in the upper water during the summer growing season. A pattern, similar to that of $PO_4^=$, was shown to exist for NO_3^-. These changes were evident in the discharge from the reservoir.

Chemical conditions at the outflow of Long Lake were the result of the complex interaction of density flow, thermal strati-fication, seasonal inflow and outflow, and summertime biologically induced epilimnetic nutrient decreases as well as hypolimnetic ion solubilization followed by dilution at fall overturn.

Cherokee Reservoir, Tennessee (Gordon and Nicholas, 1977). Cherokee Reservoir, located on the Holston River, is a deep-storage impoundment with a deep outlet. Its hypolimnetic DO depletion rate is the highest of all Tennessee Valley Authority reservoirs. The Holston River is greatly enriched by large quantities of industrial and domestic wastes.

Hypolimnetic DO depletion first occurred in the upreservoir reach and developed downreservoir to the dam as the summer progressed. Periodic heavy inflows of well-aerated water dissipated part of the anoxic conditions in the upreservoir reach. Without such inflows, the anoxic zone probably would have developed throughout the length of the reservoir in a few weeks. Soon after thermal stratification formed a more dense hypolimnion, the DO depletion rate increased.

Inflow to the Cherokee Reservoir was high in NH_4^+ and exhibited considerable DO demand. Seasonal trends showed the NH_4^+ concentrations decreased concomitantly with the time span of DO depletion due to nitrification. After the DO was completely exhausted, the NH_3 concentration increased to high levels during the anaerobic processes as a result of release from the bottom sediment, anaerobic deamination of organics, and from under- and inter-flows during the summer. Almost all of the NH_4^+ formed in the anoxic hypolimnion was oxidized during fall overturn.

Retention time, temperature, and the downreservoir progression of hypolimnetic anoxic depletion all influenced NH_4^+ concentration at the outlet. During the winter, with a short retention time and cold temperatures, nitrifying bacteria had little time to develop at any significant rate; therefore, the NH_4^+ passed through the reservoir unmodified. Outflow was halted during the spring to permit the lake to fill to summer pool levels. Longer retention time and increased temperature permitted bacteria to develop, which subsequently enhanced nitrification and DO demand. Both NH_4^+ and DO gradually reduced to zero over time at the outlet and remained depleted until NH_4^+ reappeared, generated under anoxic conditions.

Hebgen Lake and Quake Lake, Montana (Martin and Arneson, 1978). This study was to test Wright's (1967) hypothesis that water bodies with surface-water outflows tend to trap nutrients, whereas impoundments with deep-water outflows dissipate nutrients. The study impoundments were Hebgen Lake and Quake Lake which are located on the upper Madison River in southwestern Montana. The region is a mountainous pine forest on a bedrock of welded turf and rhyolite. Hebgen Lake is a deep, bottom-draining reservoir and Quake Lake is a natural, deep lake with a surface discharge.

The major cations of the impoundments were $Na^+ > Ca^{++} > K^+ > Mg^{++}$ and the major anions were $HCO_3^- > Cl^- > SO_4^=$. Silica was high in both lakes. Typical summertime clinograde temperature conditions were present in both lakes, but minimum DO concentrations were around 3 ppm near the bottom.

The importance of outflow depth on the chemical composition of downstream waters is shown in this study. There was an epilimnetic depletion and a hypolimnetic accumulation of NO_3^- in both lakes as summer progressed. This seasonal pattern was directly related to direct thermal stratification and fall overturn. Salinity increased in the hypolimnion of both lakes during this time.

The lake with the surface outflow discharged low nutrient and low salinity water, whereas the reservoir with the deep outlet discharged high nutrient and high salinity water during thermal stratification. The nutrients and increased salinity trapped in the hypolimnion of the natural lake would mix with the epilimnion during overturn and increase their concentration in the outflow.

Selected Tennessee Valley Authority Reservoirs (Higgins, 1978). This study shows the annual effect of impoundment on the chemical conditions of the parent river. Inflow-outflow trends were studied for several TVA reservoirs.

The average annual load of total dissolved solids, total nitrogen, total phosphorus, and 5-day biochemical oxygen demand leaving Cherokee Reservoir all decreased each year over a 10-year period compared to the load entering the reservoir. Total dissolved solids, total nitrogen, and total phosphorus for five TVA reservoirs for a one-year period were greater in the inflow than in the outflow.

These comprehensive studies substantiate the trends reported by other workers that reservoirs often act as sinks for many organic and inorganic chemicals and, therefore, that concentrations of these chemicals are reduced in the outflow.

Canyon Reservoir, Texas (Hannan and Young, 1974; Hannan and Broz, 1976; Hannan et al., in press, a, b; Segura, 1978; Ralph, 1978; Barrows, 1978). Canyon Reservoir is an oligo-mesotrophic deep-storage reservoir, with a bottom outlet, located on the Guadalupe River. The major cations of the impoundment were Ca^{++} > Mg^{++} > Na^+ > K^+ and the major anions were HCO_3^- > $SO_4^=$ > Cl^-. Studies comparing annual inflow data with outflow data have shown that of 21 physicochemical parameters, 17 were modified at the outlet as compared to concentrations of the inflow (Hannan and Young, 1974; Hannan and Broz, 1976; Segura, 1978). Annual spatial and temporal profiles of the length of the reservoir for selected parameters (Ca^{++}, Na^+, K^+, Mg^{++}, Mn^{++}, Fe^{++}, HCO_3^-, pH, specific conductance, DO, temperature) show the influence of impoundment on the chemical conditions in the outflow (Segura, 1978; Ralph, 1978; Hannan et al., in press, b).

Flow dynamics controlled the minor spatial and temporal differences in concentrations found in conservative cations (Na^+, K^+, Mg^{++}) within the reservoir and in the outflow at the dam (Ralph, 1978). An interaction of natural inflow, regulated outflow, thermal

stratification, pattern of hypolimnetic DO depletion, overturn, and
community metabolism caused pronounced longitudinal and vertical
chemical stratification in nonconservative (Ca^{++}, HCO_3^-, H^+, Mn^{++},
Fe^{++}, O_2) chemical conditions within the reservoir (Segura, 1978;
Hannan et al., in press, b). These chemical changes within the
reservoir were reflected in the outflow over time.

The hypolimnion becomes anoxic during the summer months.
Anoxia first occurs in the upreservoir and riverine reaches during
late spring or early summer and develops toward the dam as summer
progresses.

Only traces of dissolved Mn (DMn) and dissolved Fe (DFe) were
found in the epilimnion. DMn first appeared in the hypolimnion
during May, when DO was around 4 mg/liter. Both DMn and DFe subsequently appeared in high concentrations associated with the downreservoir progression of hypolimnetic DO depletion. DMn in low
concentrations appeared first and was attributed to the desorption
of Mn from Mn and Fe oxides. The subsequent increases in both DMn
and DFe were attributed to dissolution from the sediment. DMn
appeared in increased concentrations before the occurrence of DFe
and preceded DFe downreservoir, since it has a greater redox potential, thus allowing more time to solubilize from the sediment. It
was apparent that the appearance of DMn followed by DFe in the outflow at the dam was the result of an interaction of desorption,
relative redox potentials as related to the pattern of hypolimnetic
DO depletion and drawdown, and overturn.

Calcium, HCO_3^-, and specific conductance profiles all had about
the same shape and reflected the same biologically induced chemical
changes. A discussion of Ca dynamics was used to identify flow and
biologically induced chemical changes for all three parameters.
Calcium decreased in the epilimnion and increased in the hypolimnion
throughout the length of the reservoir during summer stratification.
Epilimnetic decalcification was attributed to photosynthetic uptake
of CO_2 and HCO_3^-, thus increasing pH, which precipitated $CaCO_3$.
Increased CO_2 from respiration and associated pH changes in the
anoxic hypolimnion enabled HCO_3^- to be solubilized.

A progressive downreservoir decrease in epilimnetic Ca from
the riverine reach to the dam was observed most months of the year.
The greatest decrease always occurred in the riverine region during
the growing season. This decrease has been attributed to sedimentation resulting from a decrease in velocity. A comparison of Ca
concentration from filtered and unfiltered samples indicated that
some suspended Ca was determined as DCa. However, the amount was
too small to account for the change in Ca between upreservoir and
downreservoir regions. Primary productivity was greatest per unit
volume in the riverine reach when the Ca concentration was most

reduced. The decrease in Ca level with distance downreservoir was, therefore, attributed to the precipitation of $CaCO_3$, as the phytoplankton used HCO_3^- as a carbon source, mostly in the riverine reach. The reservoir water, now lower in Ca, moved downreservoir while continually being influenced by a lower rate of photosynthesis and a longer residence time. At overturn, the lower concentration of Ca in the epilimnion at the dam end of the reservoir diluted the higher concentration that occurred in the hypolimnion during anoxia. The result was continued downreservoir decrease in Ca during the winter months.

Calcium, HCO_3^-, DMn, and DFe occurred in the anoxic zone of the hypolimnion during thermal stratification as the water moved downreservoir toward the dam. The maximum concentrations that occurred mid-way between the riverine reach and the dam during later summer did not reach the outflow before fall overturn. Increased concentrations of DMn occurred in the outflow about two months before the occurrence of DFe, whereas high Ca concentrations did not reach the dam before overturn. Studies by Hannan and Young (1974) on the different nitrogen species of Canyon Reservoir indicated that NO_3^- decreased but NH_3 and NO_2^- increased in the outflow if the anoxic zone developed to the dam during the thermal stratification. The change of NO_3^- to the more reduced state was due to denitrification occurring under the anoxic conditions in the hypolimnion.

RESEARCH NEEDS

Comprehensive field studies on reservoirs that will couple limnological principles developed from lake and stream studies to flow dynamics (e.g., thermal stratification, natural inflow, regulated or natural outflow, longitudinal and vertical mixing, density flow) are needed. Such studies are essential to determine spatial and temporal variability in chemical conditions within reservoirs that eventually occur in outflows. These studies need to include chemicals that have been little researched or have been ignored. Practically nothing is known about the components of dissolved organic matter, and relatively little is known about heavy metals as essential trace elements. Still less is known about the effect of allochthonous and metabolically produced organic complexing agents on chemical conditions in the outflow of reservoirs.

Reservoirs often receive high concentrations of allochthonous organic matter from inflowing streams and from a relatively high autochthonous riverine planktonic and littoral-zone macrophytic production. There is a need to determine the changes that occur in particulate organic matter and dissolved organic matter from the riverine and plant sources to the outflow, since the trophic relationships control community structure downstream (Cummins, 1975;

Simmons and Voshell, 1978). The form and chemical composition of total solids is regulated in part by such impoundment behavior as the development of anoxic zones in the metalimnion and hypolimnion. For example, degradation of detrital organic matter is normally reduced under anoxic conditions (Wetzel, 1975).

REFERENCES

Ackermann, W. C., White, G. F., and Worthington, E. G., eds., 1973, "Man-Made Lakes: Their Problems and Environmental Effects," Am. Geophys. Union, Washington, D.C., 846 p.

Anderson, E. R., and Pritchard, D. W., 1951, "Physical Limnology of Lake Mead," Rep. 258, U.S. Naval Electronics Laboratory, San Diego, California, 152 p.

Barrows, D., 1978, "The Trophic Status of a Deep-Storage Reservoir in Central Texas," M.S. Thesis, Southwest Texas State Univ., San Marcos, Texas, 49 p.

Bowmaker, A. P., 1976, The physico-chemical limnology of the Mwenda River mouth, Lake Kariba, *Arch. Hydrobiol.*, 77:66-108.

Cairns, J., Benfield, E. F., and Webster, J. R., eds., 1978, "Current Perspectives on River-Reservoir Ecosystems," Proc. Symp., N. Am. Benth. Soc., Columbia, Missouri, 85 p.

Churchill, M. A., 1947, Effect of density currents upon raw water quality, *J. Am. Water Works Assoc.*, 39:357-360.

Cummins, K. W., 1975, The ecology of running waters; theory and practice, p. 277-293, *in*: "Proc. Sandusky River Basin Symp. Int. Comm. Great Lakes," Heidelberg Coll., Tiffin, Ohio.

Dendy, J. A., 1945, Depth distribution of fish in relation to environmental factors, Norris Reservoir, *J. Tenn. Acad. Sci.*, 20:114-131.

Dendy, J. S., and Stroud, R. H., 1949, The dominating influence of Fontana Reservoir on temperature and dissolved oxygen in the Little Tennessee River and its impoundments, *J. Tenn. Acad. Sci.*, 24:41-51.

Driver, E. E., and Wunderlich, W. O., eds., In press, "Environmental Effects of Hydraulic Engineering Works, Proceedings of a Symposium," Tenn. Valley Authority, Norris, Tennessee.

Ebel, W. J., and Koski, C. H., 1968, Physical and chemical limnology of Brownlee Reservoir, *Fish. Bull.*, 67:295-335.

Elder, R. A., Krenkel, P. A., and Thackston, E. L., eds., 1968, "Proceedings of the Specialty Conference on Current Research into the Effects of Reservoirs on Water Quality," Dep. Environ. Water Resour. Eng. Rep. No. 17, Vanderbilt Univ., 390 p.

Eley, R. L., Carter, N. E., and Dorris, T. C., 1968, Physicochemical limnology and related fish distribution of Keystone Reservoir, p. 333-357, *in*: "Reservoir Fishery Resources Symposium," Am. Fish. Soc., Washington, D.C.

Fiala, L., 1966, Akinetic spaces in water supply reservoirs, *Verh. Int. Verein. Limnol.*, 16:685-692.

Fish, F. F., and Wagner, R. A., 1950, "Oxygen Block in the Mainstream of the Williamette River," U.S. Fish and Wildl. Serv. Spec. Sci. Rep. No. 41, 19 p.

Gnilka, A., 1975, Some chemical and physical aspects of Center Hill Reservoir, Tennessee, *J. Tenn. Acad. Sci.*, 50:7-11.

Golterman, H. L., 1975, "Physiological Limnology," Elsevier Sci. Publ. Co., New York, 489 p.

Gordon, J. A., 1978, "Definition of Dissolved Oxygen Depletion Mechanisms Operating in the Metalimnia of Deep Impoundments," Water Resources Research Center Res. Rep. No. 63, Univ. Tennessee, 110 p.

Gordon, J. A., and Nicholas, W. R., 1977, "Effects of Impoundments on Water Quality: Observations of Several Mechanisms of Dissolved Oxygen Depletion," Div. Environ. Plan., Tenn. Valley Authority, Chattanooga, Tennessee, 55 p.

Hall, G. E., ed., 1971, "Reservoir Fisheries and Limnology," Am. Fish. Soc., Washington, D.C., 511 p.

Hannan, H. H., Barrows, D. B., Fuchs, I. R., Segura, R. D., and Whitenberg, D. C., In press, a, Limnological and operational factors affecting water quality in Canyon Reservoir, Texas, *in*: "Environmental Effects of Hydraulic Engineering Works," E. E. Driver and W. O. Wunderlich, eds., Tenn. Valley Authority, Norris, Tennessee.

Hannan, H. H., and Broz., L., 1976, The influence of a deep-storage and an underground reservoir on the physicochemical limnology of a permanent central Texas river, *Hydrobiologia*, 51:43-63.

Hannan, H. H., Fuchs, I. R., and Whitenberg, D. C., In press, b, Spatial and temporal patterns of temperature, alkalinity, dissolved oxygen, and conductivity in an oligo-mesotrophic deep-storage reservoir in central Texas, *Hydrobiologia*.

Hannan, H. H., and Young, W. J., 1974, The influence of a deep-storage reservoir on the physicochemical limnology of a central Texas river, *Hydrobiologia*, 44:177-207.

Hannan, H. H., Young, W. C., and Mayhew, J. J., 1973, Nitrogen and phosphorous in a stretch of the Guadalupe River with five mainstream impoundments, *Hydrobiologia*, 43:419-441.

Higgins, J. M., 1978, "Water Quality Progress in the Holston River Basin," Water Quality and Ecology Branch, Div. Environ. Plan., Tenn. Valley Authority, Chattanooga, Tennessee, 43 p.

Hill, L. G., and Summerfelt, R. C., eds., 1974, "Oklahoma Reservoir Resources," *Okla. Acad. Sci.*, Stillwater, Oklahoma, 151 p.

Ingols, R. S., 1959, Effect of impoundment on downstream water quality, Catawba River, S.C., *J. Am. Water Works Assoc.*, 51:42-46.

Iwanski, M. L., Bruggink, D. J., and Shipp, J. W., 1978, "Field
    Research on the Effects of Impoundment on Water Quality,"
    Report by Water Quality and Ecology Branch, Tenn. Valley
    Authority, Chattanooga, Tennessee, 12 p.
Lane, C. E., Chairman, 1967, "Reservoir Fishery Resources Symposium,"
    Am. Fish. Soc., Washington, D.C., 569 p.
Likens, G. E., ed., 1972, "Nutrients and Eutrophication: The
    Limiting-Nutrient Controversy," Am. Soc. Limnol. Oceanogr.,
    Ann Arbor, Michigan, 328 p.
Lockett, C. L., 1976, "Classification of Seventeen Central Texas
    Reservoirs," M.S. Thesis, Southwest Texas State Univ.,
    San Marcos, Texas, 45 p.
Love, S. K., 1961, Relationship of impound to water quality, *J.
    Am. Water Works Assoc.*, 53:559-568.
Lowe-McConnel, R. H., ed., 1966, "Man-Made Lakes," Acad. Press,
    London and New York, 218 p.
Martin, D. B., and Arneson, R. D., 1978, Comparative limnology of a
    deep-discharge reservoir and a surface-discharge lake on the
    Madison River, Montana, *Freshwater Biol.*, 8:33-42.
Neel, J. K., 1963, Impact of reservoirs, p. 575-593, *in*:
    "Limnology in North America," D. G. Frey, ed., Univ. Wisconsin
    Press, Madison.
Nix, J., 1974, "Distribution of Trace Metals in a Warm Water
    Release Impoundment," Arkansas Water Resources Research
    Center, Fayetteville, Arkansas, 377 p.
Obeng, L., ed., 1969, "Man-Made Lakes: The Accra Symposium," Ghana
    Univ. Press, Accra, 398 p.
Prochazkova, L., Straskrabova, V., and Popovsky, J., 1973, Changes
    of some chemical constituents and bacterial numbers in Slapy
    Reservoir during eight years, *Hydrobiol. Stud.*, 2:83-154.
Ralph, J., 1978, "The Spatial, Temporal, and Mass Balance Dynamics
    of Magnesium, Sodium, and Potassium in a Deep-Storage Reser-
    voir in Central Texas," M.S. Thesis, Southwest Texas State
    Univ., San Marcos, Texas, 57 p.
Ridley, J. E., and Symons, J. M., 1972, New approaches to water
    quality control in impoundments, p. 389-412, *in*: "Water
    Pollution Microbiology," R. Mitchell, ed., John Wiley and
    Sons, Inc., New York.
Rohlich, G. A., Chairman, 1969, "Eutrophication: Causes, Conse-
    quences, Correctives," Nat. Acad. Sci., Washington, D.C.,
    661 p.
Segura, R. D., 1978, "Non-conservative Cation Dynamics in a Deep-
    Storage Reservoir in Central Texas," M.S. Thesis, Southwest
    Texas State Univ., San Marcos, Texas, 52 p.
Simmons, G. M., and Voshell, J. R., 1978, Pre- and post-impoundment
    benthic macroinvertebrate communities of the North Anna River,
    p. 45-61, *in*: "Current Perspectives on River-Reservoir Eco-
    systems," J. Cairns, E. F. Benfield, and J. R. Webster, eds.,
    Proc. Symp. N. Am. Benth. Soc., Columbia, Missouri.

Soltero, R. A., Gasperino, A. F., and Graham, W. G., 1974a, Chemical and physical characteristics of a eutrophic reservoir and its tributaries: Long Lake, Washington, *Water Res.*, 8:419-433.

Soltero, R. A., Wright, J. C., and Horpestad, A. A., 1973, Effects of impoundment on the water quality of the Bighorn River, *Water Res.*, 7:343-354.

Soltero, R. A., Wright, J. C., and Horpestad, A. A., 1974b, The physical limnology of Bighorn Lake-Yellowtail Dam, Montana: Internal density currents, *Northwest Sci.*, 48:107-124.

Straskraba, M., Hrbacek, J., and Javornicky, P., 1973, Effects of an upstream reservoir on the stratification conditions in Slapy Reservoir, *Hydrobiol. Stud.*, 2:7-82.

Sylvester, R. O., 1968, Discussion on effects of reservoirs on water quality, p. 285-287, *in*: "Specialty Conference on Current Research into the Effects of Reservoirs on Water Quality," R. A. Elder, P. A. Krenkel, and E. L. Thackston, eds., Vanderbilt Univ., Dep. Environ. Water Resour. Eng. Rep. No. 17.

Symons, J. M., 1969, "Water Quality Behavior in Reservoirs," U.S. Public Health Service Publ. 1930, Cincinnati, Ohio, 616 p.

Symons, J. M., Weibel, S. R., and Robeck, S. R., 1964, "Influence of Impoundments on Water Quality," Public Health Service, U.S. Dep. Health, Education, and Welfare, Cincinnati, Ohio, 78 p.

U.S. Public Health Service, 1965, "Symposium on Streamflow Regulation for Quality Control," Public Health Service Publ. No. 999-WP-30, Cincinnati, Ohio, 420 p.

Vick, H. C., Hill, D. W., Bruner, R. J., Barnwell, T. O., Raschke, R. L., and Gentry, R. E., 1976, "West Point Lake Postimpoundment Study," Surveillance and Analysis Division, Environmental Protection Agency, Region IV, Athens, Georgia, 89 p.

Westerdahl, H. E., Perrier, E. R., Thornton, K. W., and Eley, R. L., 1975, "DeGray Lake and Caddo River Field Studies for Water Quality and Ecological Model Development," U.S. Army Eng. Waterways Exp. Stn., Misc. Pap. 0-75-10, Vicksburg, Mississippi, 6 p.

Wetzel, R. G., 1975, "Limnology," W. B. Saunders Co., Philadelphia, Pennsylvania, 743 p.

Wiebe, A. H., 1938, Limnological observations on Norris Reservoir with special reference to dissolved oxygen and temperatures, p. 440-457, Trans. 3rd N. Am. Wildl. Conf.

Wright, J. C., 1967, Effect of impoundments on productivity, water chemistry, and heat budgets of rivers, p. 188-199, *in*: "Reservoir Fishery Resources Symposium," Am. Fish. Soc., Washington, D.C.

Wunderlich, W. O., 1971, The dynamics of density-stratified reservoirs, p. 188-189, *in*: "Reservoir Fisheries and Limnology," G. E. Hall, ed., Spec. Publ. No. 8, Am. Fish. Soc., Washington, D.C.

Wunderlich, W. O., and Elder, R. A., 1967, The mechanics of strati-
     fied flow in reservoirs, p. 56-58, *in*: "Reservoir Fishery
     Resources Symposium," Am. Fish. Soc., Washington, D.C.
Wunderlich, W. O., and Elder, R. A., 1973, Mechanics of flow
     through man-made lakes, p. 300-310, *in*: "Man-Made Lakes:
     Their Problems and Environmental Effects," W. C. Ackermann,
     G. F. White, and E. B. Worthington, eds., Am. Geophys. Union,
     Washington, D.C.
Young, W. C., Hannan, H. H., and Tatum, J. W., 1972, The physico-
     chemical limnology of a stretch of the Guadalupe River, Texas,
     with five main-stream impoundments, *Hydrobiologia*, 40:297-319.

EFFECTS OF STREAM REGULATION ON CHANNEL MORPHOLOGY

Daryl B. Simons

College of Engineering; Colorado State University
Fort Collins, Colorado 80523

INTRODUCTION

"The erosion cycle – detachment, transport, deposition – is
described by the Second Law of Thermodynamics, which states each
system tends to move in the direction of lowest energy. Because
of the potential energy of soil (rock) in elevated positions and
because of the kinetic energy of flowing water, all soil and water
will move, eventually, to the lowest possible level, i.e., the ocean
deeps, or to some temporary, intervening basin" (McHenry, 1964, 1969,
1974). The intervening basins, destined to be filled with sediment,
include the reservoirs that we have become so dependent upon. We
daily enjoy the benefits they provide, not fully realizing that they
are continually filling and that, in the not so distant future, many
will become useless. It is possible to accelerate or decelerate this
process, but the filling is inevitable, given current knowledge. The
reservoir's life is dependent on sediment deposition. The understanding of, and the ability to predict, both channel and reservoir
response processes are extremely important. The rate and magnitude
of changes experienced by watersheds and rivers depend on many
naturally imposed factors, including floods, changes in the duration
of specific flows, droughts, and changes in the water and sediment
supply imposed by earthquakes, vertical changes in the earth's crust,
floods, and mass wasting.

Similarily, man can impose many factors that affect water and
sediment supply and consequently the short- and long-term response
of the watershed and rivers. Some of the most important factors
causing changes are:

1)   Changing land-use practices:

    a)   Grazing
    b)   Timber harvesting
    c)   Clearing of land
    d)   Altering the surface geometry of the land (terracing)
    e)   Irrigation
    f)   Construction of roads

2)   Altering natural flood conditions in the rivers:

    a)   Diversions
    b)   Dams
    c)   River-training works
    d)   Navigation
    e)   Power releases
    f)   Modified discharge
    g)   Flood control
    h)   Land development

The foregoing factors not only change the flow conditions but also can cause great increases in the flow sediment, which in turn can

1)   Alter channel and watershed geometry
2)   Fill existing and proposed reservoirs with sediment, rendering them useless in relatively short periods of time
3)   Fill diversion systems, including desilted works, with sand and coarse materials
4)   Reduce the quality of water for irrigation, hydro-power production, municipal use, and fisheries.
5)   Cause aggradation in the river channel, which increases river stage, flood-damage potential, the likelihood of failure of bridges and roads, and the possibility of inundation of riparian lands.
6)   Cause channel degradation, which accelerates erosion and may initiate bridge and road failure.

There are many other responses, some favorable but most adverse, to the production of excess sediments that affect watersheds and rivers. However, the foregoing help identify the many problems associated with sediment and resource-development projects of all types. While there are numerous references on sediment effects and transport in reservoirs, few consider the changes in channel morphology due to stream regulation. Leopold et al. (1964) provided a classic base upon which to discuss fluvial processes. Shen (1971) includes discussion on sediment transport, aggradation, degradation, and their relations with channel changes and reservoir construction.

Other references on reservoir sedimentation are Pemberton (1978) and Brown (1958). Buma and Day (1977) detail channel morphology changes in addition to providing specific data on a regulated stream in Ontario. Keeley (1978) discusses water-quality aspects and environmental problems in control and response of reservoirs and waterways. Geomorphic and hydraulic analysis assists with the evaluation of probable problems and responses. This method of analysis is presented in detail by Simons and Senturk (1977).

DATA REQUIREMENTS

Analysis, design, and implementation require knowledge of the physical processes affecting the watersheds and river systems (Simons and Senturk, 1977). This knowledge exists as theory, experimentally derived relations, and experience. In all cases, certain data are required to validate existing theories, concepts, and experimental relations. Also, working with basic field data is the primary process by which one gains useful experience.

Because certain basic data are necessary to all forms of analysis, it is essential to identify present and long-range data requirements to ensure a basis for an acceptable analysis of watershed and river systems. It is also necessary to analyze the response of these complex, dynamic environments to alternate development and utilization schemes.

Data-Storage and -Retrieval Systems

Some governmental agencies, such as the U.S Geological Survey National Weather Data Exchange (NAWDEX), EPA's STORET, and the U.S. Weather Bureau, maintain large data-storage and -retrieval systems.

For the analysis of a river system, it would be beneficial to develop a comprehensive data-storage and -retrieval system for data management, analysis, and design. This system should be formulated to accommodate present and long-range requirements and should include at least: climatic, hydrological, hydraulic, water quality, soils, geology controls, and channel and watershed geometry data. Such a system was developed to complement the development of the Yazoo River Basin for the Vicksburg District, U.S. Army Corps of Engineers (Simons et al., 1978).

METHODS OF ANALYSIS

From the preceding discussion, note that river systems are complex and difficult to control. Water-resource development adds

to the complexity and alters the behavior of these systems. The
analysis and design of water resources development projects can be
evaluated as follows:

1)   Quantitative geomorphic, hydrologic, and hydraulic
     analysis. This method gives approximate values use-
     ful for planning and provides an important check on
     more comprehensive methods of analysis and design.
2)   Conventional analysis coupled with the input and checks
     provided by method (1).
3)   Comprehensive analysis utilizing data-storage and
     -retrieval systems coupled with mathematical models.
     This approach is identical to (2) but additionally
     provides a means of looking in more detail at a larger
     number of alternatives. Also, this method more effec-
     tively utilizes the total data base available. For a
     state of the art review of the utilization of mathe-
     matical modeling techniques, refer to Simons and Li
     (1978).

UTILIZATION OF MATHEMATICAL MODELS FOR WATER AND SEDIMENT ROUTING

Mathematical models are currently widely used for planning,
analysis, and design of water-resources projects. Many benefits
can be derived from such an approach.

First, only modular, physical process-oriented water- and
sediment-routing models are suggested. These models are based
upon selected relationships that can be modified to better model a
particular environment. The details of physical-process modeling,
as applied to the analysis of watersheds and river systems, are
given in detail by Simons and Li (1979). Specific suggested uses
are to

1)   Use a comprehensive water- and sediment-routing model
     to identify the most important variables that must be
     measured to assure accurate modeling. This is a speci-
     fic means of checking the present and long-term valid-
     ity of the data-collection program for both watersheds
     and rivers.
2)   Conduct a sensitivity analysis to determine variables
     of primary, secondary, and negligible importance. This
     method of analysis is extremely valuable in setting pri-
     orities for both short-term and long-term data-collection
     programs.

3) Incorporate statistical and time-series methods of
    analyzing the basic data as a part of the data-storage
    and -retrieval system. In fact, routing-analysis and
    -design problems can significantly reduce the time and
    expense of analysis and design.

4) Use the selected and modified water- and sediment-
    routing models to analyze various alternatives. Such
    models provide an efficient method for analyzing both
    the short- and long-term response of a part of the sys-
    tem, or even the total system, to various alternatives.
    More specifically:

    a) Evaluate rates of aggradation or degradation in
        specified reaches of channel. By routing both
        water and sediment, it is possible to identify rates
        of change as a function of time and space. This
        provides an excellent means of evaluating the
        effects of the embankments, the spurs, the diver-
        sions, and the barrage on aggradation.

    b) Evaluate various reservoir sites. By this
        procedure, it is possible to route the water and
        sediment into the reservoir and determine
            --Loss of reservoir volume as a function of
              time
            --Aggradation in the upstream channel
            --Degradation downstream from the reservoir
        By this means, single reservoirs or combinations of
        reservoirs can be evaluated to determine the best
        management practice.

    c) The watershed model can be used to evaluate the
        effects of changing land-use practices on water and
        sediment yield. By this means afforestation, irri-
        gation, and grazing can be evaluated to determine
        their impacts on water and sediment yield.

IMPACTS OF IMPOUNDMENTS

The construction of dams to impound and control the flow of
water and sediment may significantly modify the system. Important
interactions related to impoundment development and operation are
identified in Fig. 1. The impacts of impoundments can be analyzed
by considering (1) the response of the river channel and tributaries
upstream from the impoundment, (2) the physical processes within
the impoundment, and (3) the response of the river channel and
tributaries downstream from the impoundment. Each of these categories, though interdependent, are considered separately as
follows.

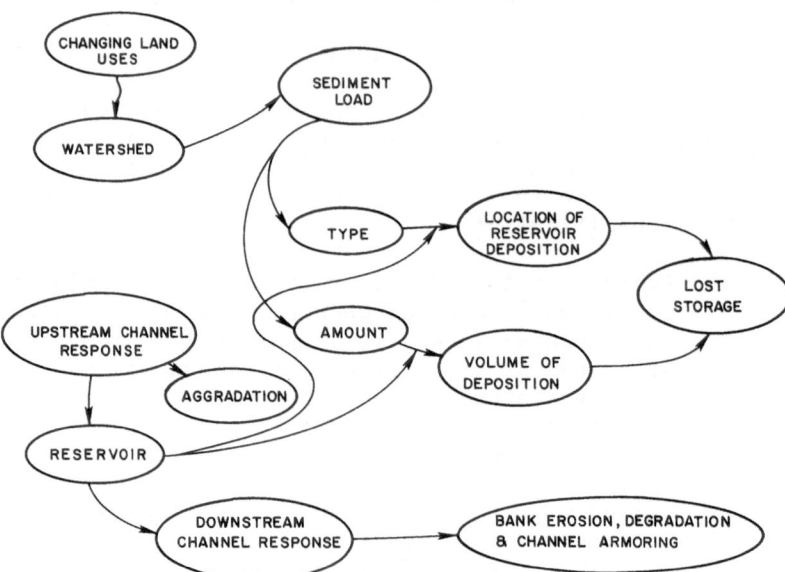

Fig. 1. Watershed/reservoir interaction

Impacts on Channels and Tributaries Upstream from Reservoirs

1)   The operation of the reservoir continuously alters
      reservoir stage and the backwater conditions imposed
      on the immediate upstream channel and its tributaries.
2)   The impoundment and the backwater effect of the im-
      poundment reduces the velocity of the flow, increases
      depth of flow, and causes deposition of bed material
      size sediments.
3)   The depositions of sediments causes general aggradation
      of the upstream channel and its tributaries. However,
      as reservoir stage is varied, the sequence of both ag-
      gradation and degradation is induced. Also, the zone of
      activity varies with water discharge.
4)   The deposition of sediment upstream from the reservoir,
      although variable with space and time, results in a
      general increase in river-bed elevation, which in turn
      increases flood stage and the potential for increased
      flood damage.

5) The increase in river-bed elevation and water level in
    the upstream channel may result in increased seepage and
    an increase in the elevation of the water table. This
    can affect riparian land use, riparian vegetation, and
    drainage requirements.
6) The increase in river stage caused by the impoundment
    will cause an increase in stage in the tributaries,
    induce aggradation, and affect riparian vegetation and
    land use.
7) The aggradation in the river and its system of tribu-
    taries will reduce the effective waterway beneath bridges
    and, in the extreme, can result in their failure.

Dynamics of Reservoirs

1) Depending upon reservoir size, volume, and geometry,
    varying percentages of the total inflowing sediment
    discharge are trapped. Large reservoirs will trap
    almost one hundred percent of the inflowing sediment.
    Smaller reservoirs will trap much smaller percentages
    of the total load, particularly the finer sediment
    sizes consisting of clays and silts. In fact, these
    clay-silt sizes can be carried significant distances
    within reservoirs in the form of density currents
    (Rice, 1979).
2) The inflowing sediment reaching a more tranquil flow
    environment will begin to deposit in the reservoir,
    (see Fig. 2). The coarsest particles will deposit in
    the channel upstream from the impoundment in the back-
    water reach and in the upper end of the reservoir,
    forming a delta. The finer sands and silt will deposit
    farther into the reservoir. The finer silts and clays
    require much less current and turbulence to keep them
    in suspension and may or may not be trapped, depending
    on the size of the reservoir, its geometry, method of
    operation, and the system of gates for release of
    water and sediment in the dam. The coarser sediments
    that deposit to form a fan-shaped delta can be periodi-
    cally gutted by the formation of a channel through the
    deposit when reservoir stage is low. In general, the
    depositional pattern is a function of the quantity and
    quality of the sediment, the discharge into the reser-
    voir, the reservoir operation, the surface area and
    geometry of the reservoir, aquatic growth, and the impact
    of boat- and wind-generated waves on suspension and
    transport of sediment. Note that most irrigation, hydro-
    power, and flood-control reservoirs are sufficiently large
    to trap almost one hundred percent of the inflowing
    sediment.

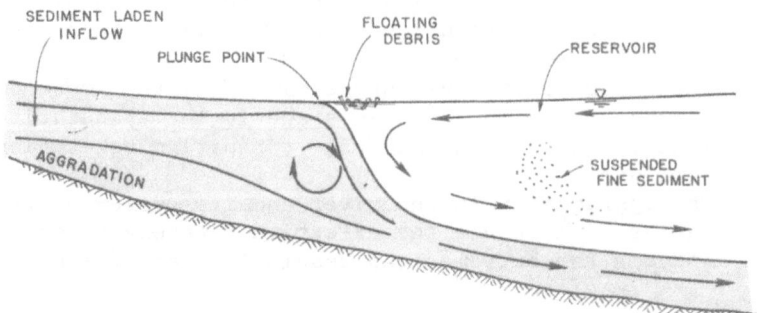

Fig. 2.   Idealized reservoir deposition.

3)   In the design of reservoirs, it is common to provide
     storage space for water that is classified as a) live
     storage and b) dead storage. In some instances a rec-
     reational pool is maintained. The dead storage is often
     assumed to provide storage space for sediment. Because
     this space is usually designated as the storage below
     some specified reservoir stage, it is essential to rec-
     ognize that it may not serve its intended purpose.
     The coarse sediment is deposited in the live-storage
     pool. The dead-storage volume is often the last part
     to fill with sediment. Hence, the determination of
     required reservoir volume and height of dam must be
     calculated from the quantity and quality of inflowing
     sediment and the rate of loss of live storage resulting
     from deposition of sediment.

4)   Increasing the water table causes water-logging and
     drainage problems.

5)   The useful life of reservoirs can, in some cases, be
     extended by using special measures, such as:

     a)   Changes in land-use practices
     b)   Stabilization of channels
     c)   Minimizing the opportunity for mass watering
     d)   Controlling peripheral development of roads,
          homes, septic tanks, and recreational use
     e)   Construction of off-channel storage
     f)   Dredging sediment
     g)   Routing sediment through a reservoir

Impacts on Channels and Tributaries Downstream from Reservoirs

1) Impoundments trap inflowing sediment, resulting in the release of essentially clear water (sediment-hungry water) into the downstream system (Leopold et al. 1964).

2) The relatively clear water derives its equilibrium load by entraining bed sediments and eroding river banks.

3) The erosion induced by the release of clear water results in degradation of the channel, which, if uncontrolled, can endanger the dam and even bridges and diversion works and cause failure of downstream structures, such as river-training works (Buma and Day 1977). Degradation is often limited by the development of an armor layer of particles too large to be transported by the existing hydraulic conditions. Finer particles, such as sand, may thus be overlain by a protective armor layer of cobbles or rubble (Fig. 3).

Fig. 3. Example of armoring in a river channel.

4)   Degradation of the channel increases the gradient of
      local tributaries.
5)   With increased gradients the tributaries are subjected
      to increases in velocities, degradation, bank erosion,
      and, possibly, head cutting. This new source of sediment
      may reduce, and even reverse, the local aggradation of the
      main channel that was induced by the release of clear
      water.
6)   Channel stabilization may be required in the main
      channel and tributaries to protect riparian land and
      adjacent developments.
7)   Degradation of the channel system may reduce the water
      table in flood plains.
8)   Changes in the river channel and improved drainage may
      cause a change in vegetation.
9)   The release of clear water may induce an enviroment
      favorable to the growth of algae and aquatic plants.
10)   The water released from the dam may be significantly
      colder than normal and devoid of oxygen, which may sig-
      nificantly alter the downstream environment for sev-
      eral kilometers.
11)   The consequences of eliminating the silts and clays
      from the downstream river system can be very signifi-
      cant. The effects include increased seepage, decreased
      channel stability, decreased moisture-holding capacity
      of newly formed banks (which may not support vegetation),
      and increased growth of aquatic plants.

MASS WASTING

     Mass wasting is an important physical process that disrupts
flow in river channels and canals, contributes to sediment yield,
destroys roads and bridges, and disrupts hillside farming activities. Mathematical models have been developed that identify
mass wasting-prone areas and indicate possible remedial actions
to reduce the sediment production and associated hazards (Ward
et al., 1979). It is suggested that

     1)   Current physical-process mass-wasting models should
           be evaluated
     2)   Selected models should be used to identify basic
           data requirements for implementation
     3)   A methodology should be formulated that relates mass-
           wasting hazards to construction of dams, roads, bridges,
           and diversions

Mass wasting can be greatly accelerated around the shoreline of reservoirs. Principal factors include saturation of the reservoir banks, drawdown, seismic activity, and changing land-use practices. Mass wasting in this environemnt needs particular attention. For example, large landslides can greatly reduce the storage capacity of a reservoir, but, even more important, it may cause failure or partial failure of the impounding structure. Such potential failures must be carefully evaluated because of the hazard to those who use the downstream river valley.

THE RESPONSE OF A RIVER SYSTEM AND INTERVENING REACHES OF CHANNEL TO THE CONSTRUCTION OF MAJOR RESERVOIRS

In the analysis of river systems, it is common to find water-resources development projects that involve the construction and use of major reservoirs in a series. Such a case is outlined in Fig. 4. The reservoir identified as A has been constructed on a major river. Reservoir B, farther downstream, is planned for subsequent development. Location C is a city that will be affected by the backwater from reservoir B. With this general background, consider the responses of the system to the development. To begin with, reservoir A is large. It has a capacity on the order of 14,000 hm^3 of water, although it stores only a small percent of the flow discharged from its watershed annually, as illustrated by the hydrograph. The hydrograph indicates that approximately five times the volume of the reservoir runs down the river system each year. All of the sediment carried into the reservoir is trapped, and the quantity of water that is desilted is equal to approximately five times the volume of the reservoir, about 62,000 hm^3 per year. The sediment load carried by the river is average to large, which results in a large volume of sediment being deposited within reservoir A each year. This deposition can significantly decrease the storage capacity of the reservoir for hydro power, irrigation, and flood control in a relatively short time. Furthermore, essentially clear water is released downstream from the dam. The channel is comprised of materials ranging from fine sand, silts, and clays to coarse gravel. The fact that there are coarser materials in the bed of the stream will tend to limit the degradation induced downstream from the dam by the release of clear water. Nevertheless, it is essential not only to evaluate the rate at which the loss of live storage will occur in the reservoir but also to document the rate and magnitude of ultimate degradation, so that the structure's safety can be evaluated.

Next, consider reservoir B, which has approximately the same storage capacity as reservoir A. The waters released into this reservoir from reservoir A will be essentially clear. However, two other river systems flow into the reservoir. These are not

desilted and, consequently, will carry significant quantities of
sediment into the reservoir because of the high sediment yield
from the upstream watershed and river channels, which are relatively unstable. Once again it is necessary to consider the
rate of loss of live storage as a consequence of inflowing sediments. Ultimately, reservoir A will lose its capacity to store
sediment, and sediments may be discharged from reservoir A to
reservoir B. There is, however, the possibility that other
reservoirs upstream from reservoir A will be constructed, which
can significantly increase the lifetime of reservoirs A and B.
This is typical of what has happened to downstream reservoirs as
a result of the construction of upstream reservoirs along the
Colorado River.

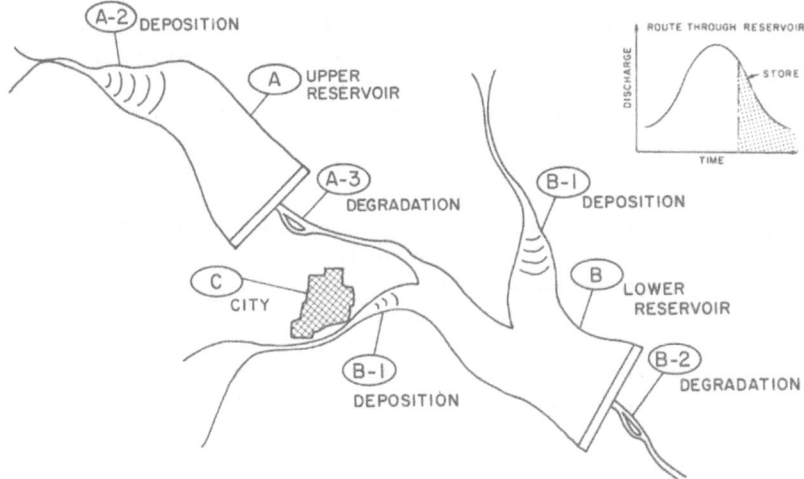

Fig. 4. The response of river system and intervening reaches of
         channel to the construction of major reservoirs given
         the hydrograph at top right.

A.   Upper Reservoir
     1. Large volume of runoff (74,000 hm^3 per year)
     2. Desilting of 5-6 times the storage capacity of the
         reservoir, hence rapid filling with sediment because
         of the large size of the reservoir
     3. Downstream degradation limited by armoring of the
         channel bed, lateral movement, and bank erosion

B.   Lower Reservoir
     1. Deposition in upper reaches, including below reservoir
         A after it has filled with sediment
     2. Degradation downstream from the structure because of
         the clearwater release

C. Effects on City
    1. Backwater of reservoir
    2. Deposition of sediment
    3. Increase in river stage
    4. Flooding
    5. Increase in groundwater level
    6. Flood protection required, designed to consider
        changing stage with time

D. Conclusion
    Analysis should consider routing sediment through the
    second major reservoir by releasing first part of runoff
    through reservoir and storing last part in order to reduce
    the rate at which the reservoir will fill with sediment.

It is possible to construct reservoir B so that relatively
large quantities of sediment can be discharged annually from it
instead of being stored within it. This would involve building a
large gated structure, so that reservoir B could essentially run
as an open river during those periods of high flow when water
could be passed through the system. Then, toward the end of the
runoff period, it would be necessary to close the gates in sufficient time to allow filling of the reservoir. With this type of
operation it would be necessary to empty the reservoir annually.
Otherwise, it would not be possible to pass excess sediments
through the reservoir system. Hence, it can be seen that there
are many trade-offs that have to be considered when determining
the economics of these systems and the response of the rivers
and impacts that may result.

If a city is located on a river discharging into an arm of
the reservoir (Fig.4), the backwater from the reservoir may require
construction of flood-protection works for the city. Determining
only water level in the vicinity of the city is not adequate.
It is also necessary to route the water and sediment down the
river system into the reservoir to determine the rates of deposition, the locations of deposition, and how these deposits of
sediment may increase river stage, aggravate flooding, increase
the groundwater level, and perhaps cause other adverse responses.
The response of these river systems is complicated, involving rates
of deposition in the reservoir, aggradation upstream, degradation
downstream, and the economic production of water for irrigation,
hydro power, and municipal use, and can best be evaluated by comprehensive model studies, such as those identified earlier.

Conducting these comprehensive model studies requires an adequate data base. In most instances, the data base is probably inadequate. Various alternatives can be used to alleviate this problem.

One is to use physical models with mathematical models when detailed
information on the performance of portions of the structure or short
river reaches is desired. It is, in fact, entirely feasible to predict with physical models how the system will respond. However,
certain field data are necessary in order to validate the physical
model. Consequently, whether one is using a physical model, a
mathematical model, or both, it is necessary to carefully overview
the system, determine those data that exist, and conduct a study
to determine how important actual data are to obtain accurate results. With this knowledge, a data base can be formulated. It is
recognized that to obtain an adequate data base of all the hydrological and climatological variables may require years. Hence, in
most instances, river development proceeds without all of the desired data. This makes the physical-process model more applicable
to the analysis. With properly constituted physical-process models,
one can minimize the amount of data required and can arrive at
logical conclusions regarding evaluation of river development.
It is possible to gather the existing data, synthesize missing
data, then formulate process models and analyze the system to gain
a better answer (in terms of the design, long-range planning and
spatial and temporal response of the system) with physical-process
models than by any other methodology presently available. Furthermore, it should be stressed that these models, once properly constituted, can be easily coupled with economic, environmental, social,
and related management models. Furthermore, these models, since
they are based upon physical processes, can be used continuously
throughout the construction of the project and, after implementation
of it, can be used to predict and adjust methodologies related to
operation of the system for optimum economic results.

REMOVAL OF CLAY AND SILTS FROM STREAMS

     In the early 1900s a major earthquake in the area caused a
large landslide to form a dam of considerable size across the
Gros Ventre River Valley in Wyoming. The resulting lake is of
sufficient size to have essentially removed the silts and clays
that normally flow in the river at high stages. Consequently, in
recent decades the banklines have changed and new banklines have
formed; but without silts and clays to hold moisture in the newly
deposited material, no new vegetation has been able to grow. The
end result has been the elimination of a significant amount of
riparian vegetation along the stream line, greatly reducing the
beauty of the stream system and in many ways deteriorating its
value for recreational use and fisheries habitat. Only recently
has the importance of silts and clays on stream and river morphology been recognized (Simons and Li, 1979).

A final example of the importance of washloads on river and canal morphology is a study conducted by Colorado State University in Venezuela for a period of years, (Simon et al. 1971a,b). This study dealt primarily with the geomorphology and hydraulics of rivers and the relationship of these river systems to the watersheds. It was found that vegetation developed along certain segments of the bankline and floodplain. In other sections or locations, vegetation sometimes started growing but died before a root system could be developed, because of an inadequate supply of water and nutrients. Extensive soil sampling along the banks and in the floodplains of these rivers showed that wherever a substantial growth of vegetation was occurring, there was a significant percentage of silts and clays. Hence, where a river system experiences natural cutoffs that subsequently fill with finer sediments, an environment is formed that can support rapid and vigorous growth of vegetation. In those segments of the floodplain that consist largely of gravels and sands, the area drains so rapidly that significant vegetation usually does not develop.

CONCLUSIONS

The design of dams, diversion structures, and river-training works, all require detailed evaluation of the impacts they impose on watersheds and river systems. An adequate analysis of river systems and the impacts of water-resources development, requires calculation of

1) The total sediment load that enters the reservoir
2) The characteristics of the sediment, including size distribution, specific weight, and density of the depositied sediment as a function of time
3) The pattern of deposition of sediment in the reservoir as a function of time and space
4) The effect of pool fluctuations on sediment storage
5) The trapping efficiency of the reservoir
6) Backwater effects on flood stage and sediment deposition
7) The rate of loss of live and dead storage
8) The useful life of the reservoir
9) Possible means of bypassing sediment
10) Aggradation rates and magnitudes upstream from impoundments
11) Rate and magnitude of channel degradation below the dam
12) Channel stabilization requirements, both upstream and downstream
13) Impacts of river development on water quality and on the biomass of the system

REFERENCES

Brown, C. B., 1958, "Factors Affecting Useful Life of Reservoirs,"
    Proc., ASCE, Vol. 84, No. IR4, Paper 1503.
Buma, P. G., and Day, J. C., 1977, Channel morphology below
    reservoir storage projects, *Environ. Conserv.* 4:279-284.
Keeley, J. W., Mahloch, J., Barko, J. W., Gunnison, D., and
    Westhoff, J. D., 1978, "Reservoirs and Waterways Identifi-
    cation and Assessment of Environmental Quality Problems and
    Research Program Development," Tech. Rep. E-78-1, Environ.
    Lab., U.S. Army Engineer Waterways Exp. Sta., Vicksburg,
    Mississippi.
Leopold, L. B., Wolman, M. G., and Miller, J. P., 1964, "Fluvial
    Processes in Geomorphology," Freeman, San Francisco, Calif.
    552p.
McHenry, J. R., and Dendy, F. E., 1964, "Measurement of Sediment
    Density by Attenuation of Transmitted Gamma Rays," p. 812-817,
    in: Proc. Soil Sci. Soc. Am., Vol. 25, No. 6, Madison, Wiscon-
    sin.
McHenry, J. R., Hawks, R. H., and Gill, A. C., 1969, "Consolidation
    of Sediments in a Small Reservoir in North Mississippi Measured
    in Situ with a Gamma Probe," p. 101-112, *in:* 4th Proc. Missis-
    sippi Water Resources Conf., Jackson, Mississippi.
McHenry, J. R., 1974, "Reservoir Sedimentation," Water Resources
    Bull., Vol. 10, USDA Sediment Lab., Oxford, Mississippi.,
    p. 329-337.
Pemberton, E. L., 1978, "Reservoir Sedimentation " Bureau of Recla-
    mation, Denver, Colorado.
Rice, T. L., 1979, "Investigation Reservoir Sedimentation." Unpubl.
    Rep.
Shen, H. W., 1971, "River Mechanics," Vol. I and II, Dep. Civil
    Eng., Colorado State Univ., Fort Collins.
Simons, D. B., and Senturk, F., 1977, "Sediment Transport Tech-
    nology," Water Resources Publ., Fort Collins, Colorado.
Simons, D. B., and Li, R. M., and Duong, N., 1978, "Sedimentation
    Study of the Yazoo River Basin - User's Manual for the Yazoo
    Data Storage and Retrieval System - Vol. II," Prepared for the
    U.S. Army Corps of Engineers, Vicksburg District, Vicksburg,
    Mississippi.
Simons, D. B., and Li, R. M., 1978, "River Mechanics, Morphology,
    Watershed Management," Paper presented at Instream Flow
    Criteria and Modeling Workshop, Colorado State Univ., Fort
    Collins.
Simons, D. B., and Li, R. M., 1979, Short Course titled, "Analysis
    of Watershed and River Systems," Colorado State Univ., Fort
    Collins.
Simons, D. B., Richardson, E. V., Stevens, M. S., Duke, J. H., and
    Duke, V. C., 1971a, "Streamflow, Groundwater, and Ground
    Response Data," VIMHEX Hydrology Rep., Vol. II, Dep. Civil
    Eng., Colorado State Univ., Fort Collins.

Simons, D. B., Richardson, E. V., Stevens, M. A., Duke, J. H., and
    Duke, V. C., 1971b, "Geometric and Hydraulic Properties of the
    Rivers," VIMHEX Hydrology Rep., Vol. III, Dep. Civil Eng.,
    Colorado State Univ., Fort Collins.
Ward, T. J., Li, R. M., and Simons, D. B., 1971, Mapping potential
    landslides in forest watersheds, *J. Geotech. Eng. Div.*, ASCE,
    Submitted.

Section II
Geographical Reviews

STREAM REGULATION IN AFRICA: A REVIEW

Bryan R. Davies

Institute for Freshwater Studies
Rhodes University
Grahamstown, 6140, R.S.A.

INTRODUCTION

Almost every river on the continent of Africa has been inter-fered with, mainly by construction of man-made lakes. However, most ecological-impact research has been directed towards the biological development of such impoundments as inundation proceeds. Only a very small proportion of research has made any attempt to elucidate downstream effects. Exceptions may be found in studies on the Nile (White and Blue), Volta, Zambezi, Vaal, and Pongolo rivers.

Table 1 lists the major dams and catchment water transfer schemes on the continent. The lack of data in certain areas is obvious, but where possible, the table attempts to list the salient features of these major regulatory systems. Discharge figures are omitted because of the great variety of presentation of such data in the literature, but where possible, relevant information regarding this parameter is in the text.

Fig. 1 illustrates the geographical distribution of African streams regulated by impoundments >100 km^2 surface area. Although large areas of the continent are still comparatively free of *major* regulatory structures, 13 countries have constructed 23 large man-made lakes on 12 major river systems, regulating 9,200 km of river (Fig. 1). The total length of stream regulated in each country by such large impoundments is indicated on the figure by cross-hatched columns. In compiling these data, I have omitted a number of major schemes, e.g., Owen Falls Dam on the Victoria Nile, because Lake Victoria is itself a natural reservoir (Rzóska, 1976). It must also be made clear that these data give a false impression of continental

Table 1.   Details of the Major Stream-Regulation Schemes of Africa
(Surface Area >100 Km^2; Fig. 1).   Discharge Data Have
Been Omitted (See Text), but Information on Two South
African Catchment Water Transfer Schemes Have Been
Included to Illustrate the Magnitude of Such Regulation
Projects in the Region

Region	Name	Country	Completion Date	Size (km^2)	Volume (m^3 × 10^6)	Affected River	Length of Regulated River (km)	Function and Comments
North Africa	Aswan	Egypt/Sudan	1964	6,217	157,000	Lower Nile	1,200	Irrigation and fisheries
	Gebel Aulia	Sudan	1937	600	3,500	White Nile	1,080	Irrigation
	Sennar	Sudan	1925	140	900	Blue Nile	375	Irrigation
	Roseires	Sudan	1966	290	3,000	Blue Nile	400	Flood control and irrigation
West Africa	Ayamé	Ivory Coast	1959	186		Bia	23	
	Kossou	Ivory Coast	1971	1,710	20,500	Bandama	250	Irrigation and hydroelectric power
	Bui	Ghana				Black Volta	152	
	Volta/Akosombo	Ghana	1964	8,727	165,000	White Volta	102[e]	Hydroelectric power, fisheries, and irrigation
	Kainji	Nigeria	1968	1,280	13,970	Niger	1,000	Hydroelectric power and fisheries
	M'Bakou	Cameroon				Tributary of Sanaga	700	
Central Africa	Mwandingushu Lualaba	Zaire				Lufira[c] Lualaba	390	
	Inga I and II	Zaire				Zaire	150	Hydroelectric power
East Africa	Seven Forks	Kenya				Tana	570	
	Numba ya Mungu	Tanzania	1965	150	9,000	Pangani/Ruvu	185	Hydroelectric power, fisheries, and irrigation
Southern Africa	Kafue Gorge	Zambia	1972	809	740	Kafue	40[d]	Hydroelectric power
	Cabora Bassa	Mocambique	1974	2,660	52,000	Middle Zambezi	520	Hydroelectric power
	Kariba	Rhodesia Zambia	1959	4,250	160,000	Middle Zambezi	250	Hydroelectric power
	H. F. Verwoerd[a]	South Africa	1971	374	6,960	Orange	124[f]	Flood control, hydroelectric, irrigation, and fisheries
	P. K. le Roux	South Africa	1975	138	3,186	Orange	1,410	Fisheries, storage, irrigation, and flood control
	Vaal Dam[b]	South Africa	1938 (raised 1956)	300	2,330	Vaal	360[g]	Domestic water and flood control
	Bloemhof	South Africa		228	1,273	Vaal	430[h]	
	J. G. Strijdom/ Pongolapoort	South Africa	1970	Indeterminate	23,366	Pongolo	180	Hydroelectric power and irrigation

[a]The H. F. Verwoerd Dam is the source supply for the catchment water transfer schemes, Orange-Fish and Fish-Sundays, which involve several
smaller dams and some 100 km of tunnel and 90 km of canal, transferring an annual flow of 1,147 × 10^6 m^3 of Orange River water to semi-arid
headwater regions of the Great Fish and Sundays rivers (Noble, 1972).

[b]The Vaal Dam is part of a complex of five dams and a diversion scheme involving water transfer from the Tugela River on the Natál side of the
Drakensburg mountain range to the upper Vaal catchment. This multipurpose project involves dams, siphons, tunnels, and pumps on an interregional scale.

[c]To the Lualaba confluence (Fig. 1).

[d]To the Zambezi confluence (Fig. 1).

[e]To the upper reaches of Lake Volta.

[f]To the upper reaches of P. K. le Roux.

[g]To the upper reaches of Bloemhof Dam.

[h]To the Orange River confluence.

stream regulation, because of the large numbers of dams of surface
area <100 km^2.   In fact, it is impossible to give any *accurate*
estimate of total length of regulated stream because of the paucity
of reliable data.

In order to partly circumvent this problem, I have attempted to
apply a correction factor to the continental figure of 9,200 km of
stream regulated by dams >100 km^2 surface area, by generating data

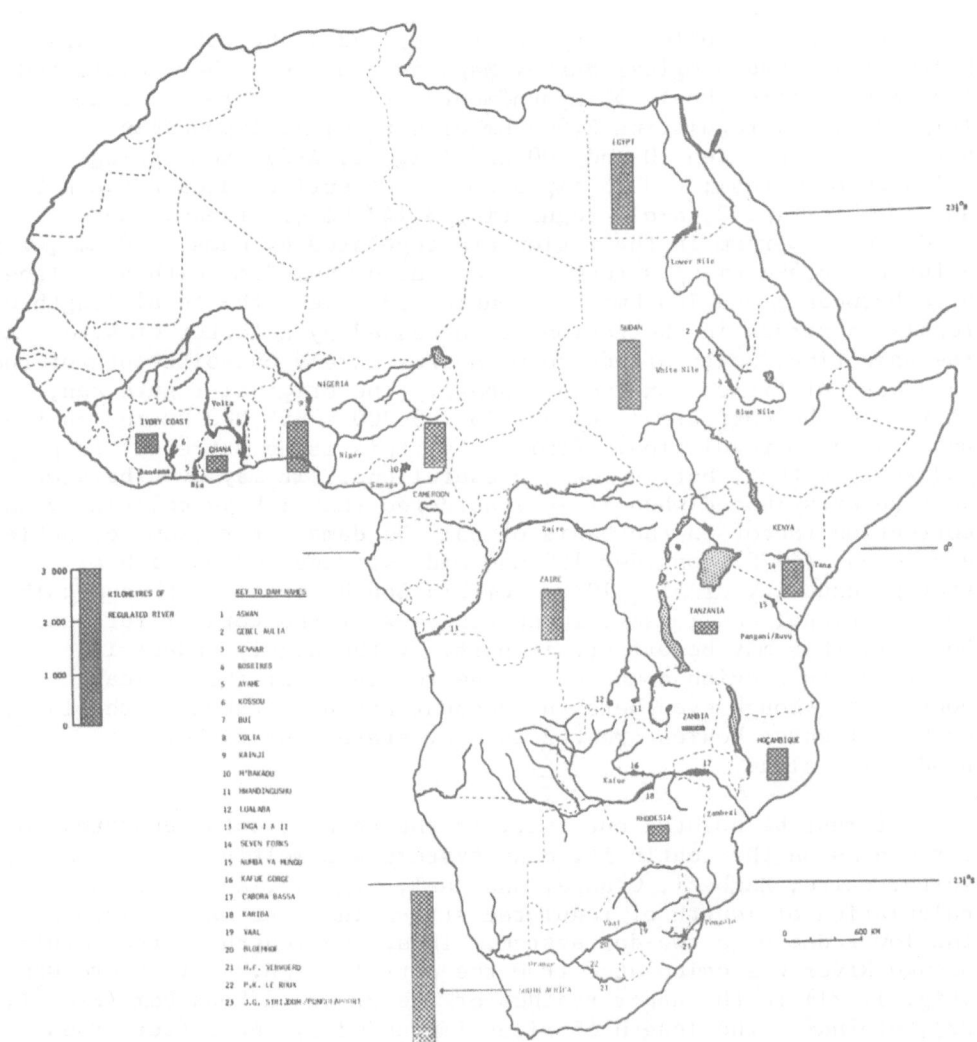

Fig. 1. Major regulation schemes of surface area >100 km^2, and
their major rivers in Africa (numbered 1-23). The length
(km) of regulated stream in each country is shown by
cross-hatched columns, and political boundaries by broken
lines; natural lakes are stipled.

on the length of stream regulated by dams with a surface area of
between 5 and 100 km^2, for a region which is fairly well documented,
South Africa.

Thus, using Noble and Hemens (1978) and 1:500,000 Union of
South Africa Hydrological Survey Maps as sources, I have estimated
that South Africa has: 5 impoundments of surface area >100 km^2
(Fig. 2, 1-5), regulating 2,504 km of stream; 15 impoundments of
surface area between 10 and 100 km^2 (Fig. 2, A-O), regulating
5,454 km of stream; and 15 impoundments of surface area between 5
and 10 km^2 (Fig. 2, a-o), regulating 3,142 km of stream. Now,
2,504 km of stream in the region are regulated by dams >100 km^2, but
a further 8,596 km of stream are regulated when dams with a surface
area between 5 and 100 km^2 are considered, i.e., the total length of
regulated stream in the region is increased by a factor of 4.43. At
the same time, South Africa forms approximately one-sixteenth of the
land mass of Africa, excluding Sahara. The original figure can,
therefore, be corrected: 4.43 × 16 × 9,200 = 652,000 km for continental stream regulation. Admittedly, this is an extremely rough
and ready method, but, I feel, a useful one. It may even be somewhat underestimated when it is considered that I have calculated the
correction factor on the basis of only 35 dams. For example, Noble
and Hemens (1978) consider 150 man-made storage bodies with a
storage capacity of >5 × 10^6 m^3 within South Africa, with a possible
300 additional small dams, impounding 40% of the total river runoff.
However, this may be offset, in part, by the highly industrial
nature of the region compared to the remainder of the African
continent, though even here, no account has been taken of the large
number of water bodies created in many states for fisheries
production alone.

It must be pointed out that, in the case of data generated for
large dams on the continent, many systems are multiple (e.g., Nile,
Volta, Zaire, Zambezi, Orange, and Vaal; Fig. 1), and that the
calculation of length of regulated stream in these cases excludes
the lower dam on a two-dam system. Thus, the length of the regulated
Orange River was calculated from the outfall of H. F. Verwoerd Dam
(Fig. 1, 21) to the upper reaches of the P. K. le Roux Dam (Fig. 1,
22), *excluding* the length of river impounded by the latter, then
from below the P. K. le Roux Dam to the sea. Again, with the Orange
River system as an example, where two major streams are linked and
both are regulated, duplication of regulated length below their
confluence has been avoided. Thus, for the Orange-Vaal, the length
of the regulated Vaal has been calculated from below the Vaal Dam to
the confluence of the Vaal with the Orange, *and no further*.

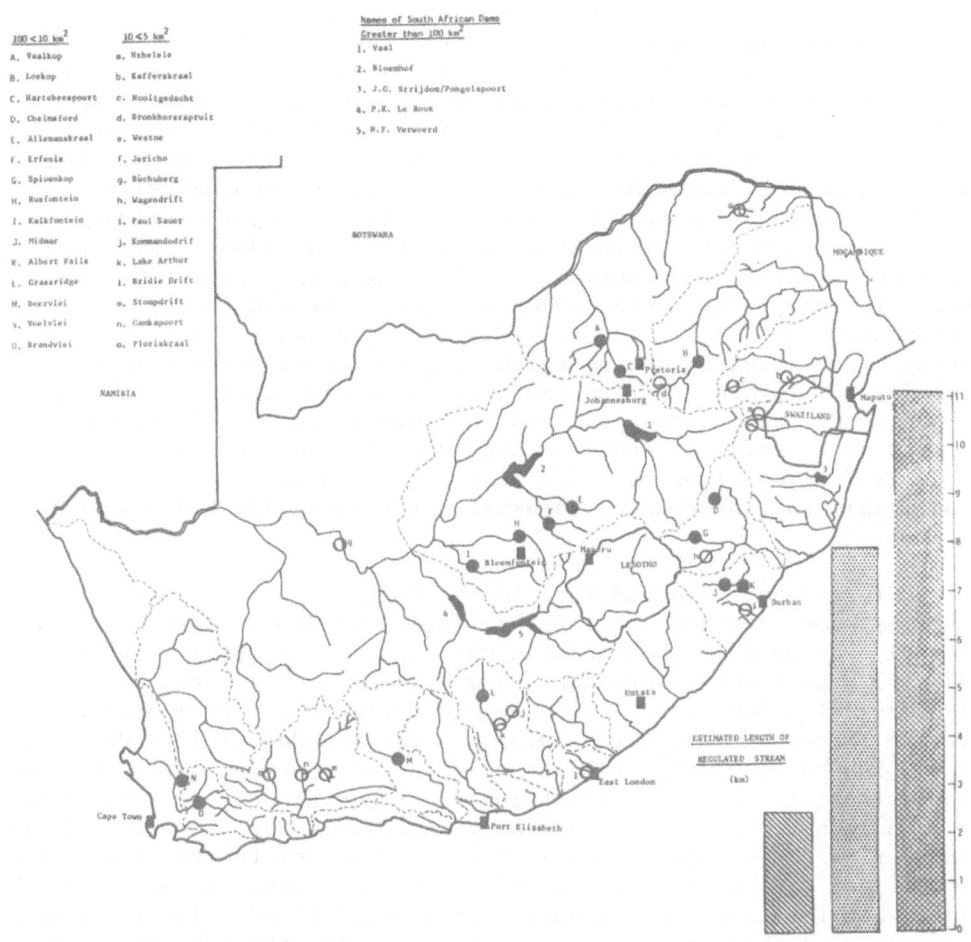

Fig. 2. Stream regulation in South Africa, showing dams with surface
         area >100 km² (1-5, closed areas), dams with surface area
         between 10 and 100 km² (A-O, closed circles), and dams with
         surface area between 5 and 10 km² (a-o, open circles)
         (Noble and Hemens, 1978). Rivers are marked together with
         major catchment boundaries (broken lines), and the lengths
         (km × 10³) of regulated stream are illustrated in histogram
         form: by dams >100 km², diagonal lines (2,504 km); by dams
         >100 km², plus those of surface area between 10 and 100 km²,
         stipled (total length = 7,958 km); and by dams >100 km²,
         plus those of surface area between 5 and 100 km², cross-
         hatched (total length = 11,110 km); see text.

PHYSICO-CHEMICAL LIMNOLOGY

Flood Regime--Some Extremes

The overriding factor in stream regulation in Africa is alteration of the flood regime, either for flood control, hydroelectric
power production, or irrigation and water storage. A most comprehensive study of the effects of alteration of flood regime has
taken place on the Pongolo River, Kwa-Zulu (Fig. 1). The driving
force behind this study has been concern for the economically
important Pongolo floodplain, which contains 25 major bodies of
water--pans--directly influenced by river floods (Jubb, 1972) and
which were, until construction of a dam, extremely productive in
terms of fish yield. Coke (1970a,b) notes that closure of
Pongolapoort Dam in 1970 caused several major pans to dry out. Some
months after closure, during which time the river dropped to exceptionally low levels, the dam was opened, causing a flood of 500-
1,000 cusecs, which later swelled to 5-6,000 cusecs. Practically
all pans were refilled, and two pans were flooded for the first time
since 1967. The flood equalled past major floods, and although
beneficial, the intervening intermittent flow regime had devastating
effects.

Unfortunately, in addition to the Pongolo River, a number of
other systems have been seriously affected by dam closure procedures.
Probably the most serious example was reported by myself (Davies,
1975a,b), where the Cabora Bassa Dam cut the Zambezi from an average
discharge of 3,000 cumecs to 60 cumecs during the four months of
filling (December 1974-March 1975). A recommended minimum discharge
of 400-500 cumecs during filling was ignored (Davies, 1975a) on the
grounds that tributary rivers below the dam would ensure river flow.
This could never have been the case. Available data showed that the
tributaries could supply a maximum of only 10% of the total Zambezi
flow at any time of the year (Davies, 1975b). Thus, drying took
place when the river should have been in full flood. The irony was
that not even short-term gain was achieved, for the lake filled so
rapidly that north bank turbine chambers were flooded. Unfortunately, this then led to emergency discharge (Jackson and Rogers,
1976) of some 13,000 cumecs via the sluices during early April and
May, in order to draw down the lake sufficiently, causing a massive
unseasonal flood at a time when the normal flood should have been
subsiding. The lower Shiré River (Malawi), itself a tributary of
the Lower Zambezi (Davies et al., 1975; Hall et al., 1977), was
similarly starved of water in its lower reaches because of closure
of the Liwonde Barrage, with a drop in flow in 1956-1957 to almost
nothing. The barrage, one of three regulating structures on the
Shiré, was built to control the level of Lake Malawi, allowing a
flow of 6,000 cusecs and retaining the rest. In fact, the lake
level has been so high that flow has been restricted only to allow

hydroelectric station construction downstream and for short periods
to draw down the level of the Elephant Marsh. This marsh and others
are important fisheries sites, and prolonged flow restriction can
adversely affect them.

Begg (1973) has described the highly erratic early discharge
from Kariba between 1961 and 1971. In 1967, discharge commenced in
February and continued to the end of May (late flood to dry season
of pre-Kariba river), while in 1968, discharge commenced in mid-
January and ended towards the end of April. Discharges during 1969
took place throughout most of the year at irregular intervals.
Thus, no pre-Kariba flow regimen were maintained during these three
years, with profound ecological disturbances downstream (Attwell,
1970; Begg, 1973).

At the other end of the scale, at least one attempt has been
made to rationalize the flow regime of a regulated river. White
(1973), working on the Kafue (Zambia), carried out an interesting
modelling experiment, using runoff and flow data for the system, and
examined the hydrological regime of the river, assuming various
characteristics of the Kafue Gorge and Itezhitezhi Dams (the latter
shortly to come into operation). The results indicated that
Itezhitezhi will have a negligible effect on the Kafue Flats between
the two systems, *other than in dry years*, as long as annual drainage
of the lower Kafue Gorge Dam takes place to reduce back flooding
caused by buildup of water from Itezhitezhi. However, even this
sytem has now become cause for concern (Rees, 1978a,b,c).

Silt

Stream regulation involves reduction of silt load in water
downstream from a barrage. This reduction, caused by velocity and
flow rate changes and therefore sediment-carrying capacity of water
(see Smith, 1975) as a river enters an impoundment, causes deposition
within the impoundment. Discharge is therefore clearer and has a
higher silt-carrying capacity, together with altered deposition and
erosion characteristics. In the case of the Vaal, when the Vaal Dam
closed in 1937, a considerable drop in silt load occurred, where an
average of 36 ppt in 1935-1936 decreased to an average of 4 ppt in
1937 (Schwartz, 1969).

Reports from the Nile system include Hammerton (1972, 1976),
who noted an increase in depth of the euphotic zone of the Blue Nile
below Roseires Dam caused by silt deposition in the impoundment.
Similarly, Entz (1976) noted that, during the first years of filling
Lake Nasser-Nubia, suspensoids were flushed through the lake, but
after 1967-1968 the current speed of the river entering the new lake
dropped so low that sediment deposition began in the basin, resulting

in a green and clear river in 1969 and drastic reduction in sediment
deposition in the Delta.

The floodplain regions of a number of African rivers have
caused concern in recent years, and the delta of the Zambezi, as
well as the Nile Delta, has generated considerable discussion in
terms of the effects of silt reduction caused by regulation of the
river (e.g., Attwell, 1970). Tinley (1971) and Tinley and Sousa-Dias
(1973) observed that one of the most important influences of rivers
along the fragile Mocambique coast is deposition of beach-building
material and alluvium. As Tinley (1971) points out, what is not
known, and there has been no attempt to study the situation, is the
effect that regulating dams (Kariba and Cabora Bassa) will have on
this aspect. Reduction in source material may well lead to a long-
term erosion of the coastline. The delta floodplain may also suffer
from reduced alluvium deposition (Tinley, 1971, 1975; Tinley and
Sousa-Dias, 1973; Hall and Davies, 1974; Davies et al., 1975), but
there are no quantitative data to support these observations.

Increase in silt-carrying capacity (and therefore erosion
potential) of regulated streams has certainly been illustrated for
systems outside Africa. Leopold et al. (1964) have shown that river
beds below dams are degraded to the order of 0.1 ft yr^{-1} or more
and that large storage dams tend to trap something in the order of
95-99% of the sediment that previously passed down river. Schwartz
(1969), discussing one of the major catchment transfer schemes of
South Africa (Orange-Fish-Sundays; Table 1), certainly has no doubt
that comparatively silt-free water from H. F. Verwoerd Dam (Figs. 1
and 2) will degrade the Thebus stream into which discharge water
will flow to reach the Great Fish River. Such suggestions are
supported by Talling (1976), who observed that from Aswan to Cairo,
there was a steady decrease in transparency, probably caused by
silt and detritus being picked up by the Nile on its way north to
the sea. This confirms Watermeyer's view (1965) that in the case of
discharge from Lake Kariba tailraces, the river can be expected to
pick up fresh silt downstream of the Gorge. Interestingly, Wafa and
Labid (1973) observe that seepage loss from Lake Nasser, though huge
(2,000 × 10^6 m^3 yr^{-1} at 180 m lake level), is advantageous in the
sense that it is a loss of silt-free water that would otherwise
cause serious scour problems downstream.

Plant Nutrients and Minerals

Ionic concentrations normally vary tremendously from river to
river and catchment to catchment and, in the case of regulated
rivers, according to the rate and volume, as well as type of discharge (hypolimnetic or epilimnetic). Some regulated systems in
Africa receive hypolimnetic discharge. In these cases, water

exiting through tailraces or sluices is cool, deoxygenated, and
laden with nutrients. In some cases, H_2S-laden water is also
released.

The effect of discharge temperature on the biota of regulated
African streams has been neglected. Rogers (1978) briefly comments
on temperature, observing that cold hypolimnetic water is discharged
from the Pongolapoort Dam, while Hall et al. (1977) have observed
that the temperature of the post-Kariba Middle and Lower Zambezi in
Moçambique increases 3°C towards the sea. However, the latter is
almost certainly caused by altitude and climatic zone changes. The
effect of Kariba on river temperature immediately below its outfall
appears to be minimal, with close similarity to observed patterns
and ranges for river water entering the lake (Coche, 1968; Balon and
Coche, 1974). Mitchell et al. (1975) have predicted that hypolimnetic discharge from the Darwendale Dam (Rhodesia) will affect
temperature regimes of the Hunyani River (tributary of the Zambezi),
so as to increase winter, and decrease summer, temperatures when the
lake is not spilling in flood. However, temperatures are expected
to equilibrate within a short distance of the wall, because of the
steep river bed. On the Orange River, Pitchford and Visser (1975)
have shown that, although the discharge from the H. F. Verwoerd Dam
did not affect early spring and autumn temperatures, 4 km below the
dam there was a persistent increase in winter, and decrease in
summer, temperatures. As a result, they felt that irrigation waters
would reflect these changes and encourage the perennial development
of Schistosomiasis host snails [*Biomphalaria* sp. and *Bulinus*
(Physopsis) sp.] and the possibility of summer transmission of
bilharzia.

Deoxygenation and hydrogen sulphide have been of slightly
greater concern. Kainji Lake, for example, discharges deoxygenated
hypolimnetic water (even though evidence suggests that twice annual
renewal time in the lake; El-Zarka, 1973), which is expected to have
a marked effect for a considerable distance below the dam. For the
Zambezi, Begg (1973) reported excessive spillage to excavate the
stilling pool, which dropped the lake level by 8 m in three months.
This enormous amount of water was deoxygenated and was detectable
some distance downstream. Also, as hypolimnetic discharge during
the unstable phase of the lake, it contained large quantities of
H_2S, which had drastic effects on the biota immediately below the
wall. In the case of Cabora Bassa, the rapid post-filling drawdown
(Davies, 1975a,b) led to the lake's becoming a "slow river" (Jackson
and Rogers, 1976; Bond et al., 1978), with no H_2S discharge or
deoxygenation of the river downstream.

Plant nutrient concentrations and ionic composition of the
Zambezi appear to have been altered by both Kariba and Kafue Gorge
dams. Hall et al. (1976, 1977) have compared figures for the

post-Kariba Middle and Lower Zambezi against figures published by
Coche (1968) and Balon and Coche (1974) for Kariba, and by Carey
et al. (1967) for the Kafue. Briefly, the water quality of the
Middle Zambezi was mainly determined by Kariba, with the Kafue Gorge
Dam playing a minor role. Transparency was low in the dry season,
possibly because of silt and unregulated inflowing rivers. However,
phytoplankton density changes cannot be excluded (studies by
J. Oliveira have remained unpublished), which may have been caused
by overturn at Kariba and subsequent increase in nutrient flow.
Kariba was the major source of orthophosphate, particularly as
turbine water was drawn from below the thermocline. Similarly, the
river registered high SO_4^{2-} because of H_2S discharge from the dam.
The most important finding, however, was that tributary rivers
carrying high nutrient loads exerted a great influence over the
nutrient regime of the Zambezi below confluence. Presumably, this
influence will increase with the addition of Cabora Bassa and
increased regulation (Hall et al., 1977). In the case of the Vaal
River, the tributaries below Vaal Dam so heavily influenced the main
river that new purification plant was required on the Zuikerbosch
upstream from its confluence with the Vaal (Schwartz, 1969).

There is very little information available for other regulated
streams. Kainji Dam draws turbine water from below the thermocline
and a substantial variability of discharge occurs because of uncertain prediction of floods (Henderson, 1973); but as the Niger River
is generally very low in minerals and plant nutrients, injection of
hypolimnetic water into the river below the dam will presumably have
a considerable effect on the regulated river. Reference to the lake
itself (where pH, HCO_3^-, Ca^{2+}, and Na^+ increase, K^+ and Cl^- decrease,
and PO_4^{3-} remains the same from inflow to lake) gives a good
indication that this will be a correct projection (Imevbore, 1970).

Salinification

Salinification has been shown to be a major problem for the
regulated Pongolo system. The floodplain lies on the Cretaceous
marine deposits of the Zululand coastal plain (du Preez, 1967) and,
as such, is susceptible to salinification during the dry winter
season. Heeg et al. (1978) have conclusively shown that salinification takes place in conditions of low flow. Sodium was the
dominant cation throughout the floodplain, while chloride was the
dominant anion of the pans. The main danger appeared to be salinification of the major pans by ground Cl^- in the absence of regular
floods, which flushed the system. Three of the four pans studied
had a TDS greater than 3,500 ppm and Cl^- in excess of 30% of the
TDS. One pan, Mholo Pan, which was not easily flooded, was markedly
diluted when flooding occurred, but during a normal winter dry
season the conductivity rose to 1,000 μS cm^{-1}, and during an

unusually protracted dry season in 1970, when no flood waters were released from the dam for almost a year, the conductivity reached 2,750 µS cm^{-1} (Heeg et al., 1978). Nothing is known of the tolerance of the floodplain biota to such conditions, but the salinification of the system must present a considerable danger, particularly when the standing crop of *Potamogeton crispus* appears to be negatively correlated to salinity (Rogers, 1978). Obviously, flushing by the dam is a priority if the preimpoundment water quality of the flood-plain is to be maintained. In fact, salinification of regulated streams seems to be a problem requiring considerable research, particularly bearing in mind that Talling (1976) has observed that, in most freshwater systems, salinity and discharge tend to be inversely related and that Cl$^-$ content of Lower Nile water has increased since Aswan altered the hydrological and silting regimes.

PHYTOPLANKTON

Brook and Rzóska (1954) have described the influence of the Gebel Aulia Dam upon seasonal phytoplankton development in the Nile. Original river populations that enter the upper reaches of the dam were so low as to be near the limits of quantitative observation (Talling, 1976) and consisted of Bacillariophyceae, which formed 25-90% of the total algal population. Chlorophyceae comprised 8-21% of the total, while Cyanophyceae made a poor third (≈5%) (Rzóska et al., 1955; Talling and Rzóska, 1967; Hammerton, 1972). At the outfall of the dam, the diatoms decreased to 31% of the total population, while Cyanophyceae became dominant, increasing to 86% of the total population. Chlorophyceae were still present but in very low numbers (Brook and Rzóska, 1954). Numerical changes were also pronounced, with an increase from river water upstream down to the dam wall. Below the dam, algal numbers continued to increase, probably because of mixing and better aeration of water passing through sluices (Brook and Rzóska, 1954).

Roseires Dam has increased the transparency of the Blue Nile, down to Khartoum, 670 km below (Hammerton, 1972, 1976), with an increase in phytoplankton density as a direct result (Hammerton, 1972; Rzóska, 1976; Talling, 1976). Thus, total algal primary production in the river increased by a factor of six from Ethiopia to Khartoum in May 1968 (Hammerton, 1976). Examination of Hammerton's data shows that the depth of the euphotic zone almost doubled in this distance, increasing from just over 2 m at the Ethiopian border to 4 m at Khartoum. As a result of the 10- 200-fold increase in phytoplankton densities within Roseires Basin and subsequent injection of algae into the river, Hammerton (1976) has compared the situation with increased productivity of other systems caused by eutrophication and believes that even a very mild degree

of industrial eutrophication would have a serious effect on the
Nile because of its normal high temperatures and radiation input.

Downstream towards Aswan, the influence of the upper dams is
slowly lost until Bacillariophyceae again dominate (Talling, 1976).
Within Lake Nasser-Nubia there is again an increase in phytoplankton
development similar to that reported for Gebel Aulia (Brook and
Rzóska, 1954), with a peak near Aswan High Dam wall (Raheja, 1973).
Although the prediction can be made that high densities near the
wall will probably greatly influence downstream phytoplankton
development, no published information is available. However, Abdin
(1948a,b), working on the river below the old Aswan Reservoir,
provides some indication that present conditions will be similar to
those recorded upstream, by noting the familiar predominance of
Cyanophyceae in the reservoir. It is, therefore, likely that this
group will again dominate the regulated river phytoplankton.
Talling (1976) comments that it is possible, though not proven, that
photosynthetic activity of Cyanophyceae, which causes high pH development (≈9.0), may be limiting Bacillariophyceae development in the
river. Thus, by injecting lacustrine Cyanophyceae, the dams of the
Nile appear to considerably influence the natural river diatom
communities.

ZOOPLANKTON

The classic work on the effect of stream regulation on zooplankton development also comes from the Nile. This work on the
influence of Nile storage dams has now become so well known that I
shall summarize only briefly the results here, and readers are
referred to Brook and Rzóska (1954), Rzóska et al. (1955), Prowse
and Talling (1958), and Rzóska (1976). Comparison of Brook and
Rzóska's (1954) quantitative data for river plankton upstream of the
lake and below the dam outfall, respectively, shows a disappearance
of adventitious forms and a change from dominance of Copepoda
upstream of the lake to a Cladocera-dominated zooplankton in the
river below the dam. Both Roseires Dam (Moghrabi, 1972) and Aswan
High Dam (Rzóska, 1976) have also added their influence. Thus, it
appears that multiple regulation of the whole Nile has completely
altered the zooplankton characteristics, by altering species composition and densities, with an emphasis towards lacustrine-adapted
groups and away from riverine types.

The zooplankton of other regulated systems go virtually
unrecorded. On the Zambezi, the only reported work is my own (Hall
et al., 1976). Here (the data are largely unpublished), the zooplankton of the Middle and Lower Zambezi, although sparse, is
dominated by Cladocera, and although there are virtually no pre-
Kariba data [the only records being those of Dr. G. Fryer, recorded

in both Bowmaker (1960) and Jackson (1961)], reference to the Nile
seems to point to some influence from Kariba.

ZOOBENTHOS

Remarkably little quantitative analysis has centered on the
effect of stream regulation on zoobenthos; remarkable in the sense
that some components of the zoobenthos are of considerable economic
importance. The Simuliidae are an excellent example; the larvae
preferring running water habitats, where the biology of immature
stages is, surprisingly, little known and where alteration of flow
regimen and other characteristics by regulation appear to have
profound and, at times, confusing ecological effects. Petr (1974)
has noted that *Simulium* spp. became a problem below Akosombo Dam
only when discharge across the spillways took place (see also,
Waddy, 1973). In the Blue Nile the Roseires Dam increased the
vector of human onchocerciasis, *Simulium damnosum* (Lewis, 1966).
The same author also stated that *S. damnosum* occurred on spillways
of the Sennar Dam, on the same river, and that both *Simulium*
griseicolle and the chironomid genus *Tanytarsus* were also favorably
affected by water-level manipulation.

Perhaps the most illuminating work in this field, however, is
that of Chutter on the Vaal River, South Africa (see Chutter, 1968,
1972, 1973 for detailed references). His work showed that recorded
fluctuations at the study site were largely governed by the Vaal
Dam, some 580 km above (Chutter, 1968). In addition, an important
influence proved to be a local, weekly manipulation of the river for
irrigation below the Vaalhartz Diversion Weir. This fluctuating
biotope was most successfully exploited by *Simulium chutteri*. The
same fluctuations in flow, however, were unfavorable to *S. damnosum*,
which, if we extrapolate to other regulated African rivers where
flow regulation does not include such wide fluctuations in water
level, points to a potentially serious problem in other uninvestigated systems. Chutter (1968) also recorded large numbers of
mammalophilic Simuliidae (*Simulium adersi, S. damnosum,* and *Simulium*
mcmahoni) in the steady discharge region below the Vaal Dam. In
addition, Hydropsychidae larvae were abundant, while *S. chutteri*
larvae were not; and it is possible that the habits of *S. chutteri*
may have led to predation by Hydropsychidae where water levels were
stable. Injection of plankton from dams on the Vaal was probably
responsible for the unusually large populations of Simuliidae,
Hydropsychidae, and *Hydra*. This was certainly the case for Vaal
Barrage (Chutter, 1969); but here, unlike the situation below
Vaalhartz Weir, the changes disappeared within 8 km. Since all of
this work, a new dam has commenced operation on the Vaal
(Oppermansdrif, between the Vaal Barrage and Vaalhartz Weir) with

a new flow regime in effect; but, as Chutter (1973) points out, the
biological consequences go unrecorded.

Obviously, further investigation of the factors influencing
simuliids and the effects of stream regulation on their development
and population biology are required. But what of their control?
Taylor (1973) envisaged the use of insecticides to control *S.*
damnosum below the Akosombo barrage, but, of course, flushing larvae
from spillways by judicious lake-level manipulation (Brown and Deom,
1973) is more desirable. Lake Kainji, a relatively small lake fed
by a large river, apparently requires little manipulation, the
natural rise and fall of the Niger below Kainji preventing *Simulium*
breeding (Waddy, 1973). In South Africa, severe outbreaks of
cattle-biting Simuliidae in areas adjacent to stretches of the lower
Vaal and, more recently, the newly regulated Orange River have
stimulated a number of control programs, mainly by experimental flow
manipulation (Noble and Hemens, 1978); and some success has already
been achieved on the Vaal by shutting down flow at weekends.
Control of simuliids by cessation of discharge from both H. F.
Verwoerd and P. K. le Roux dams on the Orange River is also being
attempted at the time of writing. There is, however, still cause
for concern: Bearing in mind that Chutter (1972) has described how
Simulium (*Eusimulium*) *nigritarse* and *Simulium* (*Eusimulium*) *ruficorne*
predominate in small trickles, large streams, and dams; that Harrison
(1966) has described the very rapid recolonization of a Rhodesian
stream after drought by *S.* (*E.*) *ruficorne*; and that, conversely,
Simulium (*Metomphalus*) *bovis, Afrosimulium gariepense*, and *Simulium*
(*Meilloniellum*) *adersi* colonize rivers at peak flow, it appears that
whichever way one regulates a southern African stream (at least),
some nuisance blackfly species seems capable of exploiting the new
conditions!

Of the other zoobenthos, the effects of regulating long distances below regulating structures are little known. However, the
lack of regular flushing of the pans on the Pongolo River floodplain
below the Pongolapoort Dam will result in severe problems of salinification (see above). Lamellibranchiata and Oligochaeta, which are
not adapted to such changes, are expected to disappear from pans
that do not receive timeous floods. At the same time, Pretorius
et al. (1975) felt that, should irrigation water become available in
the Makatini Flats area of the Pongolo floodplain, human
Schistosomiasis could become a dangerous problem, because of the
abundance of the aquatic snail host species in the pans of the
system.

Just before Lake Volta was filled, the extensive clam fishing
industry based on the lamellibranch *Egeria radiata* (Lawson, 1963;
Purchon, 1963) became the focus of attention because of fears of
dam-induced damage to the population (Ewer, 1966; Lawson et al.

1969). Instead of seasonal floods of 10,000 cumecs, the dam
discharged a steady 100-150 cumecs (Taylor, 1973). Spawning of
Egeria is triggered by a slight rise in salinity when flood is least
and conditions correct for veliger osmotic balance. The region of
1 ppt salinity spawning grounds (Pople and Rogoyska, 1969) was 30 km
upstream from the coast. Restriction of flow during impoundment
caused salinity penetration at 1 ppt to move upstream to a point
50 km from the coast, but subsequent low, but continuous, regulation
has since forced the spawning grounds down to 10 km from the coast
(Lawson et al., 1969; Beadle, 1974). So, rather than anticipating
an increase in salinity penetration as a result of regulation (as
speculated by a number of authors in the past, e.g., Davies et al.,
1975), ecologists should perhaps anticipate a decrease in salinity
penetration at river mouths because of steady, low, but higher than
dry season, flows discharged from hydroelectric dams. Although the
Egeria industry is still thriving, it is, as Beadle (1974) points
out, indeed fortunate that the gradient of the Volta was not such
that the preimpoundment salinity penetration was closer to the sea!

VEGETATION CHANGES

    Hall and Pople (1968) have noted the rapid development of
Potamogeton octandrus Poir and *Vallisneria aethiopica* Fenzl below
the Akosombo Dam. Prevention of floods, which once scoured the
river bed, together with a more or less constant flow, have been
cited as causal factors for this change. Water clarity caused by
silt deposition within the lake, however, should also be taken into
account. Changes were also apparent in riverbank vegetation,
compared to preregulation conditions, and in a study of a sandbank
in the regulated Volta, Hall and Pople (1968) recorded stable
populations for 5 species, a decrease in 3 species, the disappearance
of 7 species, and the invasion of the water's edge by 3 swamp
species.

    Lack of scouring in the Middle Zambezi, caused by flow regula-
tion by Kariba, has also had considerable effects on vegetation of
the river. Jackson (1961, 1966) described the Zambezi as a typical
"sand bank" river, having fairly low flow in winter and early summer
and rising enormously in middle to late summer (January-April).
During preimpoundment surveys (pre-Kariba), there was a more or less
complete absence of rooted aquatic macrophytes, except for some
Phragmites mauritianus. The dry seasons of 1973-1974, however,
showed a stronger, more turbid flow than the pre-Kariba river, and a
dense sandbank cover of *Ludwigia* sp., *Panicum repens*, and other
species had developed (Jackson and Rogers, 1976). Such changes in
vegetation have a profound effect upon the fish populations of
regulated rivers by providing extensive cover.

Attwell's (1970) prognosis for the grasslands of Mana Pools
128 km below Kariba wall was one of steadily increasing desiccation,
and a decrease in variety and productivity of the grasses. The
major problem until 1970 was out-of-season floods and in-season
floods of too little magnitude. He noted a replacement of rooted
aquatic macrophytes by floating species, including *Salvinia*, a
reduction of the fodder tree *Acacia albida*, which has a flood-
borne seed–dispersal mechanism, and a lack of resting period from
grazing for the grasses of the floodplain. This situation, however,
was unlike that in the Marromeu Reserve on the Lower Zambezi (Anon.,
1975; Tinley, 1975) in that there was no invasion of the grasslands
by woody species. Attwell (1970) and Begg (1973) also noted that
lack of flushing allowed the development of aquatic vegetation,
to the detriment of living space for *Hippopotamus amphibius*,
crocodile, and wild fowl. The particular offender was *Salvinia*, with
subsequent "Sudd" growth of *Scirpus cubensis*.

Similar vegetation changes were also recorded in the Pongolo
system, although for different reasons. Pooley (1978) reported that
artificially prolonged flooding was affecting a steady change in
grass and sedge species composition. In addition, the Fever Tree
Acacia xanthophloea was disappearing (Furness, 1978; Pooley, 1978),
as was the *Phragmites* community. Much of the riverine forest was
also suffering, particularly *Ficus sycamorus*. Unseasonal discharges
of the Pongolapoort Dam also caused changes in abundance and productivity of *Potamogeton crispus* (because of increased salinity),
Cynodon dactylon (detrimental effects in prolonged dry periods),
Cyperus fastigiatus (decrease in dry conditions), and *Dyschoriste
depressa* (increase in wet conditions) (Furness, 1978).

INFESTANT AQUATIC MACROPHYTES

The paucity of floating aquatic macrophytes in South African
streams has been attributed to the frequency of flash floods
(Mostert, 1958; Edwards, 1967), but with stabilization of river
floods and flow regime by dams, ideal conditions have been created
that are eminently suited to the downstream development of species
such as *Eichhornia crassipes* (Mart.) Solms (Edwards, 1969; Edwards
and Musil, 1975). For example, many of the rivers of the Natal
coastal region already infested with *E. crassipes*, *Potamogeton* spp.,
and *Lagarosiphon muscoides* will presumably become problem areas when
flooding and natural flushing is prevented and increased nutrient
input and flow stabilization is initiated by projected dams.

Problems of this nature have already become alarmingly obvious
in the Lower Zambezi delta (personal observation, 1974), where
Salvinia molesta D. S. Mitchell was developing in many of the arms
of the river to the extent that the northern arms, in particular,

were choked and secondary "Sudd" was in evidence. A similar situation existed in the preimpoundment Cabora Bassa area (Jackson and
Rogers, 1976), where change caused by Kariba (Jackson, 1961) led to
considerable development of both *S. molesta* and *E. crassipes*. Both
the Kariba and Kafue Gorge dams have had infestant weed problems,
with the result that a regular discharge of *Salvinia* from Kariba and
Eichhornia from Kafue has taken place over recent years. A short-
term survey in January 1974 estimated that 150,000 *S. molesta* mats,
averaging 20 cm diameter and in healthy condition, were passing the
Luangwa confluence on the Middle Zambezi per hour (Davies, unpubl.
data). On the Lower Zambezi the problem is likely to be compounded
by input of *E. crassipes* from the Shiré River, particularly because
the Shiré is rich in nutrients (Davies et al., 1975; Hall et al.,
1977) and will considerably influence enrichment of the Lower
Zambezi.

FISH

Many of the riverine fish species are adapted to exploit newly
inundated flood plains, the drowned vegetation of which provides
rich feeding grounds and sheltered nursery areas for fry. In
regulated rivers, total disruption of normal flood patterns and,
therefore, spawning migrations, has occurred (Jackson, 1975; Bowmaker
et al., 1978). The effect of dams as barriers to fish spawning
migrations is well documented (Jubb, 1960; Jackson, 1966; Lagler,
1969). Jackson (1966) and Lagler (1969) state that, in the case of
anadromesis, most dams without some form of by-pass system will have
disastrous effects. The eels of the Tana River in Kenya are cited
by Jackson (1966, 1975) as an example. In fact, the *Anguilla*
population of the regulated Zambezi has stimulated considerable
controversy in the literature, with polarization into two camps:
(1) Kariba will have, and is having, a drastic effect upon the eel
population of the Zambezi (e.g., Jubb, 1960, 1961, 1964, 1967),
and (2) the eel population is thriving, and will continue to thrive,
because elvers are capable of climbing Kariba wall and entering the
lake (e.g., Balon, 1973a,b, 1975). Obviously, eels are capable of
circumventing the barrier of Kariba, but in what numbers? In the
meantime, the long-term situation must have changed considerably
(reducing eel migration) now that Cabora Bassa (171 m high; Jackson
and Davies, 1976) has also added its bulk to the river.

There have been two interesting observations on the effect of
Kariba on the potamodromous behavior of Zambezi fish. The first
comes from Begg (quoted as *personal communication* by Bowmaker
et al., 1978) and describes a potamodromous response to tailrace
discharge from the dam. On one occasion, Sanyati River water
(inflow is above dam wall) penetrated the lake and "hit" the wall.
In doing so it was deflected up into the turbines and out through

the tail races, where the flow rate was more or less constant. The
result was a massive increase in accumulated fish in the river below
the wall *without* an increase in flow. Such accumulations may have
considerable impact on predation pressure. For example, Bowmaker
(1960) has described the impact of *Hydrocynus vittatus* (Castelnau)
upon fish accumulated in the Zambezi due to a 6-ft drop in water level
shortly after Kariba commenced inundation. The second observation
comes from Kenmuir (1976), who examined fish spawning under conditions of artificial flood on the Mana Pools floodplain below Kariba.
Here potamodromous fish were able to utilize floods for spawning
migrations, *provided* discharge took place at the correct season. He
noted that sexually active adults of a number of species moved into
flooded pools shortly after Kariba sluice gates opened; and later,
after sluice-gate closure, he recorded large numbers of juveniles in
the area.

A very clear picture of the effects of the Pongolapoort Dam on
Pongolo River fish has been sketched by Coke (1970a,b, 1971), Coke
and Pott (1970, 1971), and Phelines et al. (1973). In 1969-1970 a
prolonged drought, due to the dam, caused an almost total loss of
the Pongolo River fish populations (Coke, 1970a). A flood released
in 1970 allowed upstream spawning migration of the few remaining
fish. However, this flood and a subsequent strong, prolonged flow
had very few, if any, beneficial effects; and, as Coke (1970a)
points out, had the same quantity of water been delivered at a
faster rate for a shorter period, the resultant flood would have
been beneficial. Furthermore, a preliminary flow would have induced
upstream migration of adult fish in readiness to enter the pans
during the main flood. During the 1971 flood, large-scale fish
migrations were observed, mainly in an upstream direction, with
either active or passive entry into the pans (Coke, 1971). At the
same time, commencement of flooding of pans drew fish already in the
pans out into the river. In receding flood conditions, pans released
water into the river and, therefore, attracted fish back into the
pans. Thus, there was a continuing pattern of exchange from river
to pan and pan to river, during which time adults became ripe
running (Kok, 1978). Coke (1971) went on to predict that if regulated floods were such that they simply filled pans, adults will
quite probably vacate the pans and migrate upstream, where spawning
may fail completely; adults requiring quiet, flooded backwaters for
breeding, feeding, and refuge (Pott, 1968). Therefore, filling pans
above supply level so that discharge then takes place would attract
fish back into the pans for spawning.

Except for data on the regulated Niger, there is little information on the effects of stream regulation on fish of other African
rivers. Lelek and El-Zarka (1973) have recorded changes in the
landings of fresh fish from the Niger below Kainji, between 1967 and
1969, e.g., a decrease from 19.6 t in 1967 to 12.2 t in 1969. At

stations below the confluence with River Benue, drops were not so
sharp, because of the ameliorating influence of this river.
Although no commercially important fish disappeared below Kainji,
some families were definitely more sensitive to the changes in water
regime, particularly Mormyridae and Clariidae. Only the predatory
Lates niloticus increased at all stations.

One special problem that arises in regulated systems involves
the associated transfer of organisms. The classic example is the
Lake Tanganyika sardine, *Limnothrissa miodon* Boulenger, which was
introduced to Lake Kariba in the hope that it would exploit the
zooplankton resources of the lake. Although *L. miodon* is essentially
a lacustrine species favoring deep water, in 1970 it was present in
the stilling pool below Kariba wall, having presumably exited via
tailrace discharge (Kenmuir, 1975). The regulation of the Middle
Zambezi below Kariba may well enable colonization of Lake Cabora
Bassa (Kenmuir, 1975), and sardine alevins released via tailrace
discharge is the most likely method of spread. In the same vein,
old geographical barriers are slowly breaking down in southern
Africa because of catchment water transfer. Here, Cambray and Jubb
(1977) have described the dispersal of fish indigenous to the Orange
River into the Great Fish River via the Orange-Fish tunnel. Additional transfer is expected from the Great Fish to Sundays River via
a linked water transfer scheme. It is indeed fortunate that most of
the transfers will be beneficial (Cambray and Jubb, 1977).

LARGE MAMMALS

The significance of vegetation changes caused by river regulation is not completely appreciated until wild herbivores are considered. Game that exploit the river floodplains of Africa have long-
established feeding and breeding migratory habits, moving on to
floodplains early in the dry season to exploit the flush of new
growth after a flood and retreating to higher ground with relatively
poor grazing during flood periods. This pattern controls numbers of
game and allows grazing to recover. Wrongly timed floods from
Kariba have completely disrupted such long-standing patterns, and
grazing mammals, once forced off the floodplain in the wet season,
now exploit the drier conditions, with species such as *Loxodonta
africana* (elephant), *Aepyceros melampus* (impala), *Kobus
ellipsiprymnus* (waterbuck), and *Syncerus caffer* (buffalo) causing
most pressure on vegetation (Attwell, 1970; Begg, 1973).

On the Kafue River, Rees (1978 a,b,c) has shown that the Kafue
Lechwe *Kobus leche kafuensis* Hattenorth, a semi-aquatic antelope
endemic to the region, is in grave danger because of both the Kafue
Gorge and Itezhitezhi dams. For the purposes of this paper, the
Itezhitezhi will have a downstream regulatory effect upon the Kafue,

by reducing extremes and duration of flooding of the Lochinvar
Reserve grasslands (Rees, 1978b,c). It will also alter the speed of
flooding and flood withdrawal as well as the general maintenance of
the Kafue River within its channel during the dry season. Through
all of these influences, Itezhitezhi may reduce the lechwe habitat
by affecting nutritious floristic combinations through the dry
season. According to Rees (1978b), the "... outlook for the only
population of Kafue Lechwe is bleak."

Lower downstream, in the major coastal floodplains, river
regulation can also have major ecological repercussions. Tinley and
Sousa Dias (1973), Hall and Davies (1974), Tinley (1975), and Davies
et al. (1975) have voiced concern about the regulatory effects that
the twin dams Kariba and Cabora Bassa may have on the Lower Zambezi
coastal floodplain. The Zambezi delta is 18,000 km^2, with a sea
frontage 120 km long (Tinley, 1975). Within this wilderness is the
Marromeu Buffalo Reserve, which, pre-Cabora Bassa, contained 16,000
buffalo and some 6,000 head of game. As at Mana Pools Reserve,
Marromeu is dependent upon regular flooding for nutrients, alluvium,
and rest periods from grazing pressure. Lack of floods will disrupt
the reproductive and feeding cycles of game, with overgrazing and
subsequent reduction in carrying capacity a long-term outcome
(Davies et al., 1975).

COASTAL AND MARINE EFFECTS

Regulation of rivers, regardless of the form it takes, can
cause considerable environmental perturbations great distances from
source. In the case of the Nile, some 34×10^9 tons of water yr^{-1}
and 14×10^7 tons of mud and silt once flowed into the Delta and
Mediterranean (Ben-Tuvia, 1973). During the main flood (August-
December), sea temperatures increased by 1-2°C, salinity dropped
from 39.3-39.4 to 28 ppt (Ben-Tuvia, 1973) and the mainstream flow
was of the order of 6 miles per day (Liebman, 1935). Oren (1969,
1970) and Oren and Hornung (1972) have shown that the normal September minimum of surface salinity off the Israeli coast disappeared
post-Aswan (1964), with a subsequent increase in salinity for the
whole water column. Plankton blooms, which occurred during this
period, attracted large shoals of *Sardinella aurita* and *S. madarensis*,
upon which Egyptian coastal fisheries depended (El-Zarka and Koura,
1965a,b; Ben-Tuvia, 1973; Ryder and Henderson, 1974). Since Aswan,
Egyptian *Sardinella* catches have collapsed from 4,600 t (1965) to
554 t (1966) (Aleem, 1969), and Israeli catches have shown a
decrease in average length and mass (Ben-Tuvia, 1973). Recent
statistics (Ryder and Henderson, 1974) show that Egyptian catches
have continued at a reduced level.

In addition to flow regulation by Aswan High Dam, Oren (1969) has noted even more drastic regulation caused by construction of earth dams for irrigation in the Delta. Here, *no water* reaches the sea from the Delta between March–July. Only as floods approach (August–November) are the dams opened. This general reduction of flow by both Aswan and manipulation of the Delta has led to the interesting prediction by Ben-Tuvia (1973) that an increase in westerly spread of Red Sea fish species, which have entered the eastern Mediterranean via the Suez Canal, can take place now that the Nile flood, which once acted as a barrier, has been reduced.

The Nile Delta itself forms a huge underground reservoir, with the river forming the main recharger (Rzóska, 1976). An upper stratum of freshwater-saturated sediment overlies a layer of salt water, and the very existence of the Delta lakes in their present form depends entirely upon the Nile floods, which keep a balance between fresh and salt water. The drop in flow and sediment deposition is expected to lead to a gradual decrease in fish production.

In other systems, similar coastal problems have arisen, although quantitative observations are lacking. Hall and Davies (1974) and Davies et al. (1975) have, for example, speculated upon the future of the extensive shrimp fishing industry in the Zambezi delta. The effects of flood, silt, and nutrient-input reduction are completely unknown and no studies are planned. And what of sea water penetration into estuaries in conditions of lower, regulated flow and of the effects upon estuarine and riverine biota? We have already examined the *Egeria* problem on the Volta but have no knowledge of other, less directly important benthic organisms. Similarly, we have no idea of the long-term effects of stream regulation upon the coastal and estuarine swamplands. In a recent study of the Mgobezeleni estuary Kwa-Zulu, Bruton and Appleton (1975) showed how even minor interference with the flow of a stream can have devastating ecological effects. In this case, a road causeway interfered with tidal exchange and caused an increase in water level of the river, which flooded a mangrove swamp consisting of *Bruguira gymnorrhiza* (L.) Lam. This species cannot tolerate permanent flooding, and without a daily tidal exchange, seeds cannot take root and the pneumatophores of plantlings and older trees are drowned. Such changes in flow and consequent die back has led to increased space for invasion by such hydrophilic species as freshwater sedges and reeds, as well as major changes in the fauna.

Similar changes may have also taken place in the Zambezi delta, for during a reconnaissance survey of the area in 1974, the whole sea frontage of mangrove exhibited a 1–400 m die back (Davies, unpubl. data). Considerable erosion of the marshes and mud flats of this area had also taken place, though the degree of damage was impossible to ascertain because of the lack of comparable data. In

the light of the changes described for a small system like Mgobezeleni
and the effects of a small causeway, the question arises, Was regulation of the Zambezi by Kariba responsible? We do not know, but
obviously doubts remain, and if Kariba was responsible, the addition
of Cabora Bassa is of great importance to the region.

CONCLUSIONS

All major rivers in South Africa have impoundments on them,
which will, in general, alter the frequency and size of floods and
stabilize flow. Undoubtedly, such changes will considerably alter
the biota, but, unfortunately, most of the changes will be unmonitored
(but certainly not unnoticed). Some rivers will become perennial
instead of seasonal and vice versa, and, as Chutter (1973) states,
"As the optimum value of hydroelectric power in the South African
context is its use at peak demand periods, the implication is that
water for power generation will flow in daily pulses, predominantly
in winter." A similar picture can be painted for the rest of Africa,
which, as a predominantly "third world" region, is rapidly turning
to industrialization, with a concomitant increase in demand for
water resources. In this context, there appears to be a real need
for *quantitative* research in the following areas: (1) rationaliza-
tion of flood control to meet, as far as possible, the ecological
needs of floodplains, coastal regions, and fisheries, with particular
reference to aspects of herbivore grazing pressure and vegetation
changes, flushing action and nutrient replenishment, salinification,
salinity penetration, tidal exchange, and effects on fish spawning
migrations; (2) in-depth analyses of factors influencing zoobenthos
components, and (3) as African rivers appear to be generally nutrient
poor, further quantitative investigations are necessary to ascertain
the direct effects of epi- and hypolimnetic discharges in terms of
plant nutrients, minerals, and temperature. For point 3, particular
attention should be paid to the influence of tributary rivers
entering regulated streams below their point of regulation. Additional research effort should be directed towards *elucidating the
effects of small regulatory structures* on stream physico-chemistry
and ecology.

ACKNOWLEDGMENTS

I would like to sincerely acknowledge the support of the following individuals and organizations: Dr. M. N. Bruton,
Dr. C. Howard-Williams, Mr. D. Eccles, and Mr. P. B. N. Jackson for
useful discussion; Mr. W. Holleman for photo-reduction of Figs. 1 and
2 and Table 1; S. Terry, J. Pons, P.-A. King, and C. R. Fredericks
for valuable research assistance; Professor J. Heeg and Mr. K. Rogers
for information on the Pongolo River; Dr. R. C. Hart, Mr. P. B. N.

Jackson, and Mr. J. Cambray for information on recent operation
procedures at P. K. le Roux Dam; Mr. D. Eccles for information on
the Shiré River, Malawi; Mr. P. B. N. Jackson for allowing access
to his reprint collection; the Ernest Oppenheimer Memorial Trust,
the South African Council for Scientific and Industrial Research,
Rhodes University, and Coulter Electronics S.A. (Pty) Ltd., all
of whom contributed generously to my travel expenses, and
Mrs. L. Allen-Rowlandson and Mrs. P. M. Eva for their considerable
patience and typing accuracy.

REFERENCES

Abdin, G., 1948a, Physical and chemical investigations relating to
    algal growth in the River Nile, Cairo, *Bull. Inst. Egypte*,
    29:19-44.
Abdin, G., 1948b, Seasonal distribution of phytoplankton and sessile
    algae in the River Nile, Cairo, *Bull. Inst. Egypte*, 29:369-382.
Aleem, A. A., 1969, Marine resources of the United Arab Republic,
    Stud. Rev. Gen. Fish. Coun. Mediterr., 43:1-22.
Anon., 1975, What's happening to the Zambezi?, *Afr. Wildl.*,
    29(2):18-21.
Attwell, R. I. G., 1970, Some effects of Lake Kariba on the ecology
    of a floodplain of the Zambezi Valley of Rhodesia, *Biol.
    Conserv.*, 2(3):189-196.
Balon, E. K., 1973a, The eels of Siengwazi Falls (Kalomo River,
    Zambezi) and their significance, *Zambia Mus. J.*, 2:65-82.
Balon, E. K., 1973b, Results of fish population size assessments in
    Lake Kariba coves (Zambia), a decade after their creation, p.
    149-158, *in*: "Man-Made Lakes: Their Problems and Environmental
    Effects," W. C. Ackermann, G. F. White, and E. B. Worthington,
    eds., Am. Geophys. Union, Washington, D.C.
Balon, E. K., 1975, The eels of Lake Kariba: Distribution, taxonomic
    status, age, growth, and density, *J. Fish. Biol.*, 7:797-815.
Balon, E. K., and Coche, A. G., 1974, "Lake Kariba, a Man-Made
    Tropical Ecosystem in Central Africa," Monographiae Biologicae,
    24, Junk, The Hague, p. 767.
Beadle, L. C., 1974, "The Inland Waters of Tropical Africa: An
    Introduction to Tropical Limnology," Longman, Inc., New York,
    p. 365.
Begg, G. W., 1973, "The Biological Consequences of Discharge Above
    and Below Kariba Dam," Commission Internationale des Grands
    Barrages, Eleventh Congress, Madrid, p. 421-430.
Ben-Tuvia, A., 1973, Man-made changes in the eastern Mediterranean
    Sea and their effect on the fishery resources, *Mar. Biol.*,
    19(3):197-203.
Bond, W. J., Coe, N., Jackson, P. B. N., and Rogers, K. H., 1978,
    The limnology of Cabora Bassa, Mocambique, during its first
    year, *Freshwater Biol.*, 8:433-447.

Bowmaker, A. P., 1960, "A Report on the Kariba Lake Area and Zambezi
    River Prior to Inundation and the Initial Effects of Inundation
    with Particular Reference to the Fisheries," Rep. on Training
    Centre on Fishery Surveys for the Countries of Central Africa,
    F.A.O., No. 1299, Rome, p. 100-126.
Bowmaker, A. P., Jackson, P. B. N., and Jubb, R. A., 1978, Freshwater
    fisheries, p. 1183-1230, in: "Biogeography and Ecology of
    Southern Africa," M. J. A. Werger and A. C. van Bruggen, eds.,
    Junk, The Hague.
Brook, A. J., and Rzóska, J., 1954, The influence of the Gebel
    Aulyia Dam on the development of Nile plankton, J. Anim. Ecol.,
    23:101-115.
Brown, A. W. A., and Deom, J. O., 1973, Summary: Health aspects of
    man-made lakes, p. 755-764, in: "Man-Made Lakes: Their Problems
    and Environmental Effects," W. C. Ackermann, G. F. White, and
    E. B. Worthington, eds., Am. Geophys. Union, Washington, D.C.
Bruton, M. N., and Appleton, C. C., 1975, Survey of Mgobezeleni Lake
    system in Zululand, with a note on the effect of a bridge on
    the mangrove swamp, Trans. Roy. Soc. S. Afr., 41(3):283-294.
Cambray, J. A., and Jubb, R. A., 1977, Dispersal of fishes via the
    Orange-Fish tunnel, South Africa, J. Limnol. Soc. S. Afr.,
    3(1):33-35.
Carey, T. G., Bell-Cross, G., Green, J., and Tait, C. C., 1967,
    Kafue River and floodplain research, Fish. Res. Bull. Zambia,
    3:9-31.
Chutter, F. M., 1968, On the ecology of the fauna of stones in the
    current in a South African river supporting a very large
    Simulium (Diptera) population, J. Appl. Ecol., 5:531-561.
Chutter, F. M., 1969, The effects of silt and sand on the inverte-
    brate fauna of streams and rivers, Hydrobiologia, 34(1):57-76.
Chutter, F. M., 1972, Notes on the biology of South African
    Simuliidae, particularly Simulium (Eusimulium) nigritarse
    Coquillet, Limnol. Soc. S. Afr. Newsl., 18:10-18.
Chutter, F. M., 1973, An ecological account of the past and future
    of South African rivers, Limnol. Soc. S. Afr. Newsl., 21:22-34.
Coche, A. G., 1968, Description of physico-chemical aspects of Lake
    Kariba, an impoundment, in Zambia-Rhodesia, Fish. Res. Bull.
    Zambia, 5:200-267.
Coke, M. M., 1970a, "The Status of the Pongolo Floodplain Pans and
    the Effects of a 1,000 Cusec Flood in December 1976," Unpubl.
    Rep., Natal Parks, Game, and Fish Preservation Board,
    Pietermaritzburg.
Coke, M. M., 1970b, "Fish Yields in the Pongolo Flood-Plain Pans,"
    IBP Section P.F.: Freshwater productivity, Working Group Annu.
    Rep. No. 4:32-34.
Coke, M. M., 1971, "Flooding of the Pongolo Pans, January and March,
    1971," Cyclostyled Rep. to Natal Parks, Game, and Fish Preser-
    vation Board, Pietermaritzburg.

Coke, M. M., and Pott, R. McC., 1970, "The Pongolo Flood-Plain Pans, A Plan for Conservation," IBP Section P.F. Freshwater productivity, Working Group Annu. Rep., 4:19-22.

Coke, M. M., and Pott, R. McC., 1971, "The Pongolo Flood-Plain Pans, A Plan for Conservation," Natal Parks, Game, and Fish Preservation Board, Pietermaritzburg, p. 34.

Davies, B. R., 1975a, Cabora Bassa hazards, *Nature (Lond.)*, 254:477-478.

Davies, B. R., 1975b, They pulled the plug out of the Lower Zambezi, *Afr. Wildl.*, 29(2):26-27.

Davies, B. R., Hall, A., and Jackson, P. B. N., 1975, Some ecological aspects of the Cabora Bassa Dam, *Biol. Conserv.*, 8:189-201.

du Preez, J. W., 1967, "The Geology and Geohydrology of the Makatini Flats, with Special Reference to the Problems of Salinization and Waterlogging," Unpubl. Rep., Geol. Surv., Dep. Mines, Pretoria.

Edwards, D., 1967, "A Plant Ecological Survey of the Tugela River Basin, Natal," Mem. Bot. Surv. S. Afr., 36, Pretoria.

Edwards, D., 1969, Some effects of siltation upon aquatic macrophyte vegetation in rivers, *Hydrobiologia*, 34:29-37.

Edwards, D., and Musil, C. J., 1975, *Eichhornia crassipes* in South Africa--a general review, *J. Limnol. Soc. S. Afr.*, 1(1):23-27.

El-Zarka, S., 1973, Kainji Lake, Nigeria, p. 197-219, *in*: "Man-Made Lakes: Their Problems and Environmental Effects," W. C. Ackermann, G. F. White, and E. B. Worthington, eds., Am. Geophys. Union, Washington, D.C.

El-Zarka, S., and Koura, R., 1965a, Seasonal fluctuations in production of the principle edible fish in the Mediterranean Sea off the United Arab Republic, *Proc. Gen. Fish. Coun. Mediterr.*, 8:227-259.

El-Zarka, S., and Koura, R., 1965b, "Seasonal Fluctuations in the Production of the Important Food Fishes of the U.A.R. Waters of the Mediterranean Sea," Tech. Pap. 8th Session Gen. Fish. Counc. Mediterranean, No. 34/65, FAO, Rome, May 10-15, 1965.

Entz, B., 1976, Lake Nasser and Lake Nubia, p. 271-298, *in*: "The Nile. Biology of an Ancient River," J. Rzóska, ed., Monographiae Biologicae, 29, Junk, The Hague.

Ewer, D. W., 1966, Biological investigations on the Volta Lake May 1964 to May 1965, p. 21-31, *in*: "Man-Made Lakes," R. H. Lowe-McConnell, ed., Symp. Inst. Biol., 15, Acad. Press, New York and London.

Furness, H. D., 1978, "Ecological Studies on the Pongolo River Floodplain," Working Document IV, Workshop on Man and the Pongolo Floodplain, C.S.I.R., NP 14/106/7C, held at Pietermaritzburg, January 31-February 2, 1978, p. 10.

Hall, A., and Davies, B. R., 1974, Cabora Bassa: Apreciação global do seu impacto no vale do Zambeze, *Rev. Econ. Moçambique*, 11(7):15-25.

Hall, A., Davies, B. R., and Valente, I., 1976, Cabora Bassa: Some preliminary physico-chemical and zooplankton pre-impoundment survey results, *Hydrobiologia*, 50:17-25.

Hall, A., Valente, I., and Davies, B. R., 1977, The Zambezi River in Moçambique: The physico-chemical status of the Middle and Lower Zambezi prior to the closure of the Cabora Bassa Dam, *Freshwater Biol.*, 7:187-206.

Hall, J. B., and Pople, W., 1968, Recent vegetational changes in the lower Volta River, *Ghana J. Sci.*, 8:24-29.

Hammerton, D., 1972, The Nile River--a case history, *in*: "River Ecology and Man," Acad. Press, London and New York.

Hammerton, D., 1976, The Blue Nile in the Plains, p. 243-256, *in*: "The Nile. Biology of an Ancient River," J. Rzóska, ed., Monographiae Biologicae 29, Junk, The Hague.

Harrison, A. D., 1966, Recolonization of a Rhodesian stream after drought, *Arch. Hydrobiol.*, 62:405-421.

Heeg, J., Breen, C. M., Colvin, P. M., Furness, H. D., and Musil, C. F., 1978, On the dissolved solids of the Pongola flood plain pans, *J. Limnol. Soc. S. Afr.*, 4(1):59-64.

Henderson, F., 1973, Stratification and circulation in Kainji Lake, p. 489-494, *in*: "Man-Made Lakes: Their Problems and Environmental Effects," W. C. Ackermann, G. F. White, and E. B. Worthington, eds., Am. Geophys. Union, Washington, D.C.

Imevbore, A. M. A., 1970, The chemistry of the River Niger in the Kainji Reservoir area, *Arch. Hydrobiol.*, 67(3):412-431.

Jackson, P. B. N., 1961, "Ichthyology: The Fish of the Middle Zambezi, Kariba Studies, 1," Manchester Univ. Press, Manchester.

Jackson, P. B. N., 1966, The establishment of fisheries in man-made lakes in the tropics, p. 53-69, *in*: "Man-Made Lakes," R. H. Lowe-McConnell, ed., Symp. Inst. Biol., 15, Acad. Press, New York and London.

Jackson, P. B. N., 1975, Fish, p. 259-275, *in*: "Dam-Induced Ecological Disturbances Affecting Human Health," N. F. Stanley and M. P. Alpers, eds., Acad. Press, New York and London.

Jackson, P. B. N., and Davies, B. R., 1976, Cabora Bassa in its first year: Some ecological aspects and comparisons, *Rhod. Sci. News*, 10(5):128-133.

Jackson, P. B. N., and Rogers, K. H., 1976, Cabora Bassa fish populations before and during the first filling phase, *Zool. Afr.*, 11(2):373-397.

Jubb, R. A., 1960, Some Lake Kariba fish problems, *Piscator*, 47:112-119.

Jubb, R. A., 1961, The freshwater eels (*Anguilla* spp.) of southern Africa. An introduction to their identification and biology, *Ann. Cape Prov. Mus.*, 1:15-48.

Jubb, R. A., 1964, The eels of South African rivers and observations on their ecology, p. 186-205, *in*: "Ecological Studies in Southern Africa," Monographiae Biologicae, 14, Junk, The Hague.

Jubb, R. A., 1967, "Freshwater Fishes of Southern Africa," A. A.
    Balkema, Cape Town and Amsterdam.
Jubb, R. A., 1972, The J. G. Strydom Dam: Pongola River: Northern
    Zululand. The importance of the floodplain pans below it,
    Piscator, 86:104-109.
Kenmuir, D. H. S., 1975, Sardines in Cabora Bassa Lake?, *New Sci.*,
    13:379-380.
Kenmuir, D. H. S., 1976, Fish spawning under artificial flood
    conditions on the Mana flood-plain, Zambezi River, *Kariba
    Stud.*, 6:86-97.
Kok, H. M., 1978, "Ecological Studies on Some Important Fish Species
    of Pongola Floodplain, KwaZulu," Working Document X, Workshop
    on Man and the Pongola Floodplain, C.S.I.R., NP/14/106/7C,
    held at Pietermaritzburg, January 31-February 2, 1978, p. 7.
Lagler, K. F., 1969, "Man-Made Lakes Planning and Development,"
    F.A.O., Rome, p. 71.
Lawson, G. W., Petr, T., Biswas, S., Biswas, E. R. I., and
    Reynolds, J. D., 1969, "Hydrobiological Work of the Volta Basin
    Research Project, 1963-1968," *Bull. l'I.F.A.N.*, 31 *Sér* A,
    3:965-1003.
Lawson, R. M., 1963, The economic organization of the *Egeria* fishing
    industry on the River Volta, *Proc. Malac. Soc. Lond.*,
    35:273-287.
Lelek, A., and El-Zarka, S., 1973, Ecological comparison of the pre-
    impoundment and post-impoundment fish faunas of the River Niger
    and Kainji Lake, Nigeria, p. 655-660, *in*: "Man-Made Lakes:
    Their Problems and Environmental Effects," W. C. Ackermann,
    G. F. White, and E. B. Worthington, eds., Am. Geophys. Union,
    Washington, D.C.
Leopold, L. B., Wolman, M. G., and Miller, J. P., 1964, "Fluvial
    Processes in Geomorphology," Freeman, San Francisco, p. 522.
Lewis, D. J., 1966, Nile control and its effect on insects of the
    Nile, p. 43-45, *in*: "Man-Made Lakes," R. H. Lowe-McConnell,
    ed., Symp. Inst. Biol., 15, Acad. Press, New York and London.
Liebman, E., 1935, Oceanographic observations of the Palestine
    Coast, *Rapp. P-v. Comm. Int. Explor. Sci. Mar. Mediterr.*,
    9:181-185.
Mitchell, D. S., Bowmaker, A. P., Osborne, P. L., Ferreira, J., and
    Gibbs Russell, G. E., 1975, "Environmental Impacts of the
    Darwendale Dam, Rhodesia," Div. Biol. Sci. Rep., Univ. Rhodesia,
    p. 45.
Moghrabi, A., 1972, "The Zooplankton of the Blue Nile," Unpubl.
    Ph.D. Thesis, Faculty of Sci., Univ. Khartoum.
Mostert, J. W. C., 1958, "Studies of the Vegetation of Parts of the
    Bloemfontein and Brandfort District," Mem. Bot. Surv. S. Afr.,
    31, Pretoria.
Noble, R. G., 1972, Hydrobiological studies of the Orange River
    system in relation to Hendrik Verwoerd Dam, *Civ. Eng. S. Afr.*,
    14(2):79-80.

Noble, R. G., and Hemens, J., 1978, "Inland Water Ecosystems in
    South Africa--A Review of Research Needs," S. Afr. Natl. Sci.
    Prog. Rep. No. 34, C.S.I.R., Pretoria, p. 150.
Oren, O. H., 1969, Oceanographic and biological influence of the
    Suez Canal, the Nile, and the Aswan Dam on the Levant Basin,
    Prog. Oceanogr., 5:161-167.
Oren, O. H., 1970, The Suez Canal and the Aswan High Dam, their
    effect on the Mediterranean, *Underwater Sci. Tech. J.*, Dec.,
    222-229.
Oren, O. H., and Hornung, H., 1972, Temperatures and salinities off
    the Israel Mediterranean coast, *Bull. Sea Fish. Res. Stn.,
    Israel*, 59:17-31.
Petr, T., 1974, "A Pre-impoundment Limnological Study, with Special
    Emphasis on Fishes of the Great Ruaha River (Tanzania) and
    Some of its Tributaries (River Yovi and Little Ruaha) in and
    Around the Proposed Impoundment Areas," TANESCO Rep., Uppsala,
    p. 104.
Phelines, R. F., Coke, M., and Nicol, S. M., 1973, "Some Biological
    Consequences of the Damming of the Pongola River," Commission
    Internationale des Grands Barrages, Eleventh Congress, Madrid,
    p. 175-191.
Pitchford, R. J., and Visser, P. S., 1975, The effect of large dams
    on river water temperature below the dams, with special refer-
    ence to Bilharzia and the Verwoerd Dam, *S. A. J. Sci.*,
    71(7):212-213.
Pooley, E. S., 1978, "Studies on the Vegetation in and Around Ndumu
    Game Reserve, Northern Zululand," Working Document XII, Workshop
    on Man and the Pongola Floodplain, C.S.I.R., NP/14/106/7C, held
    at Pietermaritzburg, January 31-February 2, 1978, p. 5.
Pople, W., and Rogoyska, H. S., 1969, Salinity penetration up the
    Volta River during the building of the Akosombo Dam, *Ghana J.
    Sci.*, 8:16.
Pott, R. McC., 1968, "The Fish Life of the Pongola River and the
    Effect of the Erection of a Dam on the Fish Populations," IBP
    Section P.F.: Freshwater productivity, Working Group Annu.
    Rep. No. 2, p. 69-82.
Pretorius, S. J., Jennings, A. C. Coertze, D. J., and van Eeden,
    J. A., 1975, Aspects of the freshwater Mollusca of the Pongola
    River flood plain pans, *S. A. J. Sci.*, 71(7):208-212.
Prowse, G. A., and Talling, J. F., 1958, The seasonal growth and
    succession of plankton algae in the White Nile, *Limnol.
    Oceanogr.*, 3:222-238.
Purchon, R. D., 1963, A note on the biology of *Egeria radiata*
    (Bivalvia, Donacidae), *Proc. Malac. Soc. Lond.*, 35:251-271.
Raheja, P. C., 1973, Lake Nasser, p. 234-245, *in*: "Man-Made Lakes:
    Their Problems and Environmental Effects," W. C. Ackermann,
    G. F. White, and E. B. Worthington, eds., Am. Geophys. Union,
    Washington, D.C.

Rees, W. A., 1978a, The ecology of the Kafue Lechwe: Soils, water
    levels, and vegetation, *J. Appl. Ecol.*, 15:163-176.
Rees, W. A., 1978b, The ecology of the Kafue Lechwe: As affected by
    the Kafue Gorge hydroelectric scheme, *J. Appl. Ecol.*, 15:205-217.
Rees, W. A., 1978c, Do the dams spell disaster for the Kafue Lechwe?,
    Oryx, 14(3):231-234.
Rogers, K. H., 1978, "Autecological Studies on *Potamogeton crispus*
    L. with Special Reference to the Pongola River Floodplain,"
    Working Document XIX, Workshop on Man and the Pongola Flood-
    plain, C.S.I.R., NP/14/106/7C, held at Pietermaritzburg,
    January 31-February 2, 1978, p. 11.
Ryder, R. A., and Henderson, H. F., 1974, "Fish Yield Projections on
    the Nasser Reservoir (Including Lake Nubia, the Sudan)," F.A.O.
    Tech. Rep. FI:DP/EGY/66/558, p. 32.
Rzóska, J., ed., 1976, "The Nile. Biology of an Ancient River,"
    Monographiae Biologicae 29, Junk, The Hague, p. 417: Chapters
    on Nile sediments; Lake Victoria; the White Nile; Delta Lakes;
    and Zooplankton.
Rzóska, J., Brook, A. J., and Prowse, G. A., 1955, Seasonal plankton
    development in the White and Blue Nile near Khartoum, *Ver. Int.
    Verein. Theor. Angew Limnol.*, 12:327-334.
Schwartz, H. I., 1969, Hydrologic aspects of limnology in South
    Africa, *Hydrobiologia*, 34:14-28.
Smith, I. R., 1975, "Turbulence in Lakes and Rivers," Freshwater
    Biol. Assoc. Publ. No. 29.
Talling, J. F., 1976, Chapters on water characteristics, and phyto-
    plankton, p. 357-384 and 385-402, *in*: "The Nile. Biology of
    an Ancient River," J. Rzóska, ed., Monographiae Biologicae 29,
    Junk, The Hague.
Talling, J. F., and Rzóska, J., 1967, The development of plankton in
    relation to hydrological regime in the Blue Nile, *J. Ecol.*,
    55:637-662.
Taylor, E. B., 1973, People in a rapidly changing environment: The
    first six years of Volta Lake, p. 99-107, *in*: "Man-Made Lakes:
    Their Problems and Environmental Effects," W. C. Ackermann,
    G. F. White, and E. B. Worthington, eds., Am. Geophys. Union,
    Washington, D.C.
Tinley, K. L., 1971, "Determinants of Coastal Conservation: Dynamics
    and Diversity of the Environment as Exemplified by the
    Moçambique Coast," Proc. SARCCUS Symp. Nature conservation as a
    form of land use, Gov. Printer, Pretoria, p. 125-153.
Tinley, K. L., 1975, Marromeu wrecked by the big dams, *Afr. Wildl.*,
    29(2):22-25.
Tinley, K. L., and Sousa Dias, A. H. G., 1973, Wildlife reconnais-
    sance of the Mid-Zambezi Valley in Mocambique before formation
    of the Cabora Bassa Dam, **Lourenço Marques**, *Veterin. Moçambique*,
    6(2):103-131.

Waddy, B. B., 1973, Health problems of man-made lakes: Anticipation and realization, Nigeria and Kossou, Ivory Coast, p. 765-768, *in*: "Man-Made Lakes: Their Problems and Environmental Effects," W. C. Ackermann, G. F. White, and E. B. Worthington, eds., Am. Geophys. Union, Washington, D.C.

Wafa, T. A., and Labib, A. H., 1973, Seepage losses from Lake Nasser, p. 287-291, *in*: "Man-Made Lakes: Their Problems and Environmental Effects," W. C. Ackermann, G. V. White, and E. B. Worthington, eds., Am. Geophy. Union, Washington, D.C.

Watermeyer, C. F., 1965, "A General Outline of Engineering Interests Associated with Lake Kariba," Lake Kariba Fish. Res. Inst. Rep., Cyclostyled, p. 24-32.

White, E., 1973, Zambia's Kafue hydroelectric scheme and its biological problems, p. 620-628, *in*: "Man-Made Lakes: Their Problems and Environmental Effects," W. C. Ackermann, G. F. White, and E. B. Worthington, eds., Am. Geophys. Union, Washington, D.C.

REGULATED STREAMS IN AUSTRALIA: THE MURRAY-DARLING RIVER SYSTEM

K. F. Walker

Department of Zoology
University of Adelaide
South Australia 5000

INTRODUCTION

Australia has an image, both at home and abroad, as a land of
wide open spaces. The land area is nearly that of the conterminous
United States, yet there are fewer than two people for each square
kilometer, compared with 26 people/km^2 in the U.S. The disparity is
related to the comparative scarcity of Australia's water resources.
Given low rainfall and high evaporative losses, yearly runoff
averages about 45 mm, only one-sixth the North American average. In
one year Australia's rivers together discharge a volume only two-
thirds that conveyed by the Mississippi River. Australia is, in
fact, the driest of the inhabited continents.

Perspectives like these commonly are cited in the literature on
Australian water resources and have been influential in planning and
development. National statistics may be misleading, however,
because in Australia both population and water resources are unevenly
distributed. In this regard, Australia is among the world's most
highly urbanized countries: More than 80% of the population, some
13 million, lives in or nearby the southeastern coastal cities of
Melbourne and Sydney. There, the per capita domestic and industrial
consumption of water is comparable to that in the densely populated
areas of Europe and North America.

Australian attitudes to water resources appear to be changing
from one favoring untrammeled development to one admitting the
necessity for management (see, for example, Australian Water
Resources Council, 1978). Development is, or has been, guided by
the "dry continent" viewpoint, with such a proliferation of man-made

water storages that few major rivers now are not subject to flow regulation. On the other hand, it is mainly the problems of urban Australia that have determined a role for management: Pollution and other environmental changes, including stream regulation, have stimulated a broad public awareness of the implications of careless development. Even so, effective management is hampered by a profound lack of even the most basic biological information about the nation's rivers.

Following an outline of the distribution of water resources and the extent of stream regulation, this paper surveys the very limited ecological knowledge that exists to support river management in Australia. The discussion concentrates upon a particular river system, the Murray-Darling system, for reasons that will be explained shortly. In conclusion, some recommendations are made for future research.

EXTENT OF STREAM REGULATION

The Australian Water Resources Council (1976) recognizes 12 principal drainage divisions (Fig. 1), with boundaries following physical divides and embracing related river systems. The largest of these, the Western Plateau Drainage Division, receives no sustained rainfall and is an area of largely uncoordinated drainage. The Lake Eyre and Bulloo-Bancannia Divisions are basins of internal drainage, and there again rainfall is unreliable. These three divisions together account for half the continental area, yet contribute no significant runoff (Table 1).

Nearly half Australia's total stream discharge derives from the Timor Sea, Gulf of Carpentaria, and Northeast Coast Drainage Divisions (Table 1), all in the northern part of the continent where rainfall, and hence stream flow, are determined by summer monsoons. Perennial rivers occur along the lower eastern and southeastern coastal margins, in Tasmania, and in the Murray-Darling Drainage Division, which has its main contributing catchment on the western slopes of the Dividing Range. Most of these permanent streams are influenced by winter rainfall, and some also by melting snows. The streams of drainage divisions V-XII (Fig. 1) generally are seasonal or episodic.

Table 1 contains some indication of the extent of stream regulation: It shows the numbers and combined capacities of the large dams and storages in each drainage division and allows a comparison of regulated and total discharges. The information, however, needs some elaboration.

PRINCIPAL DRAINAGE DIVISIONS

I	North - East Coast
II	South - East Coast
III	Tasmanian
IV	Murray - Darling
V	South Australian Gulf
VI	South - West Coast
VII	Indian Ocean
VIII	Timor Sea
IX	Gulf of Carpentaria
X	Lake Eyre
XI	Bulloo - Bancannia
XII	Western Plateau

Fig. 1. Drainage divisions of Australia.

Although a clear majority (64%) of dams and storages has been constructed mainly for purposes of water supply (Table 1), only about 15% of all annual water diversions is to meet needs other than irrigation (Australian Water Resources Council, 1976). Irrigation, therefore, is the dominant influence in matters of stream-flow regulation. Australia has one of the largest irrigated areas, per capita, of any country, with a particular concentration along the River Murray and its tributaries.

A clearer picture of the extent of flow regulation emerges from Table 2. In this case the possible exploitable yield has been estimated (Australian Water Resources Council, 1976) for each drainage division, with regard for logistic (but not economic) factors and standards of acceptable water quality. The exploitable yield of most divisions is less than half the total discharge. The table shows also that for the nation as a whole only 24% of the possible exploitable yield is committed (now or in the foreseeable future) to consumptive use. This low figure, however, is not representative of the Murray-Darling Division, where 91% of the available water is committed for use. Given also that 83% of the total discharge is regarded as available for exploitation, it is clear that rivers in this division are under considerable pressure. It is not a coincidence that virtually all existing information on the ecology of regulated streams in Australia pertains to the Murray and its tributaries. Most of the following discussion takes this direction.

Table 1. Basic Data on Regulated Streams in Australia (Australian Water Resources Council, 1976). Runoff and Flow Data are Annual Means, Volumes are Millions m³ (I = Irrigation, H = Hydroelectric, W = Water Supply, F = Flood Control and Recreation)

Drainage Division	Area (km²)	Runoff (mm)	Number of Storages	Main Purpose				Total Capacity	Discharge	
				I	H	W	F		Regulated	Total
I Northeast Coast	450,705	183	33	12	1	20	--	4,100	1,300	82,500
II Southeast Coast	273,553	144	99	5	5	87	2	10,700	2,700	39,396
III Tasmania	68,200	730	43	1	31	11	--	19,500	8,700	49,799
IV Murray-Darling	1,062,530	21	104	34	15	53	2	20,700	10,500	22,261
V South Australian Gulf	82,300	11.9	24	--	--	23	1	240	150	980
VI Southwest Coast	314,090	23	24	8	--	15	1	870	360	7,290
VII Indian Ocean	518,570	8	1	--	--	1	--	a	a	4,160
VIII Timor Sea	547,050	136	8	5	--	3	--	6,100	1,900	74,260
IX Gulf of Carpentaria	638,460	91	4	--	--	4	--	140	10	58,230
X Lake Eyre	1,170,000	3	2	1	--	1	--	a	a	3,260
XI Bulloo-Bancannia	100,570	5.4	--	--	--	--	--	--	--	540
XII Western Plateau	2,455,000	a	--	--	--	--	--	--	--	a
Australia	7,681,028	44.6	342	66	52	218	6	62,350	25,620	342,676

a Negligible

Table 2. Exploitation of Stream Flow in Australia (Australian Water Resources Council, 1976)

	Drainage Division												
	I	II	III	IV	V	VI	VII	VIII	IX	X	XI	XII	Australia
Exploitable yield as a percentage of total discharge	34	43	71	83	31	37	13	22	21	4	?	0	39
Total commitment as a percentage of exploitable yield	13	24	5	91	61	20	15	19	2	7	?	0	24

THE MURRAY-DARLING SYSTEM

The Murray-Darling system gathers water from nearly a seventh
of Australia's land area, and is among the most extensive river
systems in the world. The main tributaries and associated dams,
weirs, and storages are represented in Fig. 2.

The Murray, the principal river, rises in the Snowy Mountains
near Mount Kosciusko (2228 m) and follows a generally westward
course to the sea in South Australia. It has a short (350 km),
well-watered mountain tract, but for the greater part of its 2570-km
length flows across semi-arid inland plains. Major flow contribu-
tions come from the Murrumbidgee and Goulburn rivers, and particularly the Murray and its tributaries upstream from the Murrumbidgee
junction. Natural flows in the uppermost reaches of the Murray and
Murrumbidgee are supplemented by diversions from the Snowy Mountains
Hydroelectric Scheme. The contribution from the Darling River is
extremely variable, but on average relatively small.

Although the average annual discharge of all river basins in
the system amounts to 22,261 million m^3, only 10,500 million m^3
passes along the mainstem of the Murray as regulated discharge
(Table 1). The discrepancy is accounted for by direct diversions
and losses to groundwater, evaporation, terminal lakes, and other
avenues (Australian Water Resources Council, 1976). The remainder,
regulated discharge, is subject to the requirements of the three
riparian states (New South Wales, Victoria, and South Australia),
and under present conditions less than half reaches the sea.

The meagre quantities of water available from the Murray-
Darling system, and the variability of river flows (see below), have
encouraged development of a complex network of dams, weirs, barrages,
diversions, and off-river storages, operated by the River Murray
Commission on behalf of the states. Flow regulation will be enhanced
when Dartmouth Dam, recently constructed on the Mitta Mitta River,
becomes operational. The new dam will impound a lake of 4000 million
m^3 and become the largest of the Murray-Darling storages. Other
large impoundments are Lake Eildon (3390 million m^3) on the Goulburn
River and Lake Hume (3038 million m^3) on the Murray, at the Mitta
Mitta confluence. Several other dams, with storage capacities
ranging between 1000 and 1700 million m^3, are indicated in Fig. 2.

Some 80 km downstream of Lake Hume, Yarrawonga Weir provides
for a gravity diversion to two large irrigation canals and impounds
a shallow but extensive lake, Mulwala. The weir does not have a
navigable pass and so is a barrier to upstream movement of boats and
river animals. Two similar concrete overflow weirs, again without
locks, are located on the lower Murrumbidgee River. Distributed
along the length of the Murray below Yarrawonga Weir is a series of

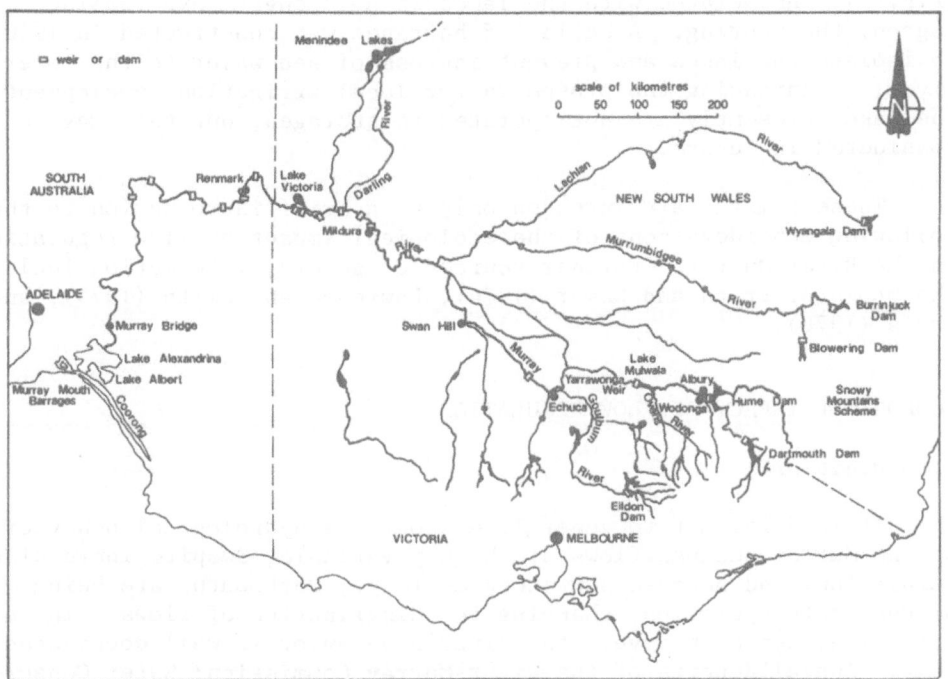

Fig. 2. Part of the Murray-Darling river system, showing main
rivers, principal dams and weirs, and some geographic
points.

13 lock-and-weir constructions, 9 of them located between the Darling junction and the Murray mouth. These were installed in the 1920s and 1930s to assist riverboat traffic but, with the demise of the riverboats, have assumed a role in irrigation flow management. In South Australia the weirs have changed the Murray to a sequence of nearly overlapping pondages.

The Murray enters the sea via two large shallow silted lakes, Alexandrina and Albert, separated from the sea by a narrow sand dune (Fig. 2). Associated with the lakes is an elongate hypersaline lagoon, the Coorong. A series of barrages was constructed in 1940 to isolate the lakes and prevent ingress of sea water to the lower river, so enhancing the prospects for local irrigation development. The lakes presently are not operated as storages, but this may be considered in future.

These remarks are intended only as a brief introduction to the following considerations of the ecological impact of flow regulation on the River Murray. Further sources of general information include the books of Frith and Sawer (1974), Lawrence and Smith (1975), and Davis (1978).

ECOLOGICAL IMPACT OF FLOW REGULATION

Flow Behavior

It is difficult to generalize about the hydrological behavior of the Murray because flows are highly variable, despite increasing regulation, and because new storages (e.g., Dartmouth) are being brought into operation, changing the distribution of flows. In an historical sense, however, the river's behavior is well documented (e.g., Annual Reports of the River Murray Commission; Water Conservation and Irrigation Commission, 1975). General accounts, appropriate to different stages of tributary storage development, are given in Frith and Sawer (1974) and Lawrence and Smith (1975). A useful recent summary is provided by Baker and Wright (1978).

Following Baker and Wright (1978), three broad effects of water resource development on river flows are recognizable. First, the seasonal distribution of flows has changed: winter flows have been decreased (as surplus water is taken into storage) and summer flows increased (as irrigation demands are met). The effect is illustrated by Fig. 3, showing river flows at Albury, immediately downstream from Hume Dam, before and after construction of the dam (first stage, 1936). Second, flow regulation has been used to provide some degree of flood amelioration. When storage is available, as in winter, surplus water may be impounded and, if necessary, released later in controlled amounts. Operations at Hume Dam have

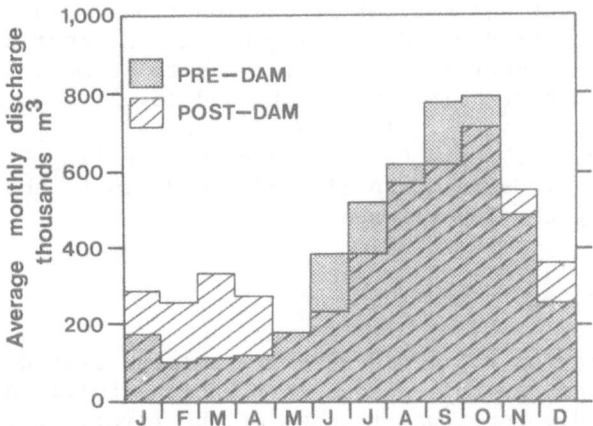

Fig. 3.   Monthly flows in the River Murray at Albury before and
          after construction of Hume Dam (Water Conservation and
          Irrigation Commission, 1975).

significantly reduced the severity and incidence of flooding along
the Murray near Albury. Third, flow regulation has set a long-term
continuing trend toward reduction of average flows in the middle
reaches of the Murray (Baker and Wright, 1978). The degree of
control presently exercised through Hume Dam will be enhanced
greatly once Dartmouth Dam assumes full operation.

Siltation and Bank Erosion

     Suspended sediment is a troublesome pollutant in Australian
water supplies (Higginson, 1973). In the River Murray watershed,
agricultural development (involving land clearing, irrigation, and
pasture management) has no doubt caused substantial increases in the
river sediment load, although comprehensive investigations have not
been made.

     Severe problems of bank erosion exist along the Murray between
Hume Dam and Yarrawonga Weir. The problems arise from the way flows
are released from Lake Hume for irrigation downstream and are
presumably compounded by the increased erosive power of water
depleted of suspended sediment by settlement in storage (relevant
data given by Walker and Hillman, 1977). Another contributing
factor is that Lake Hume is situated where historically the Murray
emerged from its mountain tract to meet the first of its plains
tracts. Downstream from the lake the river spreads out in a maze of
meandering waterways, with numerous billabongs (ox-bow lakes) across
a broad floodplain. The local character of the river is not ideally

suited to efficient delivery of irrigation flows. The River Murray
Commission has carried out desnagging operations near Albury
(removing fallen timber from the river bed to promote flow) and is
investigating bank stabilization works and levee construction as
further possible measures. Introduced willows (*Salix* spp.) have
been planted in some areas to promote bank stability. In South
Australia they have spread rapidly and now dominate the riverside
vegetation along the lower reaches.

Salinity

Tertiary marine sediments underlie a large area of the Murray-
Darling Basin and generally have contributed significant amounts of
salt to the middle and lower reaches of the river. Extensive irrigation development has accelerated, by various means, the rate of
salt accession, and steadily increasing river salinities now present
a major problem to irrigation farmers and other water consumers.
Water quality deteriorates progressively with distance downstream,
and in South Australia salinities frequently exceed the limits
tolerated by irrigated crops (ca. 500 parts per million). The
problem is complex and has been given much government attention in
the last decade (e.g., Engineering and Water Supply Department,
1978); a review is given by Collett (1978). The salinity problem is
related directly to stream regulation, as subsurface flows along
hydraulic gradients created by the locks and weirs are an important
avenue of salt input to the river. In addition, releases from
upstream storages are influenced by the occasional need for diluting
flows in the lower river. Finally, although no directly relevant
studies have been carried out, there seems to be no real evidence
that present river salinities are adversely affecting the river
plants and animals.

Floodplain Communities

Eucalypt forests, predominantly the river red gum (*Eucalyptus
camaldulensis*), occupy large tracts of the Murray's floodplain.
Although the forests have been depleted greatly by logging and
sawmilling, a number of large red gum stands remain. Red gums are
dependent upon periodic flooding for seed germination but succumb to
prolonged flooding. Flow regulation has drastically curtailed
regeneration in most areas, and most impounded areas, particularly
Lake Mulwala, typically have large numbers of standing dead trees
associated with them. Dexter (1978) discusses red gum silviculture
with regard for river flow management.

Waterfowl populations in the Murray-Darling system have been
affected, though perhaps not dramatically, by flow regulation and

its reduction of habitat and limitation of breeding opportunities.
Frith (1977) provides inroads to an extensive literature on this
subject, and Braithwaite (1975), in particular, discusses approaches
to management.

Aside from their significance as habitats for waterfowl (and
other wildlife), the billabongs, swamps, and minor streams and lakes
of the Murray's floodplain are breeding refuges and habitats for a
variety of river animals (Shiel, 1976; Walker and Hillman, 1977).
In this context, the aquatic environments of the river floodplain
have been drastically affected by flow regulation and other changes.
The absence or curtailment of flooding limits opportunities for
exchanges between the parent river and all but the most recently
formed billabongs (those nearest the river channel). The remaining
billabongs are subject to use for stock watering, waste disposal,
and other purposes, and few if any could be considered pristine
environments. Furthermore, the present controlled-flow regime is
unlikely to promote development of new billabongs and swamps. It is
important, therefore, that steps be taken to preserve an ecologically
significant sample of the remaining billabongs and other floodplain
habitats. Clearly, this will require concessions from
flow-regulation authorities.

Eutrophication and Water Weeds

Phosphate and nitrate concentrations in the Murray and associated waters generally are well in excess of algal growth requirements (Walker and Hillman, 1977), although under certain conditions
nitrate may limit phytoplankton development in the lower reaches of
the river (Falter, 1978). Algal growth problems have occurred in
the lower river, forcing shutdowns of irrigation machinery, and also
in some of the Murray-Darling reservoirs. In Burrinjuck Reservoir
on the Murrumbidgee River, for example, there are problems with
recurrent blooms of the potentially toxic blue-green algae *Anacystis
cyanea* and *Anabaena circinalis* (May, 1978). There are potential
problems at Lake Mulwala and Lake Hume, on the Murray, in regard to
planned urban development at Albury-Wodonga (Walker and Hillman,
1977).

Problems of algal growth, however, are to a large extent
checked by prevailing high turbidity, so that potential nutrient
effects are over-ridden by light limitation (Walker and Hillman,
1977). Turbidity is a key factor in the lower Murray, where the
seasonal development of phytoplankton typically involves two growth
periods (Falter, 1978). One peak occurs in spring and involves
diatoms (*Melosira* spp.), and a second peak occurs in late summer,
with development of blue-green algae (*Anacystis cyanea, Anabaena
circinalis*), given certain environmental conditions. The summer

growth peak is dependent upon adequate light and temperatures (exceeding 22°C) and upon turbidities below an apparent threshold level of 30 NTU. A further condition is that nitrate concentrations be sufficiently low to discourage continued development of the spring growth. Seasonal changes in both nutrient and turbidity levels are influenced by river flow behavior (e.g., Walker and Hillman, 1977). In South Australia, in particular, the prevalence of impounded waters along the river must favor the growth of phytoplankton.

Water weeds (hydrophytes) presently are not a serious problem in the River Murray proper, perhaps because of generally high turbidities, unstable sediments and variable flows. There are, however, two areas of concern. An extensive raft of water hyacinth (*Eichhornia crassipes*) occurs on the Gingham Watercourse, part of the Darling drainage in New South Wales, and there are fears that an extensive flood could distribute plants more widely through the Murray-Darling system. Conditions in the lower Murray and a number of impounded waters evidently are conducive to the establishment of water hyacinth and other potential problem plants, including salvinia (*Salvinia molesta*). Another problem area concerns weed growth (and hence flow retardation) in irrigation canals supplied from the Murray and Murrumbidgee rivers. The problem species include *Elodea canadensis*, *Typha* spp., *Potamogeton* spp., and *Myriophyllum* spp. An excellent review of these problems and of similar problems elsewhere in Australia is provided by Mitchell (1978).

Water Quality Effects of Dams

The water-quality effects associated with Hume Dam on the Murray are described by Walker and Hillman (1977) and Walker et al. (1979), in reports originating from an ecological survey of the Murray with regard for planned urban development at Albury-Wodonga. It seems reasonable to consider the relationship between Lake Hume and the River Murray typical of the several Murray-Darling reservoirs. Like the other principal storages (Fig. 2), Lake Hume is a deep-release reservoir, follows a "warm monomictic" pattern of seasonal thermal stratification (as do most Australian lakes), and is governed by irrigation water needs.

The thermocline in Lake Hume forms at about 10-12 m depth and gradually deepens before overturn in early autumn. As water discharged to the Murray is drawn from near the lake bottom (41.4 m), river temperatures below the dam in spring, summer, and autumn differ from those that prevailed before impoundment. Fig. 4 compares data for two sampling stations, one on the Murray above Lake Hume (and representative of the preimpoundment regime) and the other on the Murray immediately below Hume Dam. The lake has

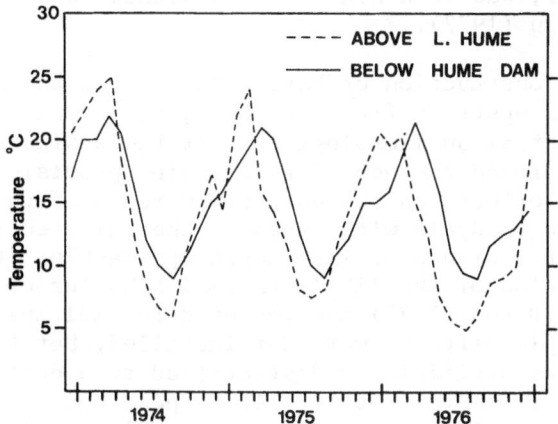

Fig. 4.   The effects of Lake Hume on temperatures in the River
          Murray (Walker and Hillman, 1977).

reduced the annual temperature range in the river by about 6°C, as
the range is 10-21°C rather than 7-24°C.   The lake also has delayed
seasonal changes in river temperature such that trends in spring,
summer, and autumn are retarded by about one month.   Other data for
stations farther downstream show that these effects are apparent,
although progressively diminished, up to 200 river-kilometers
downstream (Walker et al., 1979).

    In most summers the hypolimnion of Lake Hume develops anoxia,
so that an oxygen sag is transmitted to the Murray.   Summer oxygen
levels in the river 1 km below the dam tend to values of about 50%
saturation.   Further on, 37 km below the dam and beyond Albury-
Wodonga, values typically approach 70% saturation.   However, there
are no signs of summer oxygen depletion at stations 100 km and more
downstream from the dam.

    Associated with the summer oxygen gradient in Lake Hume are
gradients in the concentrations of heavy metals, notably iron and
manganese, and other water-quality characteristics.   Iron and,
manganese concentrations in the hypolimnion may exceed 500 and 300
mg/m^3, respectively; but the soluble fractions of both are greatly
depleted by the aeration that occurs in discharge to the river.
Hydrogen sulphide concentrations also are rapidly diminished by
aeration but remain a seasonal problem in management of fish stocks
at a nearby trout farm.   The lake is effective as a sediment trap,
as noted earlier.   Of the 535 million kilograms of suspended sediment
imported to Lake Hume in the three years 1974-1976, 83% was retained.
Over the same period, 38% of the total phosphorus load input to the
lake (842,000 kg), but less than 1% of the Kjeldahl-nitrogen load

(10.2 million kg), was retained. More explanation is provided by
Walker and Hillman (1977).

The recent construction of Dartmouth Dam, on the Mitta Mitta
River about 90 km upstream from Lake Hume, has attracted considerable
interest from Australian limnologists. At the design stage a controversy developed around the need for including provisions that would
minimize adverse effects on the downstream reach of the Mitta Mitta
River. Most concern dealt with probable thermal effects, as the dam
is to impound water with a maximum depth approaching 180 m. This
led to investigation of the likely thermal behavior of the new
reservoir (e.g., Burns, 1977) and the need for multiple-level
offtakes. Multiple offtakes were not installed, but the reservoir
presently is being artificially destratified to a depth of 60 m,
where there is a low-level outlet. It is possible that the effects
of changed flow behavior on the Mitta Mitta River will be severe
and offset any advantages gained from artificial destratification.
In addition, it is possible that chemical stratification (meromixis)
will develop in the new lake with consequences perhaps similar to
those noted for Tasmanian reservoirs by Tyler and Buckney (1974).
The downstream effects of Dartmouth Dam are now being monitored on
behalf of the River Murray Commission, and no doubt will prove
substantial and irreversible.

Fish

Apparent declines in the range and abundance of several native
fish species have caused concern, and there is general agreement
that environmental changes associated with stream regulation have
played a major role (e.g., Weatherley and Lake, 1967; Cadwallader,
1978).

The flow and temperature effects of impoundments particularly
are implicated. Following the construction of Hume Dam, Langtry
(see Cadwallader, 1977) reported substantial changes in the local
fish fauna, leading to the disappearance of two species, the callop
(*Macquaria ambigua*) and the freshwater catfish (*Tandanus tandanus*),
from the River Murray above Yarrawonga Weir (Walker et al., 1979).
Both are essentially warm-water species whose particular flow and
temperature requirements for spawning are unlikely to be met under
the changed river regime. Lake (1975) reported that the River
Murray upstream from the region of Mildura now rarely attains a
temperature high enough (23.5°C) to induce spawning of callop.
Catfish are further disadvantaged by their habit of spawning in
gravel nests in shallow water, making them vulnerable to sudden
level changes which often occur in streams controlled for irrigation (Lake, 1967). Williams (1967), referring to Lake Eildon on the
Goulburn River, reported that the river temperature in summer below

the dam is 10-15°C cooler than the inflow water. Below the dam,
trout have displaced the formerly abundant native Murray cod
(*Maccullochella peeli*). Furthermore, the decline of Macquarie perch
(*Macquaria australasica*) in the Lake Eildon area and most Victorian
tributaries of the Murray has been linked with flow regulation
(e.g., Cadwallader and Rogan, 1977). A population of Macquarie
perch, regarded as an endangered species, was discovered in the
Mitta Mitta River recently, but is unlikely to survive the effects
of Dartmouth Dam (Walker and Hillman, 1977).

     Yarrawonga Weir, Hume Dam, and the other dams on the Murray-
Darling system do not provide fishways and hence are physical
barriers to fish migration. Yarrawonga Weir undoubtedly has been an
important factor in the decline of callop and catfish (and probably
other species) in the River Murray between Lake Hume and Lake
Mulwala. Only 2 of the 13 lock-and-weir constructions on the Murray
below Yarrawonga Weir have fishways, and there is no information
regarding their effectiveness. Yet it is known that some native
fish species, notably callop, undertake extensive upstream migrations, particularly at times of rising floods. The extent and speed
of these migrations (e.g., 1000 km in 163 days: Llewellyn, 1968;
Reynolds, 1976) suggests that the locks and weirs may not be a
serious obstacle. At times of high river levels, the weirs may be
partly dismantled to allow passage of flood waters and,
incidentally, of fish.

     Lake Mulwala periodically is drained to permit maintenance of
Yarrawonga Weir, but the effects on aquatic plants and animals have
not yet been studied (see, however, Cadwallader, 1977). Flow stop-
pages at Tallowa Dam, on the Shoalhaven River in coastal New South
Wales, have caused extensive fish kills (Bishop and Bell, 1978).

     The construction of barrages across Lake Alexandrina and the
Murray mouth in 1940 brought the demise of the local estuarine
fishery and must have profoundly altered the lake environment.
However, there is virtually no ecological information about the
prebarrage conditions.

     Bank erosion and siltation, mentioned earlier, may affect fish
in several ways. For example, siltation probably has interfered
with populations of those fish that, like Macquarie perch, deposit
adhesive eggs on the stream substrate. "River improvement" schemes,
involving desnagging, bank stabilization and channelization,
generally are unsympathetic to fish and other aquatic animals and
have attracted criticism for this and other reasons (e.g., Conservation Council of Victoria, 1977). Desnagging, in particular,
deprives fish of cover and, for some species, spawning sites:
Murray cod and some other native fish lay adhesive eggs on or within
fallen timber on the river bed. More information on these and other

factors in the dynamics of Murray-Darling fish populations is
provided in a review by Cadwallader (1978).

Mention should be included of Davis' (1977) study of changes in
the diet of catfish (*Tandanus tandanus*) before and during filling of
the Copeton Dam impoundment on the Gwydir River, a tributary of the
Lachlan River. With rising water levels, the diet of decapod
crustaceans, chironomids, and other aquatic invertebrates changed to
accomodate large numbers of terrestrial organisms. In the early
stages of stabilized water levels, the diet reflected successional
populations of lacustrine species, notably pulmonate snails and
dipteran larvae associated with extensive growths of duckweed (*Lemna
minor*).

A number of introduced fish species, including trout (*Salmo*
spp.), redfin (*Perca fluviatilis*), European carp (*Cyprinus carpio*),
goldfish (*Carassius auratus*), and mosquito fish (*Gambusia affinis*),
are well established in the Murray-Darling River system. In general,
the regulated environment is more favorable to these introduced
species than to most of the native fish. An obvious means of
offsetting the disadvantages for the native species would be to
artificially stock populations, as is practiced for trout. However,
despite progress with artificial induction of spawning in Murray cod
and other species, problems remain with high fry mortalities in
hatchery populations (Bowerman, 1975). Research and governmental
support in this area are urgently needed.

Benthic Invertebrates

There is no published information that might allow an assessment
of the effects of flow regulation upon the river benthos, and clearly
this should receive urgent attention. In general, the pristine
Murray was characterized by extreme fluctuations in flow and unstable,
shifting sediments, favoring a benthic fauna of relatively few
tolerant species. Dredging and siltation must have adverse effects
in some areas, but the overriding consequence of the controlled-
flow regime presumably has been to encourage the development of
benthic communities by providing a stabilized environment.

In the South Australian section of the Murray, there is circum-
stantial evidence that the lock-and-weir constructions are implicated
in the local extinction of the River Murray crayfish (*Euastacus
armatus*) and in the displacement of a typically riverine species of
freshwater mussel (*Alathyria jacksoni*) by another species (*Velesunio
ambiguus*) typical of billabongs and minor streams (Walker, unpublished). Studies of invertebrate benthos in the Albury-Wodonga
region now are in progress (see Walker and Hillman, 1977), and some
useful results of preimpoundment surveys of the Mitta Mitta River,

in relation to Dartmouth Dam, have been reported by Smith et al.
(1977).

Zooplankton

The zooplankton fauna of the River Murray is an assemblage of
mostly lacustrine species (predominantly Cladocera and Copepoda)
derived from headwater impoundments and inoculated from backwaters
and billabongs at times of flood. However, the least regulated of
the rivers in the system, the Darling River, has a typically riverine
fauna (predominantly Rotifera). The lower Murray in South Australia
has a mixed assemblage of rotifers, cladocerans, and copepods that
remains largely unmodified in its passage from the Darling junction
to the Murray mouth. This information is drawn from reports of a
continuing study by Shiel (1976, 1978, 1979).

Conclusion

Australian ecologists are particularly sensitive to the changing
balance between the native plants and animals and those introduced
since European settlement, some 200 years ago. The decline of many,
indeed most, native species is due variously to interactions with
exotic species, to habitat changes wrought by human industry, or to
more direct forms of persecution by man. The Australian fauna has a
distinctive character compared to the faunas of other continents.
Long evolutionary isolation and the demands of an often harsh
climate have favored the development of ecological opportunists.
Many indigenous animals are ill equipped to survive in the presence
of exotic species, and man, in seeking to impose greater stability
upon the Australian environment, has further discouraged the native
species and encouraged the introductions. This seemingly inevitable
conflict is by no means peculiar to Australia, but perhaps it is
more obvious there than in some other countries.

No better illustration of the conflict exists than in the
imposition of an irrigation regimen upon an Australian river. On
the one hand is variable flow, perhaps the outstanding ecological
characteristic of Australian streams, and the legacy of a capricious
climate. On the other, the stabilized flow required of rivers
supplying irrigation needs. The environmental ramifications of
imposed irrigation controls are so far-reaching that it is difficult
to envisage how the opposing ideals might be reconciled. Inevitably,
the answer will be treated as a matter for compromise.

LOOKING AHEAD

Rivers are accorded an uncertain status in matters of environmental conservation. Some ecologists argue that rivers are not ecosystems, and similar uncertainties are apparent in the attitudes of governmental authorities. If ecologists appear undecided upon the nature of rivers, there may be little hope for displays of ecological conscience by regulating authorities. In Australia, the importance of rivers, both as resources and as ecological systems, cannot seriously be questioned (see further Lake, 1978). Even so, there is an extraordinary lack of practical knowledge about those rivers. The survey of the River Murray in relation to Albury-Wodonga (Croome et al., 1976; Walker and Hillman, 1977) will, it is hoped, mark a turning point: It represents the first stage of a comprehensive study of the Murray with regard for local urban development. Investigations continue under the direction of Dr. T. J. Hillman of the Albury-Wodonga Development Corporation.

A first requirement, in looking ahead, must be a commitment to river ecology. Commitment is needed at the level of national resource management, from state and local government authorities, and from an informed public. Commitment is an expression of attitude, and there is evidence, in the form of the Albury-Wodonga survey, that attitudes are changing.

A first commitment should be towards gathering basic ecological knowledge. In Australia such knowledge is sparse indeed, yet it is essential to effective forward planning and to the solution of existing conflicts. Such investigations are primarily a responsibility of government, although academic researchers and others have important roles as contributors and advisors.

The intense pressures of irrigation development in the Murray-Darling system suggest that this is an important focus for future studies. In Tasmania and the Snowy Mountains of southeastern Australia, hydroelectric power generation is a major influence in stream regulation, yet the impact of this development upon streams is quite unexplored. There have been several studies of irrigation and hydroelectric reservoirs that have not considered downstream effects (e.g., reviews by Williams, 1973 and Tyler, 1974). Future researchers might seek to integrate studies of this kind.

It is important that reserves of basic knowledge be used constructively. Positive proposals from river ecologists are most likely to impress stream-regulation authorities, whether the proposals relate to planning or management. It is perhaps a fault of many academic researchers that they do not make a sufficient attempt to communicate their findings to the right people. In another sense, the same accusation might be leveled at government,

because in Australia, and no doubt in other countries, an enormous
library of potentially useful knowledge is bound in reports and
unprocessed data not readily available to others. Communications
are important, and are greatly assisted by symposia like this.

REFERENCES

Australian Water Resources Council, 1975, "Review of Australia's
    Water Resources 1975," Aust. Gov. Publ. Serv., Canberra.
Australian Water Resources Council, 1978, "Proceedings of the Water
    Planning Workshop 1978," Aust. Gov. Publ. Serv., Canberra.
Baker, B. W., and Wright, G. L., 1978, The Murray Valley: Its
    hydrologic regime and the effects of water development on the
    river, *Proc. R. Soc. Vict.*, 90:103-110.
Bishop, K. A., and Bell, J. D., 1978, Observations on the fish fauna
    below Tallowa Dam (Shoalhaven River, New South Wales) during
    river flow stoppages, *Aust. J. Mar. Freshwater Res.*, 29:543-549.
Bowerman, M. R., 1975, Important discoveries about native warm water
    fish, *Aust. Fish.*, 34:3-8.
Braithwaite, L. W., 1975, Managing waterfowl in Australia, *Proc.
    Ecol. Soc. Aust.*, 8:107-128.
Burns, F. L., 1977, Localized destratification of large reservoirs
    to control discharge temperatures, *Prog. Water Tech.*, 9:53-63.
Cadwallader, P. L., 1977, J. O. Langtry's 1949-50 Murray River
    investigations, *Fish. Wildl. Pap. Vict.*, 13.
Cadwallader, P. L., 1978, Some causes of the decline in range and
    abundance of native fish in the Murray-Darling river system,
    Proc. R. Soc. Vict., 90:211-224.
Cadwallader, P. L., and Rogan, P. L., 1977, The Macquarie perch,
    Macquaria australasica (Pisces:Percichthyidae), of Lake Eildon,
    Victoria, *Aust. J. Ecol.*, 2:409-418.
Collet, K. O., 1978, The present salinity position in the River
    Murray Basin, *Proc. R. Soc. Vict.*, 90:111-123.
Conservation Council of Victoria, 1977, "River Improvement?,"
    Environment Awareness Publ. 2.
Croome, R. L., Tyler, P. A., Walker, K. F., and Williams, W. D.,
    1976, A limnological survey of the River Murray in the Albury-
    Wodonga area, *Search*, 7:14-17.
Davis, P. S., 1978, "Man and the Murray," New South Wales Univ.
    Press, Sydney.
Davis, T. L. O., 1977, Food habits of the freshwater catfish,
    Tandanus tandanus Mitchell, in the Gwydir River, Australia,
    and effects associated with impoundment of this river by the
    Copeton Dam, *Aust. J. Mar. Freshwater Res.*, 28:455-465.
Dexter, B. D., 1978, Silviculture of the river red gum forests of
    the central Murray floodplain, *Proc. R. Soc. Vict.*, 90:175-191.

Engineering and Water Supply Department, South Australia, 1978, "The
    South Australian River Murray Salinity Control Programme, Vols.
    1-3," Adelaide.
Falter, C. M., 1978, Assessment of expected impact of diversion of
    saline groundwater and irrigation drainwater on phytoplankton
    of the River Murray, South Australia, p. 95-96, in: "The South
    Australian River Murray Salinity Control Programme, Vols. 1-3,"
    Eng. Water Supply Dep., South Australia, 1978, Adelaide.
Frith, H. J., 1977, "Waterfowl in Australia," Reed, Sydney.
Frith, H. J., and Sawer, G., eds., 1974, "The Murray Waters. Man,
    Nature, and a River System," Angus and Robertson, Sydney.
Higginson, F. R., 1973, Soil erosion and siltation within the Murray
    Valley, in: "Pollution Problems of the River Murray," Dep.
    Decentralization and Development, New South Wales, Sydney.
Lake, J. S., 1967, Principal fishes of the Murray-Darling system,
    p. 192-213, in: "Australian Inland Waters and their Fauna:
    Eleven Studies," A. H. Weatherley, ed., Aust. Natl. Univ.
    Press, Canberra.
Lake, J. S., 1975, Fish of the Murray River, p. 213-224, in: "The
    Book of the Murray," G. V. Lawrence and G. K. Smith, eds.,
    Rigby, Adelaide.
Lake, P. S., 1978, On the conservation of rivers in Australia, *Aust.
    Soc. Limnol. Newsl.*, 16(2):24-26.
Lawrence, G. V., and Smith, G. K., eds., 1975, "The Book of the
    Murray," Rigby, Adelaide.
Llewellyn, L. C., 1968, Tagging gives answers to fish growth queries,
    Fisherman (N.S.W.), 3:1-5.
May, V., 1978, Areas of recurrence of toxic algae within Burrinjuck
    Dam, New South Wales, Australia, *Telopea*, 1:295-313.
Mitchell, D. S., 1978, "Aquatic Weeds in Australian Inland Waters,"
    Aust. Gov. Publ. Serv., Canberra.
Reynolds, F. R., 1976, Tagging important in River Murray fish study,
    Aust. Fish., 35:4-6; 22.
Shiel, R. J., 1976, Associations of Entomostraca with weedbed
    habitats in a billabong of the Goulburn River, Victoria, *Aust.
    J. Mar. Freshwater Res.*, 27:533-549.
Shiel, R. J., 1978, Zooplankton communities of the Murray-Darling
    system, *Proc. R. Soc. Vict.*, 90:193-202.
Shiel, R. J., 1979, Synecology of the Rotifera of the River Murray,
    South Australia, *Aust. J. Mar. Freshwater Res.*, 30:255-263.
Smith, B. J., Malcolm, H. E., and Morison, P. B., 1977, Aquatic
    invertebrate fauna of the Mitta Mitta Valley, Victoria, *Vict.
    Nat.*, 94:228-238.
Tyler, P. A., 1974, Limnological studies, p. 29-61, in: "Biogeo-
    graphy and Ecology in Tasmania," W. D. Williams, ed., Junk, The
    Hague.
Tyler, P. A., and Buckney, R. T., 1974, Stratification and biogenic
    meromixis in Tasmanian reservoirs, *Aust. J. Mar. Freshwater
    Res.*, 25:299-313.

Walker, K. F., and Hillman, T. J., 1977, "Limnological Survey of the
    River Murray in Relation to Albury-Wodonga, 1973-1976," Albury-
    Wodonga Development Corporation and Gutteridge, Haskins and
    Davey, Albury.
Walker, K. F., Hillman, T. J., and Williams, W. D., 1979, The
    effects of impoundment on rivers: An Australian case study,
    Verh. Int. Verein. Limnol., 20:1695-1701.
Water Conservation and Irrigation Commission, New South Wales, 1975,
    "Water Resources of the Murray Valley within New South Wales,"
    Sydney.
Williams, W. D., 1967, The changing limnological environment in
    Victoria, p. 240-251, *in*: "Australian Inland Waters and their
    Fauna," A. H. Weatherley, ed., Aust. Natl. Univ. Press,
    Canberra.
Williams, W. D., 1973, Man-made lakes and the changing limnological
    environment in Australia, p. 495-499, *in*: "Man-Made Lakes:
    Their Problems and Environmental Effects," W. C. Ackermann,
    G. F. White, and E. B. Worthington, eds., Am. Geophys. Union,
    Washington.

STREAM REGULATION IN GREAT BRITAIN

Patrick D. Armitage

Freshwater Biological Association
River Laboratory
East Stoke
Wareham
Dorset BH20 6BB   England

INTRODUCTION

Published, unpublished, and ongoing work concerning the effects of various flow-regulation schemes in Great Britain is reviewed, with special reference to the few more-detailed studies of specific areas. In addition, an attempt is made to link faunal communities present in sites below reservoirs with particular reservoir types. The main points to arise from this are the variety of faunal communities and the need for more work to relate cause and effect. Future research possibilities are outlined.

THE NEED FOR WATER AND ITS AVAILABILITY

The demand for water and its relation to the natural supply are major concerns in Great Britain. Totals of residual rainfall vary regionally, from less than 125 mm in the Thames estuary and the Wash to over 2500 mm in N. Wales, the Lake District, and the Highlands of Scotland. These totals vary from year to year, and rainfall distribution within a year may fluctuate widely. This variability, together with other difficulties arising from the fact that greatest population densities occur in low-rainfall areas, can give rise to problems in water-resource development. The demand for water in the United Kingdom is expected to rise from the present 16×10^6 m^3 d^{-1} to about 33×10^6 m^3 d^{-1} by the end of the century (Rodda et al., 1976). This heavy demand is being met by increasing the management of natural water resources. Existing and possible future water-resource developments in England and Wales are depicted in Fig. 1. Here the full range of flow-regulation schemes is apparent,

165

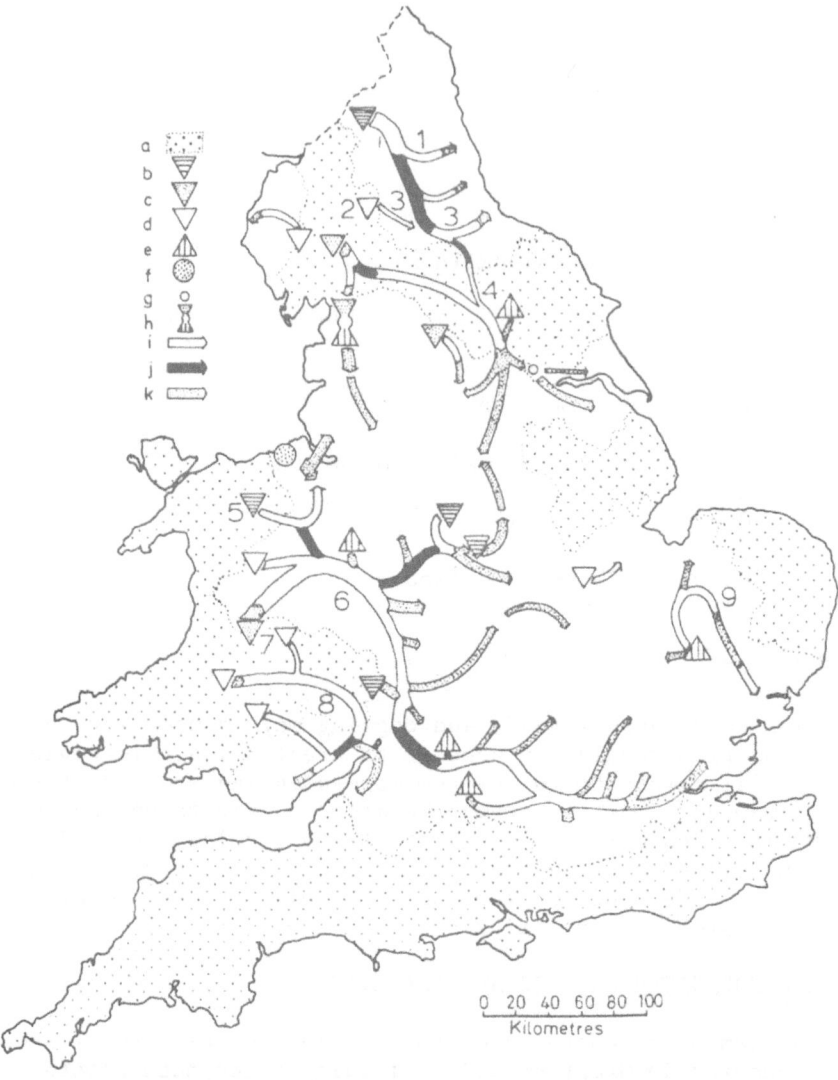

Fig. 1. Resource development in England and Wales (redrawn from a
         map prepared by former Water Resources Board). [a = self-
         sufficient areas; b = new inland reservoirs; c = existing
         reservoirs enlarged; d = existing reservoirs; e = ground-
         water development; f = estuarine storage; g = river source
         without storage; h = combined surface, groundwater, and
         river source; i = rivers used to convey supplies; j =
         river-to-river aqueducts; k = bulk-supply aqueducts.]
         Sites discussed in the text are indicated as follows:
         1 River Tyne, 2 Cow Green Reservoir, 3 River Tees, 4 River
         Swale, 5 River Dee system, 6 River Severn, 7 Craig Goch
         area, 8 River Wye, 9 Ely-Ouse to Essex transfer scheme.

including reservoir storage, inter-river transfers, and groundwater augmentation.

In Scotland there is no shortage of water resources and the chief regulatory schemes are those associated with the generation of hydroelectric power.

REVIEW OF FLOW-REGULATION WORK IN GREAT BRITAIN

Published studies are listed in Table 1. It is clear from the table that most of the work concerns management in some way and that long-term biological studies are few. In the paragraphs below examples of problems and benefits resulting from flow regulation will be illustrated with reference to published and unpublished work on particular schemes.

Regulation of the River Dee

Large-scale regulation began in the mid-fifties, when sluices were constructed at the outlet of the natural lake Bala (Fig. 1). The purposes of this regulation were flood control and the maintenance of flow downstream during periods of natural low flow. A second impoundment, Llyn Celyn, was brought into service in 1965, and a third impoundment has recently been constructed on the Afon Brenig, a tributary of the Dee.

After initial deterioration, caused by large amounts of suspended solids during construction of Llyn Celyn, water quality in the Dee improved, indicating that regulation helped to reduce the concentration of nutrients during drought by adding impounded water low in dissolved salts.

Blezzard et al. (1971) have drawn attention to the fact that many of the rivers in Great Britain that lend themselves most readily to water conservation schemes are those that produce the main weight of migratory fish. The Welsh Dee falls firmly into both of these categories, and great efforts were made to avoid endangering the fisheries. However, construction of Llyn Celyn submerged four miles of first-class spawning ground and imposed a barrier to migration. Since installation of a fish trap just downstream of the dam site, the number of salmon reaching the trap has dropped from about 100 in 1965 to about 10 in 1977. Below Llyn Celyn, salmon appear to spawn preferentially in small, unregulated tributaries about 3 km below the fish trap.

In general, therefore, regulation of the Dee system has apparently not been detrimental to salmonids, although additional schemes like Llyn Celyn, which has been shown to deter migratory fish, may radically alter the situation. Studies on invertebrates are few and unpublished.

Table 1. The Distribution of Studies on Aspects of Stream Regulation in Great Britain.

Subject	Reference
Discussion and Reviews	Alabaster, 1976; Davies and Price, 1979; Elder, 1966; Menzies, 1966; Ministry of Agriculture, Fisheries and Food and National Water Council, 1976.
General	Blezzard et al., 1971; Davies and Price, 1979; Hall, 1977; Iremonger, 1971; Isaac, 1967; Menzies, 1966; National Water Council, 1978; Severn-Trent Water Authority, 1977.
Management schemes:	
a. Hydroelectric	Aitken, 1963; Rogers, 1978.
b. Reservoirs general	Blezzard et al., 1971; Crann, 1978; Fordham, 1970; Kennard and Reader, 1975.
c. Transfers	Ash, 1966; Davies and Price, 1979; Guiver, 1976; Hall, 1977; Huntington and Armstrong, 1974.
d. Groundwater	Anon., 1977; Avon and Dorset River Authority, 1973; Great Ouse River Authority, 1974; Severn-Trent Water Authority, 1977.
e. Fisheries	Aitken et al., 1966; Alabaster, 1970; Blezzard et al., 1971; Calderwood, 1945; Guiver, 1976; Hartley and Simpson, 1967; Hellawell, 1976; Howells and Jones, 1972; Iremonger, 1971; Jones, 1968; Millichamp, 1976; Mills, 1965a,b, 1968; Mills and Shackley, 1971; North and Hickley, 1977; Pyefinch and Mills, 1963; Stuart, 1962.
Special studies:	
a. Macrophytes	Central Water Planning Unit, 1978a; Holmes et al., 1972; Holmes and Whitton, 1977a,b.
b. Fish	Alabaster, 1970; Central Water Planning Unit, 1978a; Crisp et al., 1978; Gee et al., 1978; Mills, 1968; Pyefinch and Mills, 1963; Stuart, 1962.
c. Invertebrates	Armitage, 1976, 1977, 1978; Armitage and Capper, 1976; Armitage et al., 1974; Brooker and Hemsworth, 1978; Central Water Planning Unit, 1978a; Chubb, 1977.
d. Physico-chemical	Central Water Planning Unit, 1978b; Crisp, 1977; Lavis and Smith, 1972; Northumbrian Water Authority, 1976; Truesdale and Taylor, 1978.

Regulation of the Tees by Cow Green Reservoir

The reservoir was completed in 1970, and its purpose is to regulate the flow of the Tees to supply the increasing demand of industrial complexes at the mouth of the river. The Freshwater Biological Association became involved in 1967 in a programme of studies emphasizing the effect on fish but including some monitoring of physico-chemical features and invertebrate populations.

Fish. To date, published information on the effects of regulated flow on fish below Cow Green is limited to an account of the stomach contents of fish before and after impoundment (Crisp et al., 1978). Samples were taken from an area about 600 m below the dam and from the adjacent unregulated tributary Maize Beck. Pre-impoundment observations showed a general similarity in the diet of trout (*Salmo trutta*) in both streams. Following regulation of the Tees, there was little change in the stomach contents of trout from Maize Beck, but in the Tees itself, trout ate more Ephemeroptera nymphs and Chironomidae larvae and less terrestrially derived material. Minnows (*Phoxinus phoxinus*) in the post-impoundment phase fed chiefly on chironomid larvae and microcrustaceans. Apparently, only trout and minnows were able to exploit the large numbers of microcrustaceans discharged from the reservoir.

The bulk of post-impoundment studies on fish (1970-1975) is being prepared for publication, but Dr. Crisp (pers. comm.) has kindly provided some provisional general information, which is given below.

Since regulation, there has been a substantial increase in population density and biomass of trout and bullhead (*Cottus gobio*) at the study site in the Tees, about 600 m downstream of the dam. No such increase has been demonstrated in the unregulated, adjacent tributary Maize Beck. In addition, detailed analyses show a small, but significant increase in trout growth rate as a result of regulation. No changes were detected on the growth of bullheads.

Physico-chemical studies. These have been described by Crisp (1977) and additional data are given in Armitage (1976, 1977). *Discharge*: Two important effects of regulation are: (i) the virtual elimination of daily minimum discharges <0.1 times the annual mean, and (ii) the elimination of daily maxima >8 times the annual mean and a marked decrease in the frequency of discharges >5 times the annual mean. This pattern holds throughout the year, despite the fact that 80% of this discharge in some winter months is uncontrolled flow down the spillway. *Temperature*: Diel fluctuations have been markedly reduced, and the spring rise and autumn fall in water temperature have been delayed by 20-50 and 0-20 days, respectively. *Water chemistry*: Fluctuations in ionic content of water entering the reservoir are large and can be related approximately to river discharges. Fluctuations in outflowing water are smaller and appear to be largely seasonal. Dissolved-oxygen concentration fluctuated relatively little during any 24-h period in either the regulated Tees or the unregulated tributary Maize Beck, and concentrations were generally between 90 and 103% saturation. *Suspended solids*: Apart from an increase during construction, concentrations of suspended solids have been reduced by regulation. Crisp (1966) has shown that 95% of the transport of suspended solids (mainly peat) in a small tributary of the Upper Tees occurred during brief periods when stream discharge was at least 6 times its annual mean value. That discharges >5 times the annual mean have been markedly decreased in the regulated Tees suggests that the ability of the stream to transport solids will be reduced. Peat and mineral particles make up the bulk of suspended solids in the unregulated Maize Beck but were virtually absent from the Tees, which has a similar peaty catchment. This indicates that these components are deposited in the reservoir. It is suggested (Armitage, 1977) that at least 92% of suspended matter entering the reservoir is retained.

Invertebrate studies. These have been described in a number of papers (Armitage et al., 1974; Armitage, 1976, 1977, 1978; Armitage and Capper, 1976) and deal mainly with changes in invertebrate drift and benthos below the dam following impoundment. The presence of a large unregulated tributary, Maize Beck, proved useful as a control in assessing observed changes. *Drift*: From closure of the dam in 1970 until the end of regular observations in 1973 the faunal composition of drift catches in Maize Beck remained similar. In the Tees, however, change was very marked, with microcrustaceans from the reservoir completely dominating numbers and biomass. Seasonal fluctuations in both numbers and biomass of drifting organisms originating from the benthos were similar in the two rivers, but diel fluctuations, particularly of baetid nymphs, were greater in Maize Beck, and peaks of abundance appeared to be depressed in the Tees. As noted, the discharge of microcrustaceans from the reservoir was the single most important feature affecting composition, numbers, and biomass of the drift in the Tees. Greatest losses in microcrustaceans occurred in the first 400 m below the dam, and it was estimated that the river bottom there receives about 160 mg m^{-2} day^{-1} (dry weight) in the August-September period. This supply of zooplankton, even if not fed upon directly, is likely to have considerable influence on the Tees benthos when it settles out and decomposes to produce a nutrient-rich detritus. *Benthos*: Mean faunal density increased after regulation in the Tees but remained similar in Maize Beck. The increases in the Tees were caused mainly by large numbers of *Hydra*, Naididae and Orthocladiinae. Species diversity was least at the site immediately below the dam but increased downstream, and values 500 m below the dam were not significantly different from those in Maize Beck. Faunal numbers and biomass were greatest 240 m below the dam and lowest in the unregulated Maize Beck.

Some workers (Pearson et al., 1968; Spence and Hynes, 1971; Lehmkuhl, 1972; Ward, 1976) have reported reduced numbers of species of Plecoptera, Ephemeroptera and Trichoptera below dams. However, there has been no reduction in the total number of species in any of these groups in the Tees, despite some change in the species present. In contrast, there has been an increase in the numbers and biomass of certain taxa, but generally not at the expense of the previous fauna. This may be partly because of the particular characteristics of slope, substratum, and discharge regime found below Cow Green Dam. Rapid flow over a generally heterogeneous bottom may prevent clogging of interstitial spaces by silt, thereby maintaining the variety of ecological niches necessary for a diverse fauna. High discharges during winter overflow contribute to the flushing process; but because coarse particles are retained in the reservoir, these flows will not scour the bottom. Problems caused by hypolimnial release do not occur at Cow Green, because the reservoir is well mixed; and although the water flowing out of the reservoir is colder in summer and warmer in winter than in the unregulated Maize Beck, no gross effects have been discernible, though it is possible that the timing of the life histories of certain groups will be altered.

The Craig Goch Scheme

This is a major resource scheme involving the enlargement of the existing Craig Goch reservoir to regulate the flows of the Severn and Wye (Fig. 1). A programme of sponsored research was mounted in 1975 to study the pre-regulation situation as a basis on which to assess the effects of regulation on the rivers Wye and Severn. Although a large part of the work concerns sites on the two main rivers, observations have also been made on smaller, regulated streams in the area. Unfortunately, the bulk of this Craig Goch work is not generally available; however, two relevant studies, dealing with varied effects of regulation in the area, have been published recently. Truesdale and Taylor (1978) have drawn attention to the problems of organic deposits rich in manganese or iron found below soft-water reservoirs in the Craig Goch area. These deposits are likely to be detrimental to benthic fauna by clogging interstitial spaces, but the precise effects have yet to be studied. Brooker and Hemsworth (1978) have studied the effects of an artificial discharge in the River Wye and found that the artificial-flow increases affected invertebrate drift in a manner similar to natural-flow increases.

Inter-river transfers--Tyne-Tees-Swale

Some pre-transfer studies of the possible effects of transferring water from the River Tyne into the Rivers Tees and Swale have been carried out by the Freshwater Biological Association (Fig. 1).

Hancock (1977), studying fish populations in the area, notes that the Tees/Swale transfer will introduce Tees water to the Swale at almost exactly the point where the fishery changes from trout to grayling (*Thymallus thymallus*) and course fish. The quality of the Tees water appears to be markedly different from that of the Swale, and any changes in temperature and quality occurring downstream of the transfer point are likely to have significant effects upon the coarse fish populations and fisheries. Grayling are particularly susceptible to environmental changes, and the transfer may have some adverse effects upon the population.

Holmes and Whitton (1977a,b) studied zonation of vegetation in the Tees and Swale. They note that colonization by submerged angiosperms is at present taking place in the Tees as a result of reservoir regulation and suggest that this process is likely to be accelerated by the addition of innocula from the Tyne. In the case of the Tees to Swale transfer any influence of the Tees may result from a change in water chemistry. Available data suggest that PO_4-P, NO_3-N and Cl levels in the Tees at the point of abstraction are between four and nine times greater than values for the same elements at the point of inflow in the Swale. Great changes in the flora do not seem likely, however, since any new innocula will be subject to growth restricting factors similar to those affecting the existing macrophytes.

Invertebrate studies (Armitage, unpublished data) indicated that major faunal changes resulting from water transfer are unlikely but that relative and absolute abundance may well be affected by the way transfers are introduced or withdrawn.

Operational considerations of the whole scheme are described by Hall (1977). Integral to the scheme is the transfer tunnel in which water may be stored for months or years, but there may be problems caused by the release of water whose quality has deteriorated during storage. Other factors that may bring about ecological change are the temperature differential between donor and recipient streams and the possibility of transferring "pheromones," which may confuse homing salmonids.

Studies of the effects of flow regime on the young stages of salmonids are currently being carried out in the River Tees by the Freshwater Biological Association. Field surveys and experimental channels will be used to investigate spawning success in the natural stream and under controlled conditions.

Inter-river transfers--Ely-Ouse to Essex

In the 1960s, a scheme was implemented to transfer water from the catchment of the Great Ouse (Fig. 1), where population densities were low and rainfall runoff to rivers is not fully required to meet

demands, to highly populated areas in Essex and along the north bank
of the Thames, where full use had already been made of river water
and underground chalk-water supplies. The scheme consists of the
transfer of water from near the mouth of the Ouse to reservoirs in
Essex, a distance of 145 km. About two-thirds of this transfer
route is tunnels, pipelines, and modified channels. Remote sensing
of dissolved oxygen in this system is therefore important to prevent
release of anaerobic water into the recipient river.

Some biological effects of the transfer have been described by
Guiver (1976). Problems have arisen from the introduction of the
diatom *Stephanodiscus* into the River Stour in such numbers as to
cause complaints, and small fry of unwanted predatory fish species
(zander) *Stizostedion lucioperca* have also been transferred to the
River Stour.

Another adverse effect of the scheme was observed in 1973, when
tomato growers in Essex, using the polluted public water supply,
produced stunted, malformed, unmarketable fruit. The source of this
pollution was eventually traced to a factory discharging a herbicide
into the River Cam, some 200 water km away. Despite these adverse
effects, the transfer can be of considerable benefit in improving
water quality; for example, the impact of an accidental discharge of
liquid ammonia to a sewage works on the upper Stour was greatly
reduced by using the transfer scheme to augment flows in the Stour,
thereby reducing the concentration of ammonia.

Groundwater

A number of studies of the effects of groundwater abstraction
and augmentation of flows on biota in rivers have been carried out.
However, the majority of these have not been published, and information is often not easily available. There are many groundwater
schemes proposed or in operation and details of some of these are
listed in Anon. (1977). Recent developments in ecological studies
of chalk streams are outlined in a report of the Central Water
Planning Unit (1978a). In general, definite effects caused by flow
regulation using groundwater have been difficult to demonstrate,
particularly on submerged macrophytes and invertebrates. This is
partly because of the natural variation in a river system and the
generally short periods of pumping. During the drought of 1976,
augmentation of flow in the Lambourn, a chalk stream in the south of
England, proved immediately beneficial near the perennial head by
increasing stream width and flooding marginal vegetation, providing
a greater area of substratum for colonising invertebrates.

Effects of groundwater schemes on fisheries have also been
difficult to demonstrate. Unpublished studies of a lined intermittent stream receiving borehole compensation water at a temperature of 10°C indicated that trout grew exceptionally well in their

first year because enhanced winter water temperatures, in particular, allowed growth at a time when it would normally have been
checked. Increased flow may also increase survival of eggs and
alevins by improving water circulation through the redds and making
more gravel areas available for spawning.

Hydroelectric schemes

Aitken (1963) has reviewed the state of hydroelectric-power
generation in Great Britain, and Elder (1966) has described biological effects of water utilization by hydroelectric schemes.

The bulk of biological work has been done in Scotland and has
been directed towards management (see Table 1). The reason for this
is purely economic: Salmonids are by far the most valuable freshwater fish in Scotland, and hydroelectric schemes are more common
there than in the rest of Britain. It is nevertheless very surprising that there appear to be no available studies dealing with the
effects of hydroelectric schemes on invertebrates below impoundments
in Britain. Proposed studies of the impact of the Dinorwic (North
Wales) pumped storage scheme on lakes and streams of the area are
outlined by Rogers (1978), but further intensive work below other
hydro schemes does not appear to be planned.

BIOLOGICAL MONITORING

For the purposes of water management, Great Britain has been
divided into a number of areas, each the responsibility of a particular water authority or river purification board. Most of these
authorities have staff scientists, part of whose work is to collect
physico-chemical data and to monitor bottom fauna and fisheries in
rivers in their areas. Some of the sampled rivers are regulated,
and for the present study appropriate authorities were requested to
provide biological information on benthos from sites as close to
dams as possible together with some physico-chemical data (distance
below dam, stream width, slope, substratum characteristics, water
hardness, mean summer temperature, water velocity and discharge
regime from the reservoir). It was hoped that the faunal communities below the dams would fall into groups that could be linked with
certain physico-chemical features. The data available were very
varied, with different sample methods and sample frequencies; and
the only way to compare faunas was to apply relative abundance to
faunal groups rather than to species, since levels of identification
varied from authority to authority. The faunal group/site matrix
was subjected to cluster analysis by the Czekanowski coefficient and
Group Average methods after an arcsin transformation. Some major
groupings emerged (Fig. 2), some of which can be loosely linked with
physical and chemical features. For example, Group 1 consists
basically of upland soft-water sites with variable regulated flow ($\bar{x}$
0.1–1.0 m^3 s^{-1},, max. 12.0–60 m^3 s^{-1}), Group 6 contains upland soft-

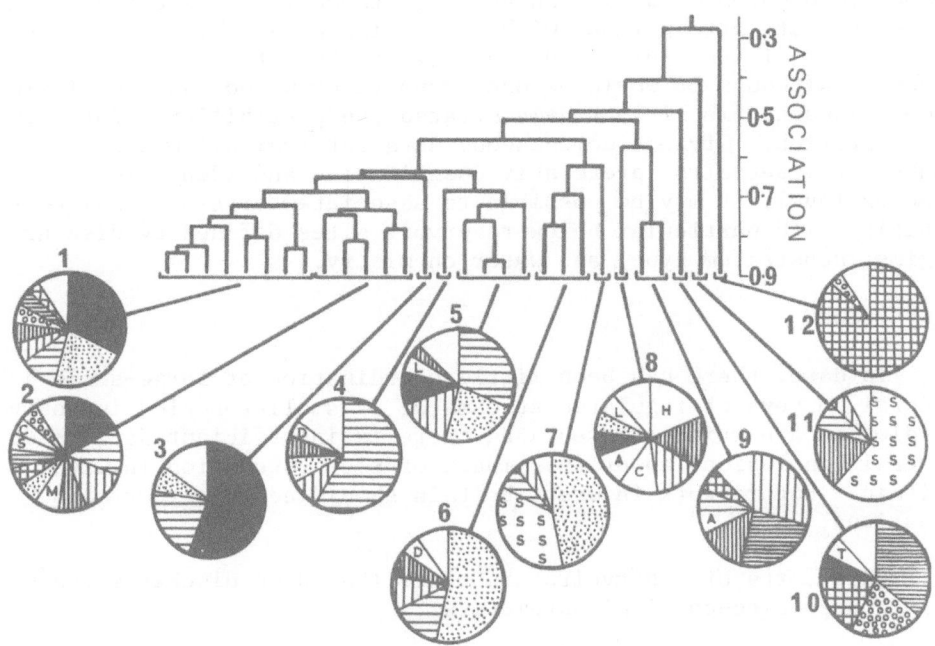

Fig. 2.   Dendrogram from a cluster analysis showing faunal composition (relative numbers) of groupings of 29 sites below 25
reservoirs in Great Britain. Groups 1-12 contain sites
below the following reservoirs: 1 Fontburn (1.7), Grass-
holme (1.8), Burnhope (1.3), Catcleugh (2.1), Alwen (8.0),
Celyn (6.4), Carron (2.4), Usk (0.2); 2 Cow Green (0.2),
Blithfield (8.0), Ogston (2.0), Ladybower (8.0), Tittes-
worth (1.6); 3 Taf Fechan (4.4); 4 Taf Fechan (0.4); 5 Taf
Fechan (2.0), Venachar (0.4), Doon (0.8), Celyn (2.0); 6
Nant-y-Moch (0.1), Dinas (0.1), Alwen (1.0); 7 Brianne
(3.0); 8 Tegid (0.1); 9 Thornton (2.5), Weir Wood (0.1); 10
Llandegfedd (0.4); 11 Carno (0.1); 12 Bewl Bridge (5.0).
[Bracketed numbers following a reservoir name indicate
distance of site (km) from the dam.]
Key to pie charts: ⊤ Tricladida, ▥ Oligochaeta, ▤
Gammarus, Ⓐ Asellus, ⊡ Plecoptera, ■ Ephemerop-
tera, ▭ Caseless caddis, Ⓒ Cased caddis, ▨
Coleoptera, ▥ Chironomidae, Ⓢ Simuliidae, Ⓓ Other
Diptera, Ⓗ Hemiptera, Ⓜ Mites, ▦ Gastropoda,
Ⓛ Bivalvia, ▢ Others.

water sites with relatively constant discharge ($\bar{x}$ 0.1-0.5 m^3 s^{-1}),
and Group 9 consists of lowland sites with very low constant discharge (0.01-0.05 m^3 s^{-1}). In general, though, it was not possible
to demonstrate strong links with particular physico-chemical features. This is not surprising in view of the data available.
However, although it would be dangerous to draw too many conclusions
from this analysis, it has demonstrated the possibilities for future
investigation. Given a homogeneous data set from below a large
number of reservoirs, preferably quantitative and identified to
species level, it may be possible to associate certain faunal communities with particular below-reservoir sites defined by discharge
regime, substratum type, and water chemistry.

CONCLUSIONS

     To date, there has been little coordination of large-scale
general surveys of regulated streams or of smaller-scaled intensive
studies in a particular area, and there is insufficient data to draw
general conclusions about the impact of flow regulation in Great
Britain. Future work in Great Britain should address four main
questions:

1. What sorts of reservoirs are there (based on discharge regime
    and physico-chemical parameters)?

2. What faunal communities are found below them?

3. How do these communities compare with those in similar unreg-
    ulated streams?

4. What factors determine the presence of a particular community?

     The first three questions could be answered in part by general
surveys of a large number of sites, which would provide much useful
information. However, the fourth requires intensive study in particular areas and should clarify some points arising from existing
studies. Of special interest would be (i) the relationship between
discharge regime and the substratum, especially its effect on interstitial spaces and particle entrainment and movement; and (ii) the
fate and nature of reservoir-derived seston and its contribution to
the nutrient budget of the benthos, particularly filter-feeding
forms.

     Where possible, use should be made of existing reservoirs to
test the effects of particular releases, as was recently carried out
in the Craig Goch scheme (Brooker and Hemsworth, 1978) and by the
Northumbrian Water Authority below Cow Green Reservoir. Such
studies would be greatly enhanced by supporting smaller-scale experimental work involving observations of the behaviour of particular

species and substratum types under differing conditions in artificial channels.

The prime aims of flow-regulation studies should be to gather information that will assist in the optimal management of our water resources, that is, management having minimal effect on the natural environment. At the same time this should provide much useful scientific data. This combination of management and scientific objectives may help to generate research funds from sponsors, which would allow a wider, yet more intensive, coverage of flow-regulation systems.

ACKNOWLEDGMENTS

I am most grateful to staff from the following organisations for providing me with data: Welsh National Water Development Authority, Central Water Planning Unit, Severn-Trent Water Authority, Northumbrian Water Authority, Southern Water Authority, Central Electricity Generating Board, North of Scotland Hydro-electric Board, Freshwater Fisheries Laboratory at Pitlochry, and the Clyde and Forth River Purification Boards. In addition, I am indebted to Dr. Bob Abel for access to his literature collection, to Dr. D. T. Crisp for permission to use unpublished Cow Green fish data, and to Dave Cooling for supplying the program for the Cluster Analysis.

REFERENCES

Aitken, P. L., 1963, Hydro-electric power generation, *Proc. Symp. Civil Eng. London,* 34-42.

Aitken, P. L., Dickerson, L. H., and Menzies, W. J. M., 1966, Fish passes and screens at water power works, *Proc. Inst. Civil Eng.,* 35:29-57.

Alabaster, J. S., 1970, River flow and upstream movement and catch of migratory salmonids, *J. Fish Biol.,* 2:1-13.

Alabaster, J. S., 1976, The water quality aspects of water transfers, *Chem. Ind.* (London), 4:138-142.

Anon., 1977, British groundwater schemes utilizing aquifer storage for river regulation or conjunctive use, *Consult. Eng. London,* 41(6):42-43.

Armitage, P. D., 1976, A quantitative study of the invertebrate fauna of the River Tees below Cow Green Reservoir, *Freshwater Biol.,* 6:229-240.

Armitage, P. D., 1977, Invertebrate drift in the regulated River Tees and an unregulated tributary Maize Beck, below Cow Green dam, *Freshwater Biol.,* 7:167-183.

Armitage, P. D., 1978, Downstream changes in the composition, numbers and biomass of bottom fauna in the Tees below Cow Green reservoir and in an unregulated tributary Maize Beck, in the first five years after impoundment, *Hydrobiologia,* 58:145-156.

Armitage, P. D., and Capper, M. H., 1976, The numbers, biomass and
    transport downstream of microcrustaceans and *Hydra* from Cow
    Green Reservoir (Upper Teesdale), *Freshwater Biol.*, 6:425-432.
Armitage, P. D., MacHale, A. M., and Crisp, D. C., 1974, A survey of
    stream invertebrates in the Cow Green Basin (Upper Teesdale)
    before inundation, *Freshwater Biol.*, 4:369-398.
Ash, R. V., 1966, The Great Eau Scheme: North-east Lincolnshire
    Water Board, *J. Inst. Water Eng.*, 20:435-458.
Avon and Dorset River Authority, 1973, "Upper Wylye investigation,"
    Detailed report by Avon and Dorset River Authority in coopera-
    tion with the Water Resources Board and the West Wilts Water
    Board, Poole, 120p.
Blezzard, N., Crann, H. H., Iremonger, D. J., and Jackson, E., 1971,
    Conservation of the environment by river regulation, *Yb. Assoc.
    River Auth. 1971*, 70-115.
Brooker, M. P., and Hemsworth, R. J., 1978, The effect of the re-
    lease of an artificial discharge of water on invertebrate drift
    in the River Wye, Wales, *Hydrobiologia*, 59:155-164.
Calderwood, W. L., 1945, Passage of smolts through turbines: Effect
    of high pressures, *Salmon Trout Mag.*, 115:214-221.
Central Water Planning Unit, 1978a, Recent developments in ecologi-
    cal studies of chalk streams, *Cent. Water Plan. Unit Annu. Rep.
    1977/78*, 40-48.
Central Water Planning Unit, 1978b, Projects--river regulation
    losses, *Cent. Water Plan. Unit Annu. Rep. 1977/78*, 19-21.
Chubb, J. C., 1977, "A review of the parasite fauna of the fishes of
    the R. Dee system," *Cent. Water Plan. Unit, Reading*, 107p.
Crann, H. H., 1978, Llyn Brenig, Part I: The concept and its promo-
    tion, *J. Inst. Water Eng. Sci.*, 32:279-287.
Crisp, D. T., 1966, Input and output of minerals for an area of
    Pennine moorland; the importance of precipitation, drainage,
    peat erosion and animals, *J. Appl. Ecol.*, 3:327-348.
Crisp, D. T., 1977, Some physical and chemical effects of the Cow
    Green (Upper Teesdale) impoundment, *Freshwater Biol.*, 7:109-120.
Crisp, D. T., Mann, R. H. K., and McCormack, J. C., 1974, The popu-
    lations of fish at Cow Green, Upper Teesdale, before impound-
    ment, *J. Appl. Ecol.*, 11:969-996.
Crisp, D. T., Mann, R. H. K., and McCormack, J. C., 1978, The
    effects of impoundment and regulation upon the stomach contents
    of fish at Cow Green, Upper Teesdale, *J. Fish Biol.*, 12:287-301.
Davies, A. V., and Price, D. R. H., 1979, Problems of water quality
    protection in inter-river transfers, (Anglo-Soviet seminar on
    River Basin Management, June 1978), Department of the Environ-
    ment, London (in press).
Elder, H. Y., 1966, Biological effects of water utilization by
    hydro-electric schemes in relation to fisheries, with special
    reference to Scotland, *Proc. R. Soc. Edinb.* Sect. B (Nat.
    Environ.), LXIX, Pt. III/IV:246-271.
Fordham, A. E., 1970, The Clywedog reservoir project, *J. Instn.
    Water Eng.*, 24:17-76.

Gee, A. S., Milner, N. J., and Hemsworth, R. J., 1978, The effect of
    density on mortality in juvenile Atlantic salmon (*Salmo salar*)
    J. Anim. Ecol., 47:497-505.

Great Ouse River Authority, 1974, Some aspects of the ecology of the
    River Thet, *Great Ouse River Authority Report*, 27p.

Guiver, K.,1976, Implications of large-scale water transfers in the
    UK, The Ely Ouse to Essex transfer Scheme, *Chem. Ind.* (Lond.)
    4:132-135.

Hall, D. G., 1977, Operational aspects of the Kielder water scheme,
    Chem. Ind. (Lond.), 9:338-340.

Hancock, R. S., 1977, The impact of water transfers and associated
    regulation reservoirs on the fish population of the Tyne, Tees
    and Swale, *Proc. 8th Br Coarse Fish Conf.*, 137-158.

Hartley, W. G., and Simpson, D., 1967, Electric fish screens in the
    United Kingdom, *in*: "Fishing with Electricity--its application
    to biology and management," R. Vibert, ed., Fishing News
    (Books), Ltd., London, 276p.

Hellawell, J. M., 1976, River management and the migratory behaviour
    of salmonids, *Fish Manage.*, 7(3):57-60.

Holmes, N. T. H., Lloyd, E. J. H., Potts, M., and Whitton, B. A.,
    1972, Plants of the River Tyne and future water transfer
    scheme, *Vasculum*, 57:56-78.

Holmes, N. T. H., and Whitton, B. A., 1977a, The macrophytic vegeta-
    tion of the River Tees in 1975: Observed and predicted
    changes, *Freshwater Biol.*, 7:43-60.

Holmes, N. T. H., and Whitton, B. A., 1977b, Macrophytic vegetation
    of the River Swale, Yorkshire, *Freshwater Biol.*, 7:545-558.

Howells, W. R., and Jones, A. N., 1972, The River Towy regulating
    reservoir and fishery protection scheme, *J. Inst. Fish Manage.*,
    3:5-19.

Huntingdon, R., and Armstrong, R. B., 1974, Operation of the Ely-
    Ouse-Essex scheme, *J. Inst. Water Eng.*, 28(8):387-401.

Iremonger, D. J., 1971, River regulation and fisheries, *Salmon Trout
    Mag.*, 191:64-79.

Isaac, P. C. G., 1967, "River Management," Maclaren & Sons, Ltd.,
    London, 258p.

Jones, A. N., 1968, The relationship of river flow and salmon ang-
    ling success in the River Towy, *Rep. South West Wales River
    Authority 1968*, 48-49.

Kennard, M. F., and Reader, R. A., 1975, Cow Green dam and reser-
    voir, *Proc. Inst. Civil Eng.*, 58:147-175.

Lavis, M. E., and Smith, K., 1972, Reservoir storage and the thermal
    regime of rivers, with special reference to the River Lune,
    Yorkshire, *Sci. Total Environ.*, 1:81-90.

Lehmkuhl, D. M., 1972, Change in thermal regime as a cause of reduc-
    tion of benthic fauna downstream of a reservoir, *J. Fish. Res.
    Board Can.*, 29:1329-1332.

Menzies, W. J. M., 1966, Salmon fisheries and the development of
    hydro-electric power, *Salmon Trout Assoc. London Conf.*, 1966:1-15.

Millichamp, R. I., 1976, Some thoughts on water abstraction on
    migratory fish rivers, *Fish. Manage.*, 7:1-3.
Mills, D. H., 1965a, Smolt production and hydro-electric schemes,
    Int. Council Explor. Sea ICES/CM No. 31:1-3.
Mills, D. H., 1965b, Observations on the effects of hydro-electric
    developments on salmon migration in a river system, *Int.
    Council Explor. Sea ICES/CM* No. 32:1-5.
Mills, D. H., 1968, Some observations on the upstream movements of
    adult Atlantic salmon in the River Canon and the River Meig,
    Rossshire, *Int. Council Explor. Sea ICES/CM* No. 10:1-5.
Mills, D. H., 1971, "Salmon and trout, a resource, its ecology,
    conservation and management," Oliver and Boyd, Edinburgh. 351p.
Mills, D. H., and Shackley, P. E., 1971, Salmon smolt transportation
    experiments in the Conon River system, Ross-shire, *Freshwater
    Salmon Fish. Res.*, 40:1-8.
Ministry of Agriculture, Fisheries and Food and National Water
    Council, 1976, "The fisheries implications of water transfers
    between catchments," MAFF and NWC, London 54p.
National Water Council, 1978, "Water industry review 1978," National
    Water Council, London 99p.
North, E., and Hickley, P., 1977, The effects of reservoir releases
    upon angling success in the River Severn, *Fish. Manage.*, 8:86-91.
Northumbrian Water Authority, 1976, "Report on experimental releases
    from Cow Green reservoir on the River Tees (23-28 June 1976),"
    Northumbrian Water Authority, Newcastle-upon-Tyne 46p.
Pearson, W. D., Kramer, R. H., and Franklin, D. R., 1968, Macro-
    invertebrates in the Green River below Flaming Gorge Dam
    1964-65 and 1967, *Proc. Utah Acad. Sci. Arts Lett.*, 45:148-167.
Pyefinch, K. A., and Mills, D. H., 1963, Observations on the move-
    ments of Atlantic salmon (*Salmo salar* L.) in the River Conon
    and the River Meig, Ross-shire, 1, *Freshwater Salmon Fish. Res.*,
    31:1-24.
Rodda, J. C., Downing, R. A., and Law, F. M., 1976, "Systematic
    Hydrology," Newnes-Butterworths, London 399p.
Rogers, A., 1978, Dinorwic pumped storage scheme aquatic environ-
    mental studies programme, 1, Introduction and general apprai-
    sal, *Central Electricity Generating Board North Western Region,
    Scientific Services Department, Rep. No. NW/SSD/82/77*, 10p.
Severn-Trent Water Authority, 1977, "Developing assets--Regulation
    of the River Severn by underground water, The Shropshire ground-
    water scheme," Severn-Trent Water Authority, Birmingham. 10p.
Spence, J. A., and Hynes, H. B. N., 1971, Differences in the benthos
    upstream and downstream of an impoundment, *J. Fish. Res. Board
    Can.*, 28:35-43.
Stuart, T. A., 1962, The leaping behaviour of salmon and trout at
    falls and obstructions, *Freshwater Salmon Fish. Res.*, 28:1-46.

Truesdale, G., and Taylor, G., 1978, Quality implications in reser-
    voirs filled from surface water sources, *Prog. Water Tech.*,
    10:289-300.
Ward, J. V., 1976, Comparative limnology of differentially regulated
    sections of a Colorado mountain river, *Arch. Hydrobiol.*,
    78:319-342.

STREAM REGULATION IN SWEDEN WITH SOME EXAMPLES

FROM CENTRAL EUROPE

Jan Henricson* and Karl Müller

University of Umeå
Department of Animal Ecology
S-901 87 Umeå, Sweden

INTRODUCTION

In Sweden about 60% (54 TWh) of the economically feasible water power has already been developed. For Europe (excluding the Soviet Union) the figure is 55%, and for the whole world, some 15%. Only Norway (64%) shows a higher degree of water-power utilization than Sweden. The construction of dams has caused a range of environmental effects in regions with differing climate and geology. Compared to other regions of the world, the environmental effects in Sweden are rather moderate. The Swedish rivers are comparatively small, as are the water-power projects, compared to those in the Soviet Union, Africa, and North America.

The exploitation of water power during the last 40 years has involved the disappearance of about 1500 km of streams in Swedish rivers. Today about 125 water-power stations ($\geq$ 20 MW) have been built on the large rivers, from the River Lule älv in the north to the River Klarälven in the south. Only two large rivers, the River Kalix älv and the River Torne älv, remain unregulated. Fig. 1 shows the geographical position of Swedish water-power stations.

Studies of the biological effects of water-power exploitation in Sweden have concentrated on the lake reservoirs in the mountain area; for a review see Nilsson (1973). The biological conditions in river reservoirs have been studied by Grimås and Nilsson (1965).

---

*Present address: Fishery Board of Sweden, FÅK, St. Torget 3, S-871 00 Härnösand, Sweden

Fig. 1. Key map of water-power stations in Sweden (1 = River
         Lule älv, 2 = River Indalsälven).

CHANGES IN ENVIRONMENTAL FACTORS CAUSED BY REGULATION

     The height-of-fall in the rivers has most often been utilized
to a maximum for production of hydroelectricity by the construction
of dams combined with water-power stations. Upstream from a dam, a
river reservoir extends up to the outlet of the next water-power
station. As a consequence, long stream sections below dams, which
have not been inundated, are comparatively rare. The river has been
converted into a ladder of rather small reservoirs. For the water
to flow, a certain inclination is required, and the inclination is
largest in the narrowest upstream part of the reservoir. Here a new
kind of lotic ecosystem has been created.

The river reservoirs are regulated through changes in discharge, not water-level fluctuations, in accordance with the demand for electricity. The amplitude is small, most often 0.2-0.6 m. The regulation is short term, so the discharge and consequently the water velocity may change considerably during a day, for example, from 160 to 400 m^3/s. Zero discharge (0 m^3/s) is permitted at some power stations. The rivers are also influenced by long-term regulation created by the lake reservoirs in the mountains. This regulation results in a higher discharge during the period from late summer to spring and a lower discharge from May to August, compared to natural conditions. The most apparent effects of the regulation on discharge are thus a more balanced yearly regime and large daily variations.

In addition to discharge and water velocity, temperature is an important environmental factor in the regulated rivers. The yearly fluctuations may change considerably compared to the situation before regulation, depending upon the position of the outlets from the dams and the flow regime (Lehmann, 1927; Müller, 1955, 1962a,b; Penáz, 1966, 1967, 1968; Penáz et al., 1968; Ward, 1974). Changes in ice conditions brought about by regulation (Fremling, 1973) may have negative or positive effects on the zoobenthos (Müller, 1962a; Penáz et al., 1968). Sediment transport has decreased 2-3 times in the regulated Swedish rivers (Nilsson, 1976). The transport of colloidal particles and organic material has not decreased to the same extent. Changes in water chemistry downstream from dams are most important during the 5-10 years after impoundment. Later on, water chemistry returns to normal values (Sundborg, 1977).

EFFECTS OF RIVER REGULATION ON BOTTOM AND DRIFT FAUNA

Generally, the animal community in lotic biotopes in Sweden is dominated by insects. Accordingly, the following discussion will deal mainly with the effects of river regulation on mayflies (Ephemeroptera), stoneflies (Plecoptera), and caddisflies (Trichoptera).

There are few Swedish studies of the effects of regulation on stream benthos. Müller (1955, 1962a,b) made comparative studies of regulated and unregulated parts of the River Lule älv. Recently, the Fishery Board has started the Research Group FÅK, dealing with fishery management in regulated rivers. Studies on stream benthos are included in the program, and since 1978, work has been in progress in the River Indalsälven (Henricson and Sjöberg, unpublished).

Müller (1962a,b) found that the development of the bottom fauna was strongly influenced in a regulated river. Because of a delay in the increase in water temperature in the spring (Fig. 2) and a poor

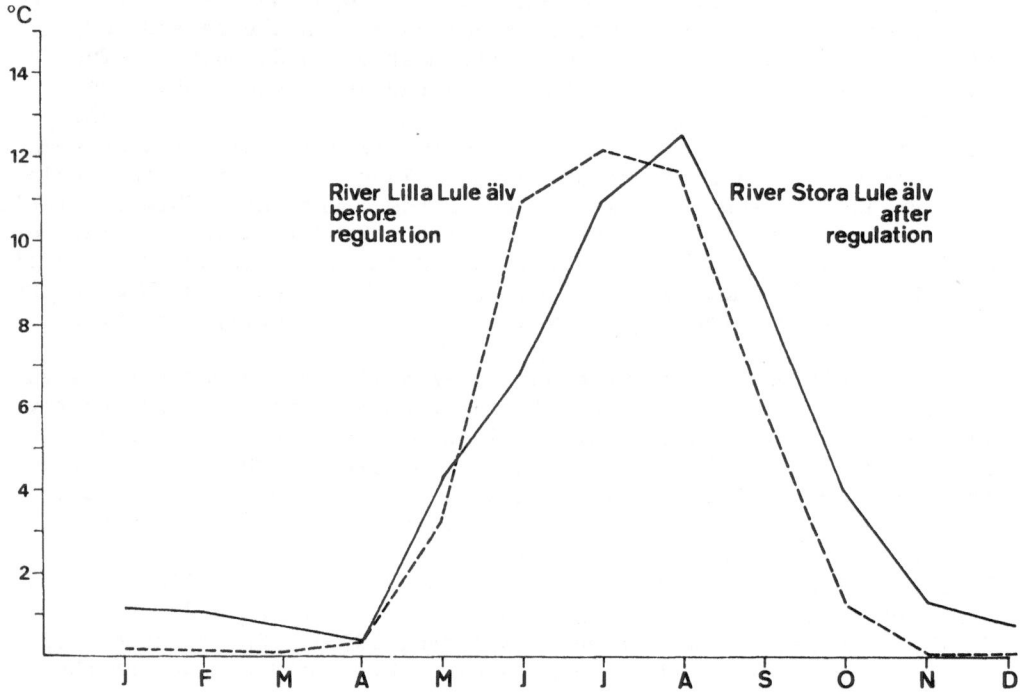

Fig. 2.   Monthly mean water temperature in the River Lilla Lule älv
          before regulation and in the River Stora Lule älv after
          regulation.

plankton drift, caused by the bottom outlet in the dam, the typical
early summer maxima of *Simulium* species were absent.   The "active"
feeders dominated the "passive" feeders (Fig. 3).

     The total biomass was significantly lower in the regulated
river during the period June to September.   Several factors were
significant for the poor development of the zoobenthos in the
regulated river: (1) the short-term regulation, (2) the changed
temperature regime, (3) the poor plankton drift, and (4) probably
the increased sediment transport connected with the construction of
the power stations, which is especially detrimental to "passive"
feeders.

     Numbers and biomass may thus be negatively influenced by
regulation, but this is not necessarily the outcome.   In the River
Indalsälven, biomass values comparable to natural conditions (5-
20 g/m^2) have been found.   Compared to the River Lule älv, however,
several factors, both natural and those produced by regulation, are
advantageous to the zoobenthos in the River Indalsälven: (1) The

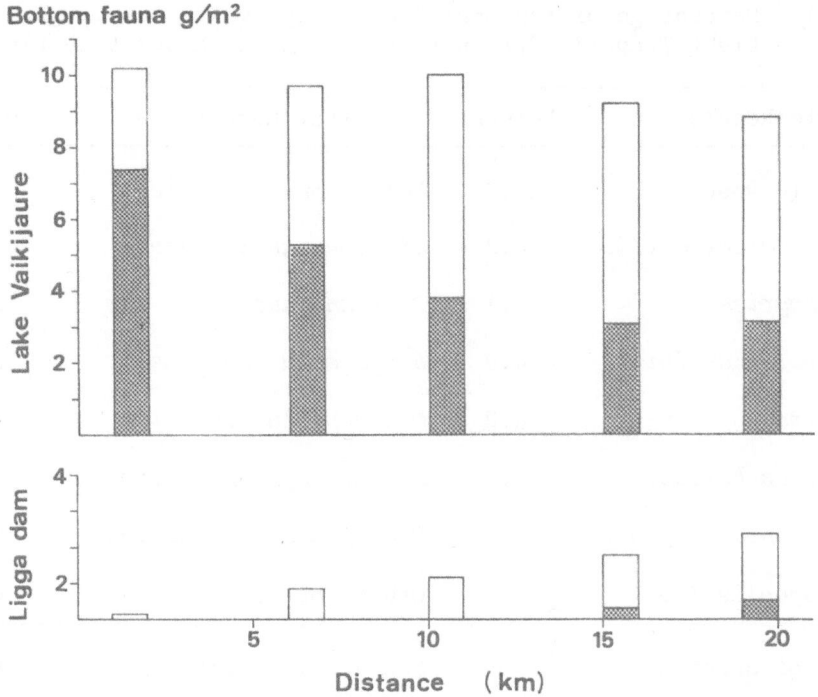

Fig. 3. Mean standing crop of total benthic fauna below the
         unregulated Lake Vaikijaure, sample period June-September
         1954-1956, and below the Ligga Dam, June-September 1957
         (shaded area = "passive" feeders).

minimum discharge at the investigated site in the River Indalsälven
is 100 m^3/s; in the River Lule älv, 0 m^3/s. (2) The water intake to
the power station is at the surface; in the River Lule älv, water is
taken from the hypolimnion. (3) The lake area in the River
Indalsälven is larger, which positively affects the nutritional
status of the river (Müller, 1956). (4) The Ca content is higher,
which increases productivity.

     Biomass values may thus be normal in the regulated river, but
the fauna composition is changed and the diversity decreases.

     In north European streams, mayflies (Ephemeroptera) are present
in large numbers. In unregulated lotic reaches, the genus *Baetis* is
most abundant. The changed environmental conditions after regulation may cause an alteration in the composition of species, which is
shown in Table 1 for unregulated parts of the River Lilla Lule älv
below Lake Vaikijaure (Q_{mean} = 177 m^3/s) and for the regulated
River Lule älv below Messaure Dam (Q_{mean} = 290 m^3/s). Before

Table 1. Percentage of the Most Common Mayfly Species Caught in
          Light Traps in Stream Biotopes in the River Lule älv

Before Regulation	Percent	After Regulation	Percent
Baetis (4 species)	44.7	*Metretopus borealis*	41.1
Heptagenia dalecarlica	8.2	*Siphlonurus lacustris*	43.2
H. fuscogrisea	5.1	*S. linneanus*	2.8
Ephemerella aurivilli	4.9	*Heptagenia joermensis*	2.5
E. ignita	4.8	*Centroptilum luteolum*	1.6
Metretopus borealis	3.2	*Paraleptophlebia strandii*	1.4
Leptophlebia marginata	2.9	*Heptagenia dalecarlica*	0.9
Other species (15)	26.2	Other species (9)	6.5
Number of species	25	Number of species	16

regulation, the Baetidae, represented by *B. rhodani*, *B. vernus*, *B. macani*, and *B. subalpinus*, were the dominating mayflies in the light-trap catches. The total number of mayfly species in the natural river was significantly higher than after regulation. The disappearance of Baetiids, which are typical for all north Swedish rivers during summer, may be explained by their extreme adaptation to lotic biotopes. Reduced current speed is a probable proximate stimulus for *Baetis* spp. to react to falling water level by increased activity (Ulfstrand, 1968a). The fluctuations in discharge caused the nymphs to drift downstream, out of their suitable environments, and thereby to perish.

In the River Indalsälven the mayflies were present in low mean densities only (Table 3). *Ephemerella mucronata* and *Heptagenia sulphurea*, the two most common species, have a period of intensive growth just before emergence. The winter is spent as eggs and/or small quiescent nymphs deep in the substrate (Ulfstrand, 1968b). This means that they spend only a short time as active nymphs on the river bottom exposed to the variations in water flow, and this may be of survival value.

Stonefly nymphs (Plecoptera) constitute up to 15% of the total biomass in northern unregulated rivers. In the unregulated River

Lilla Lule älv, there was a high diversity of stoneflies, and 21
species were caught in light traps. This may be compared to the
conditions below the Messaure Dam, where 16 species occurred
(Table 2). The stoneflies are good indicators of the environmental
disturbances caused by regulation. Changes in the frequencies of
different species occurred. After regulation, two species, *Leuctra
fusca* and *Amphinemura standfussi*, constituted 81% of the total
number captured. Species belonging to the hiemal growth type
(Brinck, 1949), i.e., *Capnia atra, Nemurella picteti, Nemoura
cinerea, Leuctra hippopus, Taeniopteryx nebulosa,* and *Diura nanseni*,
were negatively influenced; whereas species belonging to the estival
growth type, i.e., *A. standfussi* and *L. fusca*, obviously coped
better with the changed environmental conditions. The nymphs of the
hiemal type have their growth period during autumn, winter, and
spring and spend a long time active on the river bottom exposed to
the adverse effects of the short-term regulation. The estival
species, on the other hand, emerge in summer and autumn; and, after
oviposition, the eggs stay in diapause into late spring. The period
of larval growth is short, which should be an advantage in the
regulated river. In the River Indalsälven the stoneflies were very
poorly represented. The active, predatory species were practically
absent. The only species, found in low numbers, was *Isoperla*

Table 2. Percentage of the Most Common Stonefly Species Caught in
          Light Traps in Stream Biotopes in the River Lule älv

Before Regulation	Percent	After Regulation	Percent
Capnia atra	17.9	*Leuctra fusca*	41.2
Nemurella picteti	17.2	*Amphinemura standfussi*	39.8
Nemoura cinerea	14.0	*Nemoura cinerea*	8.1
Leuctra hippopus	12.7	*Nemurella picteti*	2.1
Taeniopteryx nebulosa	11.0	*Leuctra hippopus*	1.4
Amphinemura standfussi	7.9	*Capnia atra*	0.8
Diura nanseni	4.1	*Diura nanseni*	0.6
Other species (14)	15.2	Other species (9)	6.0
Number of species	21	Number of species	16

Table 3.   The Mean Numbers of Animals Per m^2,
           Based on Data Collected in May, July,
           August, and October 1978, Below the
           Krångede Dam in the River Indalsälven
           (+ = <1)

Hydra	Abundant
Naididae	60
Isoperla obscura	14
Diura sp.	+
Taeniopteryx nebulosa	+
Heptagenia sulphurea	27
Ephemerella mucronata	55
Ephemerella spp. undetermined	10
Baetis spp.	25
Ephemeroptera, other	13
Neureclipsis bimaculata	499
Polycentropus flavomaculatus	3
Hydropsyche spp.	117
Hydroptila sp.	2247
Rhyacophila sp.	5
Trichoptera, other	37
Chironomidae	1050
Simuliidae	+
Lymnaea peregra	174
Others	67
Total	4403

obscura, which, besides being a predator, ingests vegetable matter (Ulfstrand, 1967). Plecoptera are considered to be rather specialized in their habitat requirements (Ulfstrand, 1975) and are sensitive to lake regulation (Grimås, 1961).

The total number of species of caddisflies (Trichoptera) in the regulated rivers is most often quite comparable to that under natural conditions, but the species composition is changed and the diversity lowered. A total of 78 species were caught in light traps below the Messaure Dam in 1973-1974, which probably reflects a rather stabilized condition 10 years after the construction of the dam (Table 4). One species, *Hydroptila tineoides*, makes up 56% of the total number. The abundance of *Hydroptila* sp. in the River Indalsälven was also high (Table 3), peaking in July, when filamentous algae, which are their food item, were abundant. The genus *Hydroptila* is well adapted to variations in waterflow. In their first stages the nymphs live without cases deep in the substrate. Later on, when carrying cases, they attach them to the substrate and are able to survive temporary drainage by remaining inside the cases.

Rhyacophila nubila, a predator, has been found to be more abundant in unregulated streams, probably because of a greater abundance of prey species, such as simuliids and chironomids. Like the Baetiids, they react to reduced current speed by increased activity (Ulfstrand, 1968a).

Filter-feeding Trichopterans are often found in great numbers below lake outlets (Müller, 1955, 1956). This is an effect of the increased organic drift in the receiving stream. The same "lake outlet effect" may be present below dams. In the River Indalsälven a major part of the numbers and biomass were made up of net-spinning Trichopterans (Table 3). The highest number were found in August and September, correlated with the quantitative development of the plankton drift fauna. The share of filter-feeding caddisflies in the light-trap catches in regulated parts of the River Lule älv was also greater than that in natural non-outlet biotopes. These caddisfly species are obviously able to withstand the variations in water flow, and their food demands are thereby ensured. Ulfstrand (1967) considers food and substrate to be more important environmental factors than current velocity for many species.

Blackflies (Simuliidae), which are normally typical at lake outlets in northern Sweden, were absent from regulated parts of the rivers, and they have been reported to be very sensitive to changes in water flow (Carlsson, 1962). The depth, in the deepened stream channel below a dam, may also be too large, since simuliids prefer shallow waters (Ulfstrand, 1967).

Table 4.   The Most Common Caddisfly Species Caught in Light
           Traps Below the Messaure Dam in the River Lule älv

	Number	Percent
Hydroptila tineoides	24,994	56.0
Apatania stigmatella	5,048	11.3
Rhyacophila nubila	2,188	4.9
Athripsodes dissimile	2,137	4.8
Lepidostoma hirtum	1,858	4.2
Neureclipsis bimaculata	1,141	2.6
Hydropsychae nevae	1,108	2.5
Athripsodes annulicornis	995	2.2
Hydropsychae ornatula	749	1.7
Polycentropus flavomaculatus	743	1.7
Other species (68)	3,700	8.3
Total	44,661	

Certain gastropods are apparently favored by the conditions
below dams, probably by the increased abundance of periphyton and
filamentous algae. Below the Krångede Dam in the River Indalsälven,
Lymnaea peregra constituted a considerable amount of the biomass.
Hydra also occurred in great numbers. Oligochaets (family Naididae)
were most abundant in July. Armitage (1978) also found great
numbers of these three taxa below the Cow Green Dam in England.

Penáz et al. (1968) reported increased abundance and biomass of
zoobenthos below the Vir Dam in the River Svratka, Czechoslovakia,
caused by the favorable influence of the dam on the discharge
balance. The greatest part of the biomass was larvae and pupae of
caddisflies, especially *Rhyacophila nubila* and *Hydropsyche* sp. The
dam, however, caused qualitative changes. The lowering of the
annual mean water temperature caused a decrease in the thermophilous
species. The rich aquatic vegetation below the dam increased the
abundance of small animals, especially minute mayfly larvae (*Baetis*)

and chironomids. Some species were not affected at all, for example, *Ephemerella ignita*, which occurred in high numbers both before and after the construction of the dam. Penáz et al. (1968) also report a technical failure at the Vir Dam, which caused considerable quanti- ties of sediment to flow downstream. This event had a very negative influence on the zoobenthos, and the biomass decreased. The filter- feeding trichopterans were most severely affected. This accords with observations in Sweden in connection with the construction of dams.

The qualitative composition of the macroinvertebrate drift fauna in a regulated river is very different from that in unregulated conditions. In an unregulated tributary of the River Indalsälven the insect nymphs were heavily dominant and constituted between 76 to 92% of the numbers on different sampling occasions. In the regulated river the lake drift was significant as fish food (Table 5). A heavy drift of zooplankton and *Hydra* also occurred during the summer and autumn. The diversity and total amount of macroinvertebrate drift was lower in the regulated river (Fig. 4). The most important factors causing these differences are most likely the reduction of areas producing stream benthos and the daily variations in discharge. A study of the drift fauna in the River Svratka downstream from the Vír Dam has been reported by Penáz et al. (1968). They found that the animal component of the drift had increased in spring, summer, and autumn but was dominated by zooplankton. Rheobenthic forms, providing valuable fish food, were scarcely represented.

EFFECTS OF STREAM REGULATION ON THE FISH FAUNA

The influence of water-power exploitation on sport and commer- cial fishing has long been considered the most important environ- mental effect. Accordingly, changes in biological conditions and in development of the catches have been carefully studied by fishery authorities.

Before the development of water power, both Swedish and Finnish rivers had a rich stock of anadromous atlantic salmon (*Salmo salar*) and sea trout (*S. trutta*). Today the dams in the rivers prevent these economically valuable species from reaching their spawning grounds. The reproduction of salmon and sea trout is now based on hatcheries in both countries.

The balance between the resident fish species is changed in the regulated river because of the reduction of lotic reaches and the increase in reservoir area. The salmonids decrease and are replaced by lentic, economically less valuable, species, i.e., perch (*Perca fluviatilis*), ruffe (*Acerina cernua*), roach (*Rutilus rutilus*), and

Table 5. Percentage of Different Types of Macroinvertebrate
          Drift Below Two Dams in the Regulated River Indalsälven
          and in the Unregulated Tributary Ammerån

	Lotic	Lentic	Emerging	Terrestrial
Ammerån	87	0	1	12
Krångede Dam	2	90	4	4
Gammelänge Dam	33	59	3	5

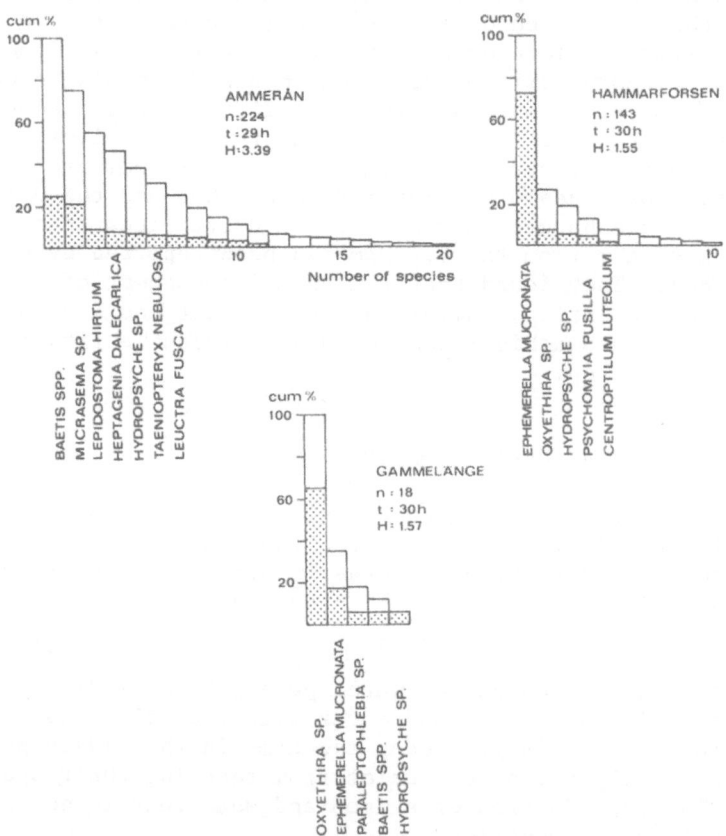

Fig. 4. Percentage distribution (shaded) and accumulated percentage
         of species of Ephemeroptera, Plecoptera, and Trichoptera
         in drift samples taken at night in August, September, and
         October 1978 in the unregulated River Ammerån and in the
         regulated River Indalsälven (H and G); n = number of
         animals, t = sampling time, H = Shannon-Weaver index.

pike (*Esox lucius*), which flourish in river reservoirs. Whitefish (*Coregonus* spp.) may also be abundant in the reservoirs, but their growth is frequently impaired, and as a rule they become heavily infected with the cestode larvae, *Triaenophorus crassus*. The lentic species also migrate into streaming parts of the reservoirs and interfere with the remaining salmonids. This circumstance was also observed by Penáz et al. (1968) above the Vír Reservoir.

The populations of grayling (*Thymallus thymallus*) and especially brown trout (*Salmo trutta*) are decreased or eliminated because of increased competition and predation. Other factors responsible for the outcome are the poor drift of stream benthos and the short-term regulation, which interfere with the territorial behavior of young trout. Despite the general trend, however, populations of brown trout and particularly grayling have been able to survive and reproduce below some dams, but compared to the situation before the regulation the catches are not economically important.

Changes in the yearly temperature cycle below dams have been reported to have great influence on the fish fauna. Lehmann (1927) found that below a dam in the River Eder in central Europe the original fish fauna (*Barbus fluviatilis, Perca fluviatilis, Esox lucius*, and *Leusiscus rutilus*) was replaced by trout, grayling, and *Cottus gobio*. Changed yearly temperature cycle and decreased annual mean temperature below the Vír Dam in Czechoslovakia (Fig. 5) favored the brown trout and *Cottus gobio*, whereas *Chondrostoma nasus* and *Barbus barbus* decreased significantly (Penáz et al., 1968). The brown trout also showed increased growth below the dam (Penáz, 1968).

Because of the drainage of bottom water lacking in oxygen, extensive fish mortality has been reported below some Czechoslovakian dams (Brádka and Rehácková, 1964). Penáz et al. (1968) also point out the potential risk of fry mortality below the Vír Dam because of oxygen super-saturation combined with intensive formation of air bubbles on submerged objects.

CONCLUSIONS

The numbers and biomass of stream benthos may be reduced or increased by regulations, depending upon the prevailing regulation conditions. The faunal composition, however, is changed and the diversity decreased. The abundance, distribution, and production of benthic animals in lotic ecosystems are determined by a complex set of factors that are variously associated with each other. Water discharge is a key factor, as it has an influence on several other factors, such as current velocity, temperature, composition of the substrate and availability of food; and manipulation of discharge thus has extensive consequences for the total lotic ecosystem.

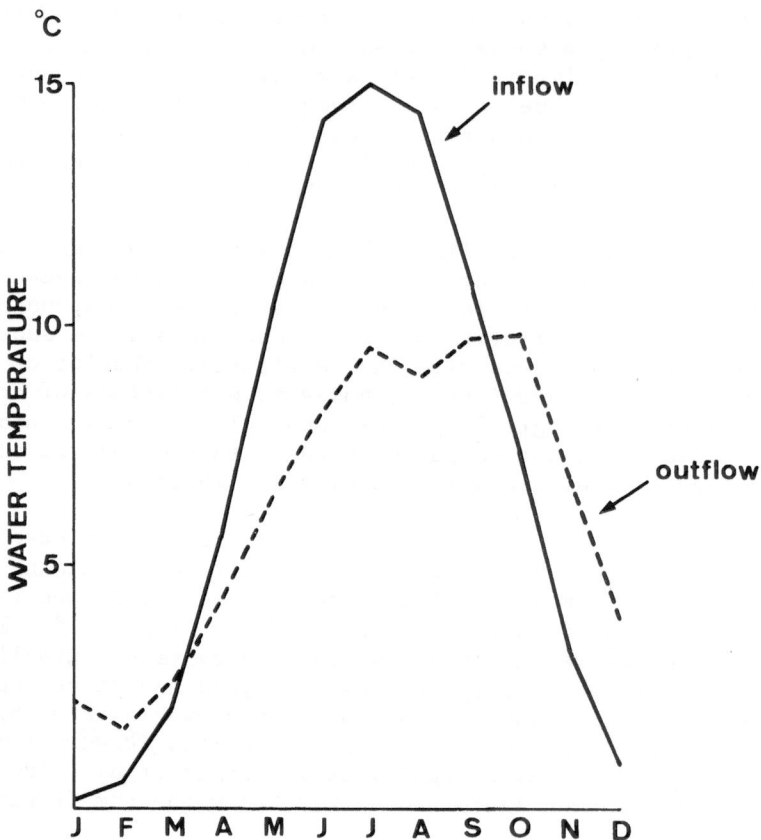

Fig. 5.   Monthly mean water temperature in the River Svratka above
          and below the Vír Reservoir [data from Penáz et al. (1968)].

        It is hard to point to one single factor as being responsible
for the decrease or increase of a species, as the animals do not
react to single factors but to factor combinations; however, food
is a basic requirement (Ulfstrand, 1967). The possibility for
individual species to survive in the regulated stream thus depends
on the nature of the environmental changes, which, in turn, depends
on the design and operation of the regulation. The specific ecological requirements, behavior, and type of life cycle are of vital
importance; for example, (1) the food preference of different species
related to the supply of different food materials in the regulated
stream, (2) the ability to withstand rapid fluctuations in water
velocity, active species facing the potential risk of being swept
away and transported out of their preferred environment, and (3) the
length of the period spent as active, full grown nymphs. Species

having only a short period are better preadapted to regulated conditions.

The fish fauna is directly influenced by changes in environmental factors caused by regulation. Fishes are also indirectly affected by the changes occurring in the plankton, drift, and bottom faunas. The typical outcome is a shift in the species composition, whose direction depends on the local ecological conditions and the type of regulation. In Swedish rivers the populations of salmonid species are reduced or destroyed and the lentic species take over.

In recent years the economic motive for exploiting the economically feasible water power remaining (40 TWh) in Sweden has increased, depending on, among other things, the increase in oil prices. In this connection, further studies on the already regulated rivers are urgently needed to increase the knowledge of these artificial ecosystems and to enable use of this knowledge in the planning of future regulations. The influence of different designs and differences in operation of the power plants on the fauna (e.g., discharge amplitude cycle and position of outlets) is especially important, partly because so-called power regulation, with daily variations in discharge from 0 to 1000 m^3/s, is planned on some Swedish rivers. The factors responsible for the survival of trout and grayling in certain streams below dams should be investigated. The necessary basic research should be the basis of fishery management measures.

ACKNOWLEDGMENTS

The investigations on which this paper is based have been supported by the Fishery Board of Sweden, the Wallenberg and Kempe Foundations, the State Power Board, and the Max-Planck-Gesellschaft zur Förderung der Wissenschaften. Some of the investigations have been accomplished by the Research Group FÅK financed by the State Power Board and the private power companies.

REFERENCES

Armitage, P. D., 1978, Downstream changes in the composition, numbers, and biomass of bottom fauna in the Tees below Cow Green Reservoir and in an unregulated tributary Maize Beck, in the first five years after impoundment, *Hydrobiologia*, 58:145-156.

Brádka, J., and Reháčková, V., 1964, Mass destruction of fish in the Slapy Reservoir in winter 1962-63 [in Czechoslovakian], *Vod. Hospodár.*, 14:451.

Brinck, P., 1949, Studies on Swedish stoneflies (Plecoptera),
    Opusc. Entomol. (Lund), 11:1-250.
Carlsson, G., 1962, Studies on Scandinavian black flies, *Opusc.
    Entomol.* (Lund), 21:1-280.
Fremling, S., 1973, "Changes of the Ice Regime in Swedish Rivers Due
    to the Development of the Hydroelectric Power," Commission
    Internationale des Grands Barrages, Madrid.
Grimås, U., 1961, The bottom fauna of natural and impounded lakes in
    northern Sweden, *Rep. Inst. Freshwater Res., Drottningholm*,
    42:183-237.
Grimås, U., and Nilsson, N.-A., 1965, On the food chain in some
    north Swedish river reservoirs, *Rep. Inst. Freshwater Res.,
    Drottningholm*, 46:31-48.
Lehmann, C., 1927, Uber den Einfluss der Talsperren auf die unterhalb
    liegende Bach- and Flussfischerei, *Zt. Fischerei Hilfswiss.*,
    25:467-476.
Müller, K., 1955, Produktionsbiologische Untersuchungen in
    Nordschwedischen Fliessgewässern. Teil 3. Die Bedeutung der
    Seen und Stillwasserzonen für die Produktion in Fliessgewässern,
    Rep. Inst. Freshwater Res., Drottningholm, 36:148-162.
Müller, K., 1956, Das Produktionsbiologische Zusammenspiel zwischen
    See und Fluss, *Ber. Limnol. Flusstn Freudenthal*, 7:1-8.
Müller, K., 1962a, Limnologisch-Fischereibiologische Untersuchungen
    in regulierten Gewässern Schwedish Lapplands, *Oikos*, 13:125-154.
Müller, K., 1962b, Limnologie und Wasserbau. Studien an
    Wasserkraftanlagen in Schwedish-Lappland, *Die Umschau in
    Wissenschaft und Technik*, 13:401-404.
Nilsson, B., 1976, "The Influence of Man's Activities in Rivers on
    Sediment Transport," Nordic Hydrology, Lyngby.
Nilsson, N.-A., 1973, "Biological Effects of Water-Power Exploita-
    tion in Sweden, and Means of Compensation for Damage,"
    Commission Internationale des Grands Barrages, Madrid.
Penáz, M., 1966, Contribution towards the knowledge of the biomass
    of zoobenthos in the Svratka River above and below the Vír
    River dam basin, *Zool. Listy*, 15:363-372.
Penáz, M., 1967, Einfluss der Talsperren auf die Ichtyofauna der
    unterhalb und oberhalb des Stausees liegenden Flussabschnitte,
    Verh. Int. Verein. Limnol., 16:1223-1227.
Penáz, M., 1968, Das Wachstum einiger Fischarten im Svratka-Fluss
    und seine Änderungen unter dem Einfluss der Talsperre Vír, *Acta
    Sci. Nat. Brno*, 2(11):1-50.
Penáz, M., Kubícek, F., Marvan, P., and Zelinka, M., 1968, Influence
    of the Vír River valley reservoir on the hydrobiological and
    ichtyological conditions in the River Svratka, *Acta Sci. Nat.
    Brno*, 2(1):1-60.
Sundborg, Å, 1977, "River, Hydropower, Environment. Environmental
    Effects of Hydropower Development" [in Swedish, English
    summary], SNV, Stockholm.

Ulfstrand, S., 1967, Microdistribution of benthic species
     (Ephemeroptera, Plecoptera, Trichoptera, Diptera:Simuliidae) in
     Lapland streams, *Oikos*, 18:293-310.
Ulfstrand, S., 1968a, Benthic animal communities in Lapland streams,
     Oikos (Suppl.), 10:1-120.
Ulfstrand, S., 1968b, Life cycles of benthic insects in Lapland
     streams (Ephemeroptera, Plecoptera, Trichoptera,
     Diptera:Simuliidae), *Oikos*, 19:167-190.
Ulfstrand, S., 1975, Diversity and some other parameters of
     Ephemeroptera and Plecoptera communities in subarctic running
     waters, *Arch. Hydrobiol.*, 76:499-520.
Ward, J. V., 1974, A temperature-stressed stream ecosystem below a
     hypolimnial release mountain reservoir, *Arch. Hydrobiol.*,
     74:247-275.

STREAM REGULATION IN NORWAY

Albert Lillehammer and Svein Jakob Saltveit

Zoological Museum, Sars gt. 1
Oslo 5, Norway

INTRODUCTION

Norway, situated in northern Europe, is characterized by mountainous area in the north and west with relatively short and fast
running streams, while in east and northeastern Norway the largest
watercourses are found. Watercourses in Norway have been used for
various purposes, such as timber transport, mills and sawmills, as
well as drinking water. However, the greatest use is connected with
the great demand for hydroelectric power, and this paper will
consider mainly the effects of such regulation.

Nearly 100% of the electric energy in Norway is based on water
power, which provides 81.2 Twh/y. Norway is the largest user of
hydroelectric power per inhabitant in the world. Over 50% is used
in industry, of which energy-intensive industries (aluminium, iron,
steel, ferroalloys) use 45%. For this purpose about 700 reservoirs
have been constructed. Only 50% of the watercourses that can be
used for hydroelectric power have been regulated, but many new
reservoirs are under construction or are being considered. Except
for the county of Nordland, the highest waterpower potential is in
the counties of southern Norway.

Norwegian streams are characterized by large variation in water
flow throughout the year. Two flow peaks are most common; one in
connection with the snow melt in early summer, and one in connection
with high precipitation in the autumn. This normally gives considerable variation during the year in the extent of stream bed covered
by water.

High altitude reservoirs, the most common type of regulation in Norway, are used to store water from melting snows and autumnal precipitation, mostly for generating electricity during winter, when demand is greatest. The reservoirs are mainly made by damming natural lakes. The power stations often receive water from many connected reservoirs, which in some cases are situated on neighboring watercourses (Fig. 1). Between the reservoir (R1) and the power station (P1) the stream (S1) is partly dry. Depending on power production, the stream (S2) below the power station will have either a high winter flow/low summer flow or a more even year-round flow compared with natural conditions. If there is a low-altitude reservoir (R2) and the water is transferred to the second power station (P2), the stream (S2) will also be partly dry. If all the water from (R1) is transferred to another watercourse, both streams (S1 and S2) will be partly dry.

A few river reservoirs have been built in Norway and are on the largest rivers, such as the Glåma, Gudbrandsdalslågen, Numedalslågen, Nidelva og Pasvikelva. Compared with regulated lakes, few studies have been carried out on affected streams in Norway; however, we will try to provide a general view of some of the consequences of stream regulation.

EFFECT OF STREAM REGULATION IN NORWAY

Constant Flow

A more constant year-round water flow has on some occasions been reported to have a positive effect both on algal development and on benthos and fish, as this involves a higher water flow in the winter and a more stable flow in the summer. The production area is increased in the winter, as more of the riverbed is covered.

In 1966-1967, regulation in the catchment area of the River Suldalslågen, western Norway (Fig. 2), increased winter flow and reduced summer flow. No significant differences have been recorded in physical/chemical factors and turbidity, except for the winter temperature, which increased in the lower part of the river by about one degree centigrade.

Before regulation, insect larvae, mainly chironomids and stoneflies, dominated both benthos and the food of small salmon fry in May and June (Lillehammer, 1973a,b; Fig. 3). In July, August, and September, there were large amounts of planktonic crustacea from the upstream lake in fry stomachs.

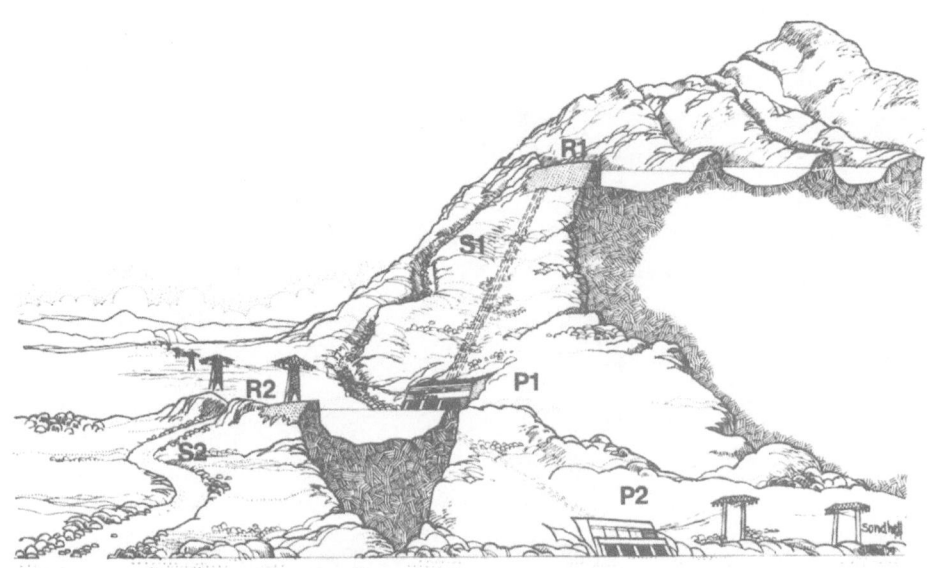

Fig. 1.   Stream regulation: R1 = high-altitude reservoirs, P1 =
first power station, S1 = stream between R1 and P1, R2 =
low-altitude reservoir, P2 = second power station, S2 =
stream below R2.

Extensive feeding of salmon fry has been reported to start at
temperatures between 4 and 5°C (Jones, 1959), while the best growth
of salmon parr occurs at temperatures above 7°C (Allen, 1940).
Temperatures above 7°C occur mainly in July, August, and September
(i.e., while the most important food items are planktonic crustacea).
The investigation period covers the whole feeding season.

Regulation does not seem to have affected either the fry's food
supply or growth. The same planktonic species (*Eudiaptomus gracilis,
Cyclops scutifer, Bosmina longispina,* and *Holopedium gibberum*) were
present in the same amount and frequency. The amount of planktonic
crustacea taken by each fry averaged 4.3 mg dry weight in 1962 and
1964 and 6.7 mg dry weight in 1976 and 1977. The mean length of fry
at the end of the first summer before regulation was 3.95 ± 0.31 cm
(SD) in 1964 and 4.13 ± 0.25 cm (SD) in 1966, while the mean length
after regulation was 4.23 ± 0.33 cm (SD) in 1977 and 4.24 ± 0.48 cm
(SD) in 1978 (Lillehammer et al., 1978). After regulation, recruitment also seems to be good. In September, fry densities ranged
between 18 and 80 per 100 m^2; parr densities, between 15 and 17 per
100 m^2. The highest density of pre smolt was 7.5 per 100 m^2 in

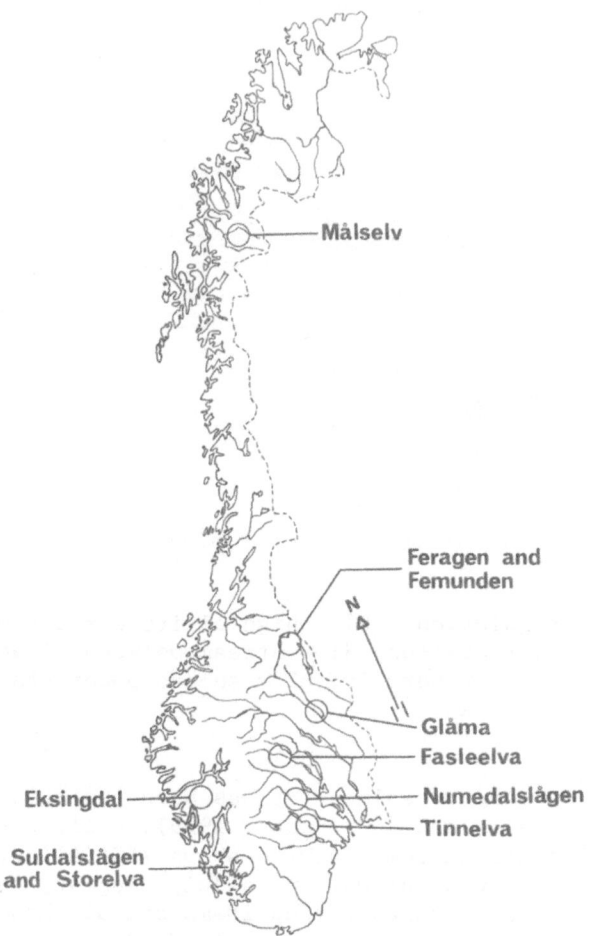

Fig. 2.   Norwegian sites referred to in this paper.

1976.   The ratio of salmon to trout fry and parr was between 3:1 and
4:1 (Lillehammer et al., 1976).

     In the upper reaches of a somewhat similarly regulated river,
Tinnelva, in the county of Telemark (Fig. 2), zooplankton also
constituted over 90% of the juvenile trout diet in June, July, and
August (Borgstrøm, 1976b).

     In the stream of Fasleelva between the lakes Strandefjorden and
Fløafjorden in the Begna watercourse (Fig. 2), where regulation
caused a more constant flow along a 0.5 km stretch, the yearly yield

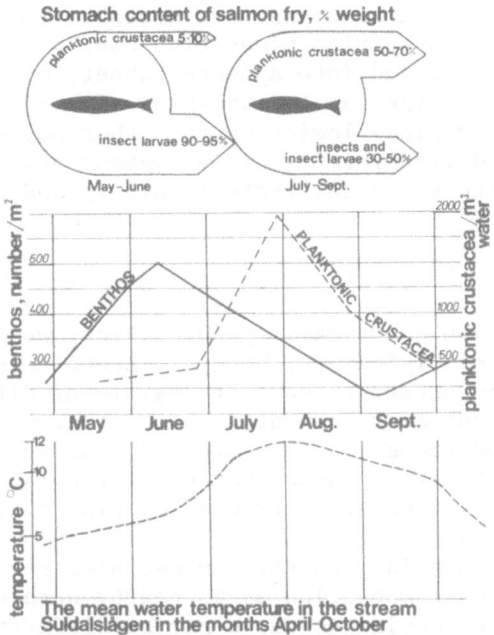

Fig. 3. Average stomach content of fry, abundance of benthic
         animals and planktonic crustacea, and the mean water
         temperature in the stream Suldalslågen from April to
         October 1961-1964.

from trout angling was estimated to be several hundred kg/ha
(Borgstrøm, 1976a; Jensen, 1976), an extraordinary high yield for
trout in Norwegian streams. The benthic fauna, dominated by gastropods, mussels, mayflies, and net-spinning caddisflies, had a mean
June biomass of from 33 to 51 g/m^2 wet weight (Borgstrøm, 1976c).

     High algal development has been reported in many different
rivers having more constant flows [e.g., Glåma, Fasleelva, Tinnelva,
Otra (NIVA 1976a,b; Skulberg and Kotai, 1978)]. Dense algal carpets
and drifting filaments make the water less suitable for drinking,
bathing, and fishing. The reasons for the high algal growth in this
type of regulated stream are difficult to explain because the
regulation has changed many factors that are important for algal
development (temperature, flow conditions, water chemistry). In the
River Glåma (Fig. 2) less flow in the summer increased the summer
temperature (Skulberg and Kotai, 1978). This together with a less

turbulent flow and a reduction in floods, which stabilized water
level and substrate and increased the ratio between growth surface
and water volume, made conditions more suitable for algal growth. A
very small increase in nutrients was also found. In this river the
algal vegetation developed into a dense carpet, mainly consisting of
single-celled benthic species (*Scenedesmus, Chlamydomonas, Tabellaria,
Pinnularia*), in very slow-flowing parts with sandy substrate and a
community of thread-like forms (mainly *Zygnema* and *Ulothrix*), in
turbulent parts with stony substrate (Skulberg and Kotai, 1978).

Reduced Flow

For streams with drastically reduced flow, we have little
information. A general feature, however, seems to be that the
vegetation on stony substrate consists mainly of different
desiccation-resistant moss communities, together with different
kinds of bushes and trees. On clay and mud, the vegetation consists
mainly of resistant marsh plants, because irregular water flow
prevents terrestrial vegetation from developing (Skulberg, 1974).

The benthic fauna in such streams may have the same densities
and the same species composition as in nearby unregulated rivers.
However, the species-dominance ratio may be different, and the total
number of benthic animals along a given stretch of stream will of
course be far less. This was the case in two nearby rivers in
western Norway studied in 1960-62 (Lillehammer, 1964). Both were
populated by the same benthic species and had nearly the same total
density per m^2 (Table 1). However, the River Suldalslågen (unregulated at that time) was dominated by Plecoptera, while the regulated
River Storelva had the greater number of chironomids. This is most
probably caused by a different temperature regime, for whereas
Suldalslågen is cold throughout the summer, Storelva is warm because
of extremely low summer flow.

High benthic densities may also result from lack of fish
predation. In salmon and sea trout streams the spawning population
is more or less prevented from entering by the dams and/or lack of
water. In addition, the possibility of survival of offspring is
very low (Møkkelgjerd and Gunnerød, 1975).

The effect of a river reservoir with a reduced flow downstream
has been studied in the River Glåma (Borgstrøm et al., 1976). The
abundance of the filter-feeding Trichoptera fauna, mainly *Hydropsyche*
spp. (Table 2), increased greatly below the Barkal Reservoir and
was detectable at station 5, 3 km below the power plant. Below
Rånåfoss Reservoir the same effect was present, although the flow
was not reduced (St. 10).

Table 1. Total Mean Densities of Benthic Fauna (Number per m^2) in
          the Regulated River Storelva, and the Unregulated River
          Suldalslågen, Based on Studies in 1960, 1961, and 1962
          (Lillehammer, 1964)

	Sudalslågen	Storelva
Oligochaeta	18.67	10.78
Mollusca		
Limnaea peregra	1.44	
Pisidium spp.	12.33	
Hydracarina	10.33	11.89
Ephemeroptera	16.78	5.56
Plecoptera	110.56	56.44
Trichoptera	61.85	66.56
Diptera		
Chironomidae	71.56	175.56
Simuliidae	22.22	17.89
Tipulidae	6.22	4.11
Other Diptera	0.44	2.11
Total	331.40	350.90

Turbid Flow

        Technical encroachment in a watercourse commonly produces
turbid water in both lakes and streams, and this has been reported
on many occasions from Norway (Borgstrøm, 1973; Aass, 1979). In the
River Målselv (Fig. 2), turbid water caused by severe bank erosion
in a lake was found to affect only salmon angling in the river below
(due to reduced transparency). In subsequent years no reduction in
the catches that could be related to the erosion was found. None of
the variations in hatching of salmon eggs or in growth or density of
juveniles and smolts could be ascribed to the increase in turbidity
(Anderson, 1979).

Table 2.   The Number of Filter-Feeding
            Trichoptera *Hydropsyche* spp. at
            10 Stations of the River Glåma,
            Where Two River Reservoirs are
            Located.

Station	*Hydropsyche* spp. (Number per m^2)
1	0.5
2	3.5
3	2.0
4	77.0
Barkal Reservoir	
5	450.0
6	100.0
7	50.0
8	190.5
9	10.0
Rånåfoss Reservoir	
10	750.0

RESTORATION MEASURES

     Norway has about 34 freshwater fish species. Atlantic salmon
(*Salmo salar*), brown trout (*Salmo trutta*), and eel (*Anquilla
anquilla*) are the most important in southern and western streams.
Northern streams also contain migratory char (*Salvelinus alpinus*).
In addition to these, the eastern and northeastern streams have
several species, such as grayling (*Thymallus thymallus*), whitefish
(*Coregonus lavaretus*), pike (*Esox lucius*), perch (*Perca fluviatilis*),
and various cyprinids.

Because many regulation projects have reduced the stream's spawning and growth capability, both brown trout and Atlantic salmon have been stocked. In 1977, 1.9 million trout were stocked. Both stocking and water transfer between systems have introduced fish species to new areas. The first documented case was in 1716, when pike came to the upper part of the River Glåma through a channel, built for timber transport, from the Lake Femunden to Feragen (Fig. 2) (Huitfeldt-Kaas, 1918). Although the transfer of water from one watercourse to another has become more common, the fish species introduced (except for one case involving grayling) are those, such as char and whitefish, that spend most or all of their lives in lakes (Aass, 1968, 1970). During the last few years, the minnow (*Phoxinus phoxinus*) has been found far outside its known distribution area (Fig. 4), because of either its use as bait or its being mixed with the stocking material, but no evidence exists for the latter. However, many trout hatcheries take in water from streams populated with minnows, and many of new habitats for trout are regulated streams and lakes annually receiving stocking material. Impurity of stocking material is also indicated by the introduction of Atlantic salmon instead of brown trout to mountain lakes (Borgstrøm, personal communication; Saltveit, unpublished). Minnows were found to take the same food items as small trout in some Norwegian lakes (Borgstrøm and Saltveit, 1975). According to Løkensgard and Borgstrøm (1976), minnow diets were very similar to those of brown trout and grayling in the River Glåma.

In some of the streams, weirs have been built to retain the water at some sites in the partly dry river and to raise the ground-water level in some areas. Biological studies have been made of production in the part of the regulated river occupied by weirs. The main part of the study Terskelprosjektet is located in the valley of Eksingdalen, western Norway. The project investigated the input of allochthonous material, primary production, and metabolism in the weir and the output to the stream below. The study also deals with trout (population) migration patterns, feeding, energy consumption, and reproduction (Larsen, 1977; Bekken et al., 1979). The study of the effect of weirs on benthic stream fauna below weirs was conducted in different watercourses, but the main study was in Numedalen (Raastad, 1979). In Eksingdal, wiers seem to have a positive effect on the trout population in the weir area (Larsen, 1977; Bekken et al., 1979). Weirs also seem to increase the filter-feeding component of the fauna below weirs, and in Numedalen a row of four weirs had a positive cumulative effect on the black fly fauna below the weirs (Raastad, 1979). However, in the main part of the river, between the weirs, the effect of the reduced flow has not been investigated.

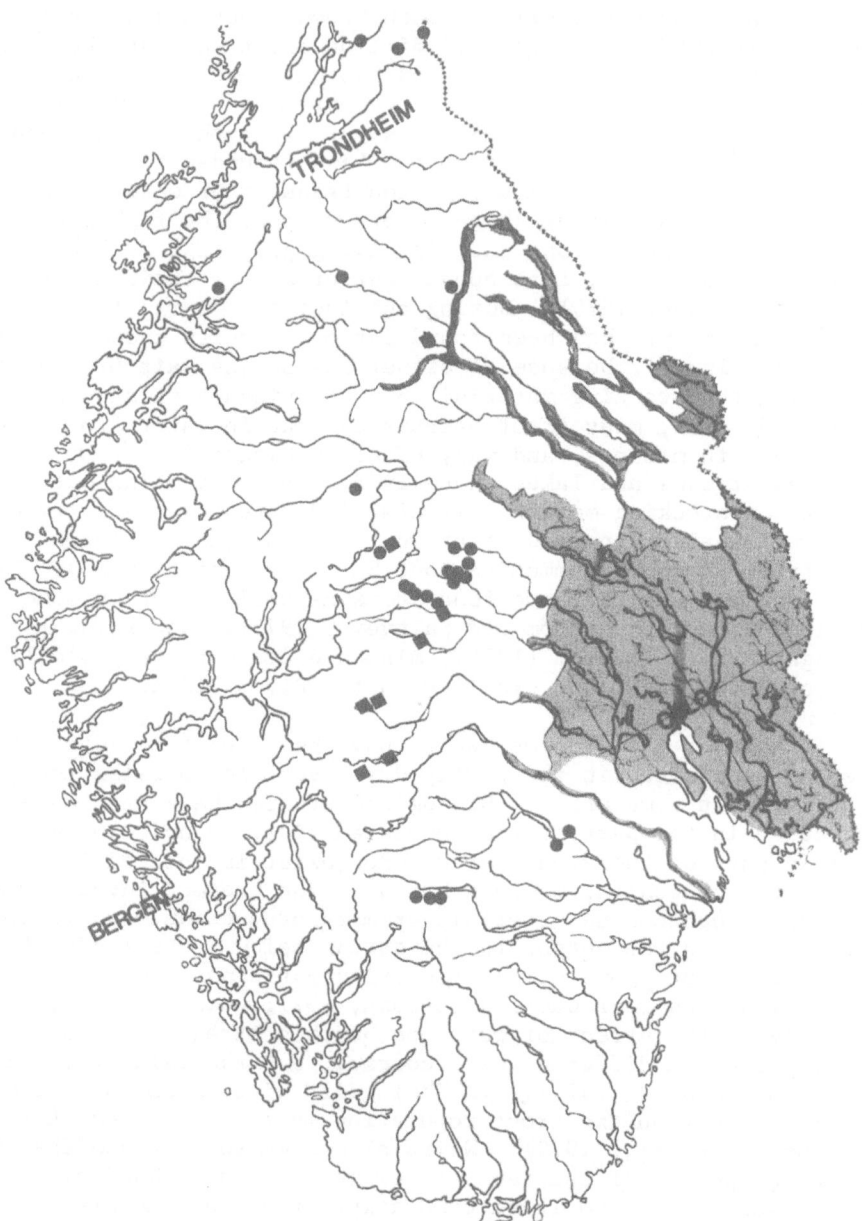

Fig. 4. The distribution of the minnow (*Phoxinus phoxinus*) in
         southern Norway [adapted from Borgstrøm (1973)], with
         additional data from Koksvik og Langeland (1975) and
         Saltveit (unpublished). Distribution area is shaded
         (Huitfeldt-Kaas, 1918), ● unregulated, ■ regulated.

CONCLUSION

The regulation of Norwegian watercourses started towards the end of the last century, but it was not until after World War II that the pace of development accelerated, to provide more employment. During recent years public debate concerning the different uses of Norwegian watercourses has led to a 10-year protection of several watercourses in order to undertake further investigations. However, many of the investigations of regulated streams are paid for by the regulating authorities. This often leads to brief investigations and superficial treatment of the material. In addition, many different institutions are working on problems connected with regulation, and the results are usually published in less-accessible reports in Norwegian; few studies of regulated streams in Norway have been published in English. Some studies from Norwegian streams have estimated production, biomass, and yield of fish (e.g. Grande, 1964; Rosseland, 1969; Power, 1973; Borgstrøm, 1976d; Heggberget, 1976; Aass, 1978), but only Borgstrøm and Aass deal with the effects of regulation. However, for the River Suldalslågen we have the advantage of data from the period before initial regulation, which, coupled with our present investigation, should enable us to determine the effect of the second regulation on salmon production. Thus, only when we have detailed comparative data from long-term studies conducted before and after regulation will we be able to precisely determine the ecological effects of stream regulation.

ACKNOWLEDGMENTS

We are indebted to Dr. John Brittain for his criticism of the manuscript and for correcting the language.

REFERENCES

Allen, K. R., 1940, Studies on the biology of the early stages of the salmon (*Salmo salar*). Growth in the river Eden, *J. Anim. Ecol.*, 9:1-23.

Aass, P., 1968, Vassdragsregulering, *in:* "Sportsfiskerens Leksikon 2," K. W. Jensen, ed.,

Aass, P., 1970, The winter migration of char, *Salvelinus alpinus* L., in ·the hydroelectric reservoirs Tunhovdfjord and Pålsbufjord, Norway, *Rep. Inst. Freshwater Res. Drottningholm*, 50:5-44.

Aass, P., 1978, Ørret og ørretfiske i Hallingdalselva ved Gol, *Terskelprosjektet*, 7:39.

Aass, P., 1979, Tilslamming i Hallingdalselva 1966-67, "Fisket i Ustedalsfjorden og Strandafjorden," Symp. om Vassdragsregule-ringers virkning på fisket, Leangkollen, 29-31 mars 1978.

Anderson, C., 1979, "Reguleringer og Utvaskinger i Målselvvassdraget,"
    Symp. om vassdragsreguleringers virkning på fisket, Leangkollen,
    29-31 mars 1978.
Bekken, T., Fjellheim, A., Larsen, R., and Otto, C., 1979, "Input
    and Output of Organic Material to the Weir Dam at Ekse,
    Eksingdal," Informasjon fra Terskelprosjekte, nr. 10, in
    press.
Borgstrøm, R., 1973a, "Spredning av Ørekyt," Jakt-Fiske-Friluftsliv,
    102.
Borgstrøm, R., 1973b, The effect of increased water level fluctuation
    upon the brown trout population of Mårvann, a Norwegian reser-
    voir, Norw. J. Zool., 21:101-112.
Borgstrøm, R., 1976a, Utbyggingsplaner for Faslefoss kraftverk.
    Virkninger på fisket, Rapp. Lab. Ferskv. Økol. Innlandsfiske,
    Oslo, 26:23.
Borgstrøm, R., 1976b, Ørretbestanden i Tinnelva. Virkninger på
    fisket ved utbygging av fallet mellom Tinnsjøen og Årlifoss,
    Rapp. Lab. Ferskv. Økol. Innlandfiske, Oslo, 30:22.
Borgstrøm, R., 1976c, "Faslefoss Kraftverk. En Vurdering av Alter-
    native Utbygginger," Notat. Lab. Ferskv. Økol. Innlandfiske,
    Oslo, 12 p.
Borgstrøm, R., 1976d, Ørretbestanden i Mørkedøla før bygging av
    terskler, Terkelprosjektet, 4:29.
Borgstrøm, R., and Saltveit, S. J., 1975, Skjoldkreps, Lepiderus
    articus Pallas, i regulerte vann II. Ørekyt og ørrets beiting
    på skjoldkrepslarver, Rapp. Lab. Ferskv. Økol. Innlandsfiske,
    Oslo, 22:12.
Borgstrøm, R., Brittain, J. E. and Lillehammer, A., 1976, Øster-
    dalsskjønne, Glåma mellom Auma og Høyegga. Virkning på
    fisket, Rapp. Lab. Ferskv. Økol. Innlandsfiske, 25:1-16.
Grande, M., 1964, En undersøkelse av bekkerøya i Øyfjell i Telemark,
    Fauna, 17:17-33.
Heggberget, T. G., 1975, "Produksjon og Habitatvalg hos Laks og
    Ørretyngel i Stjørdalselva og Forra 1971-1974," K. norske
    Vidensk. Selsk. Mus. Rapp. Zool. Ser., 1975-4.
Huitfeldt-Kass, H., 1918, "Ferskvandsfiskenes Utbredelse og
    Indvandring i Norge, Med et Tillaeg om Krebsen,"
    Centraltrykkeriet, Kristiania.
Jensen, J. W., 1976, "Planer om Nyutbygging av Faslefoss og Virk-
    ninger på Fisket," Stensil, 24 p.
Jones, J. W., 1969, "The Salmon," Collins, London, 192 p.
Koksvik, J. I., and Langeland, A., 1975, Nye funn av ørekyt,
    Phoxinus phoxinus L., i Tallsjøen (Nord-Østerdal) og
    Neavassdraget (Tydal) sommeren 1974, Fauna, 28:20-22.
Larsen, R., 1977, "Tersklenes Innvirkning på Biologiske Forhold i
    Regulerte Vassdrag - Terskelprosjektet: Ørretens Vandringer,
    Bestandsstørrelse, Vekst og Føde i Øvre delav Eksingedalselven
    før Reguleringen," Informasjon fra Terskelprosjektet, nr. 6, 31 p.

Lillehammer, A., 1964, "Benthos and Driftfauna as Food of Young
    Salmon (*S. salar* L.) and Trout (*S. trutta* L.) in the Rivers
    Suldalslågen and Storelva," Unpubl. Thesis, Univ. Oslo, 75 p.
Lillehammer, A., 1973a, An investigation of the food of one-to-four
    month old salmon fry (*Salmo salar* L.) in the River
    Suldalslågen, west Norway, *Norw. J. Zool.*, 21(1):17-24.
Lillehammer, A., 1973b, Notes on the feeding relationships of trout
    (*Salmo trutta* L.) and salmon (*Salmo salar* L.) in the River
    Suldalslågen, west Norway, *Norw. J. Zool.*, 21(1):25-28.
Lillehammer, A., Borgstrøm, R., and Saltveit, S. J., 1976, "Veskt og
    Ernaering hos Laksunger i Suldalslågen før og Etter Regulering
    av Ovenforliggende Vann," Rapp. Zool. Mus. Univ. Oslo, 26 p.
Lillehammer, A., Saltveit, S. J., and Borgstrøm, R., 1978, "En
    Sammenlikning av Oppvekstmulighetene for Laks i Sudalslågen
    før og Etter Vassdragsreguleringen," Rapp. Zool. Mus. Univ.
    Oslo, 19 p.
Møkkelgjerd, I., and Gunnerød, T. B., 1975, "Fiskeribiologiske
    Undersøkelser i Sauda 1974," Direktoratet for vilt og fersk-
    vannsfisk. Rapp., nr. 3, 1975, 30 p.
NIVA, 1976a, Tinnelva. Vassdragsundersøkelse 1976/1976 i forbin-
    delse med planlagt kraftutbygging, 0-1/76, 57 p.
NIVA, 1976b, Uttalelse om Faslefoss kraftverk, 0-42/75, 6 p.
Power, G., 1973, Estimates of age, growth, standing crop and produc-
    tion of salmonids in some north Norwegian rivers and streams,
    Rep. Inst. Freshwater Res. Drottningholm, 53:78-111.
Raastad, J. E., 1979, "Tersklenes Innvirkning på Biologiske Forhold
    i Regulerte Vassdrag - Terskelprosjektet: Bunndyrundersøkelser
    i regulerte Elver - Med Hovedvekt på insektgruppen Knott
    (Diptera, Simuliidae)," Informasjon fra Terskelprosjektet, nr.
    11, 56 p., in press.
Rosseland, L., 1969, "Om Virksomheten til Direktoratet for Jakt,
    Viltell og Ferskvannsfiske 1969," St. meld., nr. 76 (1969-
    1970).
Skulberg, O. M., 1974, Begroing i norske vassdrag. Virkninger av
    regulering, *Norsk Inst. Vannforskning, Årsrapport*, 1973:27-37.
Skulberg, O. M., and Kotai, J., 1978, Miljøfaktorer og algeutvikling
    i strømmende vann. Noen observasjoner av innvirkningene av
    vassdragsreguleringer på begroingsforhold i Glåma i Østerdalen,
    Norsk Inst. Vannforskning, Årbok, 1977:63-73.

STREAM REGULATION IN NORTH AMERICA

Jack A. Stanford and James V. Ward

Dept. Biological Sciences, North Texas State University,
Denton, Texas 76203; Dept. Zoology and Entomology,
Colorado State University, Fort Collins, Colorado 80523

INTRODUCTION

The mainstream rivers of North America once guided steamships
and fishes inland from the oceans to the confluence of major tributaries. These side-channels presented the travelers with profound
alternatives for maximizing (or minimizing) their fitness. Some
routes gave way to prosperity, others terminated at insurmountable
barriers.

Today most of the mainstream rivers are effectively manipulated
for anthropogenic purposes, manifested in complete flow control.
Until recently, flow regulation via controlled releases from impoundments and diversions has proceeded without due consideration of the
ecological consequences in the downstream environment.

Stream regulation in North America is epitomized by the Columbia
River System, which once supported a great run of anadromous salmon
and trout into more than 80 per cent of the drainage basin. Today
less than 30 per cent of the original salmon habitat remains accessible to sea-run spawners, due to the construction of mainstream dams
(Fig. 1; Robinson, 1978). Several of the structures contain no fish
passage facilities (e.g., Grand Coulee and John Day). Only concerted
management efforts involving downstream hatcheries, physical transportation of adults and fry around dams, and protection of available
upstream spawning sites from cultural developments have sustained
a portion of a much larger fishery. The Columbia salmonid fishery
is a focal point in the annals of aquatic biology, and its decline
in the face of water development projects is well documented (see
Netboy, 1974; Bently and Raymond, 1976; Robinson, 1978). Other
aspects of the river's biology were not documented at all, and it

Fig. 1.   Columbia Basin anadromous salmon and steelhead habitat
         (from Robinson, 1978, by permission of *Oregon Wildlife*).

is unfortunate that we will never know the true structure and function of the big river ecosystem embodied in the Columbia River Basin.
In North America and elsewhere, limnologists interested in problems
dealing with the ecology of regulated or unregulated streams will
always be plagued by the fact that ecosystem integrity in the lower
portions of the stream continuum (see Cummins, this volume) of big
rivers was destroyed before we knew very much about them.

     This review attempts to summarize the extent of stream regu-
lation in North America and to make geographic comparisons of the
ways in which riverine communities have been influenced by diversions
and on-channel impoundments.   Much of the North American literature
concerning the effects of stream regulation on riverine hydrology,
chemistry, and biology is summarized in the various subjects reviews
included in this volume.   Therefore, only important generalizations
will be repeated here and we will focus on several well documented
case histories, in addition to proposing ideas on the direction of
ecological research in North America rivers regulated by dams.

MAJOR DAMS AND REGULATED STREAMS

   The U.S. Federal Power Commission (1976) inventoried 1,426 conventional hydropower plants and 27 pumped storage facilities. The
fold-out map (Fig. 2) is a compilation of only those hydroelectric
facilities on mainstream rivers which utilize more than 6.2 hm^3
(5,000 acre-ft.) of water storage. Another recent inventory (U.S.
Fish and Wildlife Service, 1976) lists 1,493 reservoirs [defined as
"an impoundment with a mean annual pool of 202 ha (500 acres) or
more"] in the U.S.; at average water level these reservoirs impound
3,948,695 ha. Efford (1975) presented a similar compilation of
major reservoir sites in Canada (Fig. 3). Figures on the length of
stream channels affected by discharges from reservoirs are not readily
available, but it is clear that most major rivers in North America
are totally or partly regulated; exceptions are the Yukon and other
far northern rivers, and sections of the Upper Mississippi and Ohio
Rivers. The U.S. Public Law 95-625, as amended 10 November 1978,
provided for designation of free-flowing rivers or river segments
for inclusion (28) or study (72) in the Wild and Scenic Rivers
System. Most of these streams are, in fact, regulated. We show the
remaining free-flowing rivers in the U.S. in Figure 4 based upon
the following criteria: mainstream rivers of at least 4th order
with continuous flow; no significant impoundments or diversions;
and a length exceeding 100 km. All of Alaska's larger rivers remain
free-flowing; however, 89 hydrogeneration dam sites have been identified, including large volume storage reservoirs on the Kobuk, Noatak,
Nuyakek(Nushagak), Naknek(Kvichak), Susistna, Copper, Tanana(Yukon),
and Upper Yukon Rivers (U.S. Fed. Power Comm. 1976). Several other
important rivers retain significant free-flowing segments near the
headwaters [e.g., Rio Grande (Colorado-New Mexico); Mississippi (Minne-
sota); Green (Wyoming); Toulumne (California); New (West Virginia);
Wolf and St. Croix (Wisconsin); Iowa (Iowa); Guadalupe (Texas); and
St. John (Maine)]. There are also some large free-flowing rivers
which are less than 100 km long (e.g., the Blackwater in Florida, the
Gallatin in Montana, and the Eagle in Colorado). Efford (1975)
reported that development plans exist for all of the remaining "wild"
rivers in Canada; more than 50 are presently regulated.

Fig. 2. Major drainage basins in the U.S.A. Closed circles define
         locations of major mainstream hydroelectric reservoirs
         (see text).

1. W.A.C. Bennett Dam
2. Alcan Kemano Project
3. Proposed Moran Dam
4. Skagit Valley
5. Columbia River Projects
6. Bighorn Dam
7. Diefenbaker Lake
8. Churchill-Nelson Diversion
9. James Bay Projects
10. Rivière Jacques Cartier
11. Manicouagan Outardes Complex
12. Saint John River Projects
13. Churchill Falls Project
14. Bay d'Espoir Development

Fig. 3. Major hydroelectric projects in Canada (from Efford 1975;
reprinted by permission, *J. Fish. Res. Board Can.*).

Stream regulation in North America has resulted primarily in
response to demand for electrical power generation, flood control,
and irrigation or water supply diversions. Practically every feasible site for hydropower development on the Continent's mainstem
rivers has been utilized with the exception of Alaskan and North
Canadian rivers. Developers are now concentrating on impounding
and diverting tributaries and headwater streams. Present hydro-
operations cycle more than twice the average annual runoff in the
U.S. (Geraughty et al., 1973).

Although the U. S. Fed. Power Comm. (1976) estimated that only
about 50 per cent of the available conventional hydro-electric power
has been developed (including essentially none of Alaska's 33
million KW potential), Hayes (1979) pointed out that it has taken
the U.S. 50 years to reach a level of 57 GW of installed generating
capacity. Since the best dam sites have been utilized, hydropower
cannot be expected to relieve much of the present, chronic energy
shortage (i.e. <1 per cent of the projected need in the year 2000).
Similarly, all of the Canadian hydro-power sites in reasonably close
proximity to population centers have been utilized (Efford, 1975).
From this one might conclude that additional, significant regulation
would be unlikely. On the contrary, more and greater amplitude
regulation is probable because of the propensity of the power net-
works in North America to depend on hydro-units to satisfy peak loads.
Nuclear and coal-fired generation facilities cannot efficiently fluctuate output, as can hydro-systems. Therefore, existing and new,
smaller hydro-plants represent the only readily available mechanism
of meeting power needs above base load. Pumped storage facilities
(see Berkowitz, 1971; Krizek, et al., 1971) are an alternative,

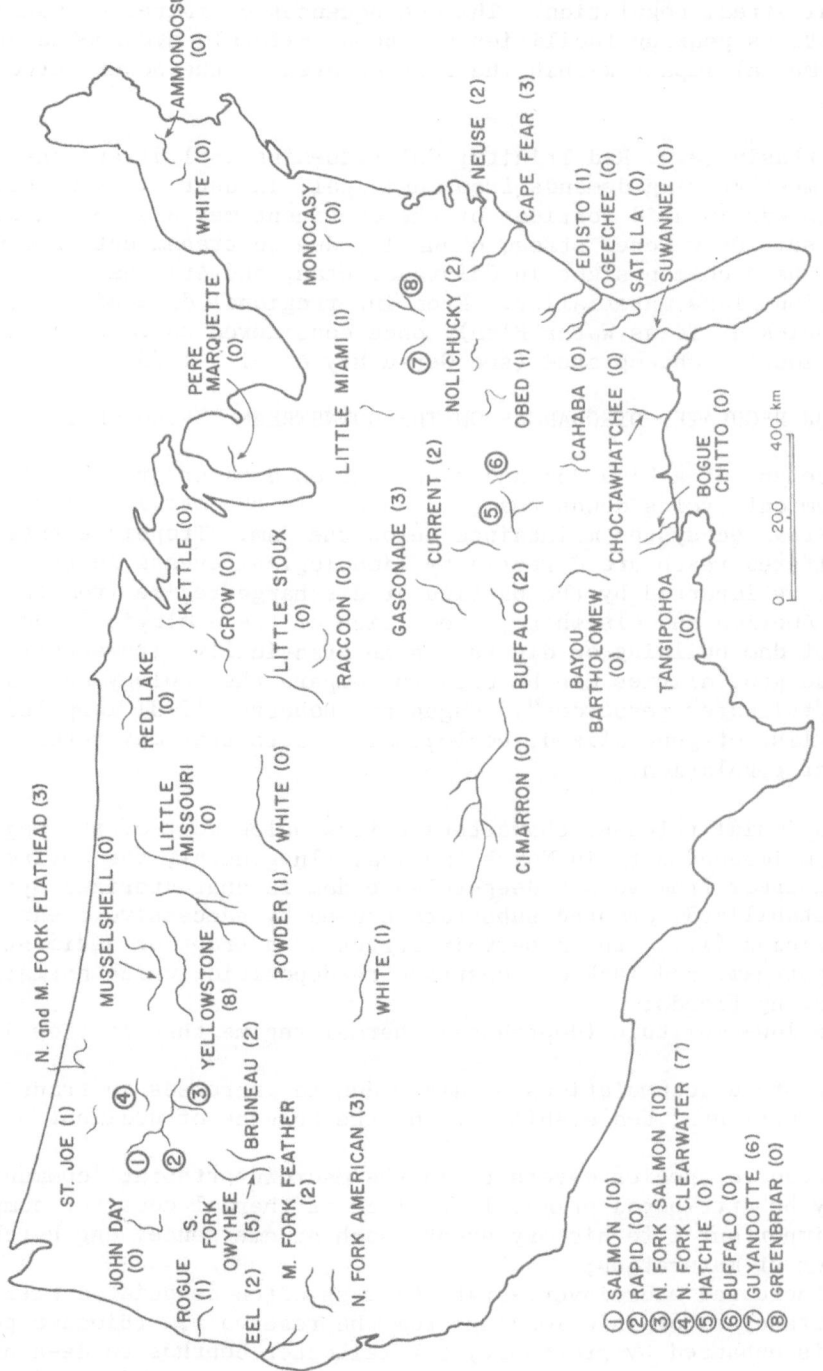

Fig. 4. Major free-flowing rivers (see text) in the U.S. Numbers in parentheses represent feasible dam sites based upon data from the Fed. Power Comm. (1976).

and many of the units are being constructed. Either case amounts to
additional stream regulation. The consequences of increased use of
hydro-units as peaking facilities has been uniformly overlooked as
an enviromental impact within the service area of the power network in
question.

Interbasin (e.g. Red-Trinity; Colorado-Missouri) diversions of
water to meet growing demands from municipal, industrial, and agricultural needs in arid portions of the continent may also be expected
to increase. De-watered stream channels, due to transmountain diversion.
will soon be a common sight in Colorado, Utah, and Arizona as authorized
construction plans materialize. Even interregional diversions (e.g.
Mississippi-Red, Texas Water Plan), once considered unrealistic are now
being seriously contemplated (see Geraughty et al., 1973).

EFFECTS OF REGULATED DISCHARGES ON THE DOWNSTREAM ENVIRONMENT

Placement of a high dam and reservoir on a river greatly alters
biogeochemical cycles downstream; in no way is the original integrity
of the stream ecosystem maintained below the dam. Trophic events
in the altered reach are directed by limnological events in the
reservoir as imparted by the particular discharge regime from the dam.
In North America and elsewhere, prediction of the ecological consequences of dam building or diversions is practically site-specific.
Only broad generalities can be used to compare the ecology and management of "tailwater resources". Hagen and Roberts (1973) compiled
a useful list of generalized, ecological impacts that may result
from river regulation.

Hypolimnial releases characterize flow below most of the major
mainstream impoundments in North America. In general, the environment downstream from such a deep-release dam is characterized by:
    a) stabilized, armored substrata caused by successive clear-
water sluicing (i.e., the reservoir serves as a trap for sediments
eroded upstream) and lack of substrate re-deposition which normally
occurs during floods;
    b) a low-amplitude (depressed) thermal regime that is flow de-
pendent;
    c) profuse accumulations of algae due to increases in trans-
parency, nutrients, bed stability, and the absence of sediment or
ice scour;
    d) reduced species diversity in the macroinvertebrate community,
which may be attributed primarily to lack of thermal cues for completion of important life history events such as emergence, egg hatching,
and timing of maturation;
    e) increased macroinvertebrate biomass often associated with
flow constancy or organic loading from the reservoir. (Biomass per
species is enhanced by propensity for tailwater benthos to de-synchro-
nize reproduction in a manner similar to that occurring in spring-fed
streams, see Ward and Short, 1978; Ward and Stanford this volume).

Viable trout fisheries may exist in tailwaters [e.g., Fontanelle Dam on the Green River in Wyoming; Morrow Point and Crystal Dams on the Gunnison River in Colorado (Wiltzius, 1978); Bull Shoals Dam on the White River in Arkansas (Hoffman and Kilambi, 1970)]. Such fisheries may produce trophy-sized trout, but growth rates and condition factors seem to wax and wane over the long term. Reproductive success is not good in many regulated streams, and factors influencing reproduction have not been adequately studied at most sites.

Less information exists for non-salmonids in tailwaters (see Holden, this volume). Dudley and Golden (1974) observed no significant differences in growth rates of bluegill sunfish (*Lepomis macrochirus*) populations in hypolimnial tailwaters compared to populations below surface release dams. Edwards (1978) found 22 species of fishes above Canyon Reservoir on the Comal River in Texas, but only 18 were residents in the stream below the reservoir. Paddlefish (*Polydon spathula*) distributions in the Missouri River have been greatly influenced by mainstream impoundments (Boehmer, 1973) and some harvested populations lack significant recruitment (Friberg, 1972). Distribution problems manifest in most river systems after impoundment are simply due to the physical barriers created by the dam. Hubbs and Pigg (1976) attributed degradation of rare species in most Oklahoma streams to distribution barriers created by impoundments.

A central question relative to the effects of stream regulation is that of minimum discharge necessary to support a biotic community. Generally, discussion is limited to flows necessary to maintain fish and invertebrates. Recent symposia have addressed the problem of instream flow requirements in regulated streams (see Pacific Northwest River Basin Commission, 1972; Fraser, 1972; Orsborn and Allman, 1976). It must be recognized that high flows will carry tailwater temperatures far downstream from the dam and that insolation and air temperature will modify water temperature as flow decreases and distance increases from the discharge source. Thermal management by stream flow may be as important to the biotic community in tailwaters as is availability of watered habitat, because of the close correlation between temperatures and metabolic rate. In addition, for most invertebrates and fish of temperate regions, timing of reproduction and sequencing of growth is thermally cued. Within this context minimum flows can be devised relative to the specific requirements of indicator or marker species (see Wesche, 1974; Ward, 1976a, 1976b). Distributions of fishes and drift rates in aquatic insects may also be related to magnitude and intensity of flow (see Ward, 1976c; Wade et al., 1978). Techniques are presently being developed that integrate these considerations into a format for predicting flow requirements of invertebrates and fish inhabiting regulated streams (Gore, 1978; Stalnaker, this volume).

Downstream manifestations of stream regulation also include
what has been referred to as loss of pulse-stimulated responses at
the water/land interface of the riverine system (Gill, 1971). All
free-flowing rivers exhibit cyclic flood events that periodically
submerge portions of the river bank or flood plain. These cyclic,
spate events re-arrange sediment deposits and riparian vegetation.
These pulse-stimulated processes may be key events in maintaining
certain wildlife habitat. For example, lush growths of willows
(*Salix* spp.) and horsetails (*Equisetum* spp.) along the MacKenzie
River in the Northwest Territories occur on sediment ridges deposited
during the floods. It had been shown that moose and ptarmigan are
dependent on the lush shoots that grow on new sediment ridges each
spring and summer. Damming of the MacKenzie would stabilize flow,
and this pulsating regrowth phenomenon would cease (see also the
Peace-Athabaska case history below). Dolan et al. (1974, 1977)
have shown that stabilized flow in the Grand Canyon of the Colorado
River has eroded the white sand beaches that were deposited annually
by flooding. Shoreline vegetation has invaded similar beaches in the
Black Canyon of the Gunnison River, Colorado,because spring floods no
longer redeposit sand bars. The result is an almost impenetrable growth
of riparian vegetation along the river course (Stanford, personal
observation; see also Simons, this volume).

CASE HISTORIES OF RIVER REGULATION IN NORTH AMERICA

Peace-Athabaska Rivers (MacKenzie Basin), Canada

Construction of Bennett Dam and subsequent filling of Williston
Reservoir created a stream regulation scenario in Canada not too
unlike the Aswan problem on the Nile River in Egypt (see Ehrlich
et al., 1978 for a review on the latter). The Peace River flows
into the shallow, northwestern end of Lake Athabaska. Prior to
construction of the upstream dam, water levels in Lake Athabaska
fluctuated 2-4 m per year. Spring flooding by the Peace River prevented outflow from the lake via the Rochers River, thereby flooding
more than 0.5 million ha of delta wetlands. This was a classic
example of a pulse-stimulated phenomenon in which perched wetlands
were rewatered and fertilized annually by flooding. After completion of the dam, flow in the Peace River was stabilized at 1000 m^3
sec^{-1}, whereas pre-regulation flow varied from 140-8680 m^3 sec^{-1}
on a mean annual cycle (Geen, 1974). The perched wetlands of the
delta immediately began succession toward a meadow system laden with
willows. Populations of muskrats, waterfowl, northern pike, and
golden eye began to decline, and the future of the wood buffalo, a
popular resident of the delta marshes, was questioned (Townsend, 1975).

A study group was formed to simulate the long-term effects of
the altered hydrological regime on the delta, and remedial alternatives were forthcoming. It was decided to place a fixed-crest

weir on Rochers River outlet to compensate for the decline in lake levels. This restored average summer water levels on the delta (Townsend, 1975). The long-term effectiveness of the solution is presently under investigation.

It is not known how far down the MacKenzie River system the effects of Peace River regulation extended. The hydrology of Great Slave Lake possibly masks most of the effects. More water development projects are proposed on the MacKenzie, and the potential effects on this pulse-stimulated system are not yet fully elucidated (Gill, 1971).

Flathead Rivers (Columbia Basin), Montana

The Flathead River system contains both free-flowing and regulated rivers (Fig. 5). While plans exist for total regulation of the system, additional upstream dams are not feasible at this time due to inclusion of the North and Middle Forks in the Wild and Scenic Rivers System. Lower river dam and pumped storage sites along the east shore of Flathead Lake, however, are being seriously considered. Also, the Bureau of Reclamation recently completed a study (unpublished) of Columbia River hydro-units to determine feasibility for additional generation capacity. The Hungry Horse unit was judged high priority, because there is room for an additional turbine in the present housing, and construction of a re-regulation (and/or pumped storage) unit may be feasible a short distance downstream from the present dam. A study of the re-regulation dam in relation to the known effects of regulated flows from Hungry Horse Reservoir on downstream biota is presently being conducted.

Productivity in oligotrophic Flathead Lake is greatly influenced by discharges from the Flathead River, which is heavily laden with clay sediments during spring runoff. These sediments are primarily derived from natural stream bank erosion in the unregulated segments of the upstream rivers, although some accelerated erosion due to road building activities has been documented. Most years an overflow turbidity plume is carried through Flathead Lake by flood discharge from the river (Stanford et al. unpubl.). These fine sediments gradually sink through the water column and have been implicated as a flocculation agent that strips nutrients and organic matter. It is thought that this process creates flocs of clay-organics that are colonized by bacteria and other microbes, producing a heterotrophically-driven primary food source for zooplankton (Stanford and Potter, 1976). A similar process is thought to occur in the unregulated rivers. Accumulations of microbially-infested sediment-detritus provide a food base for a phenomenal diversity of riverine insect species (Stanford and Gaufin, in press).

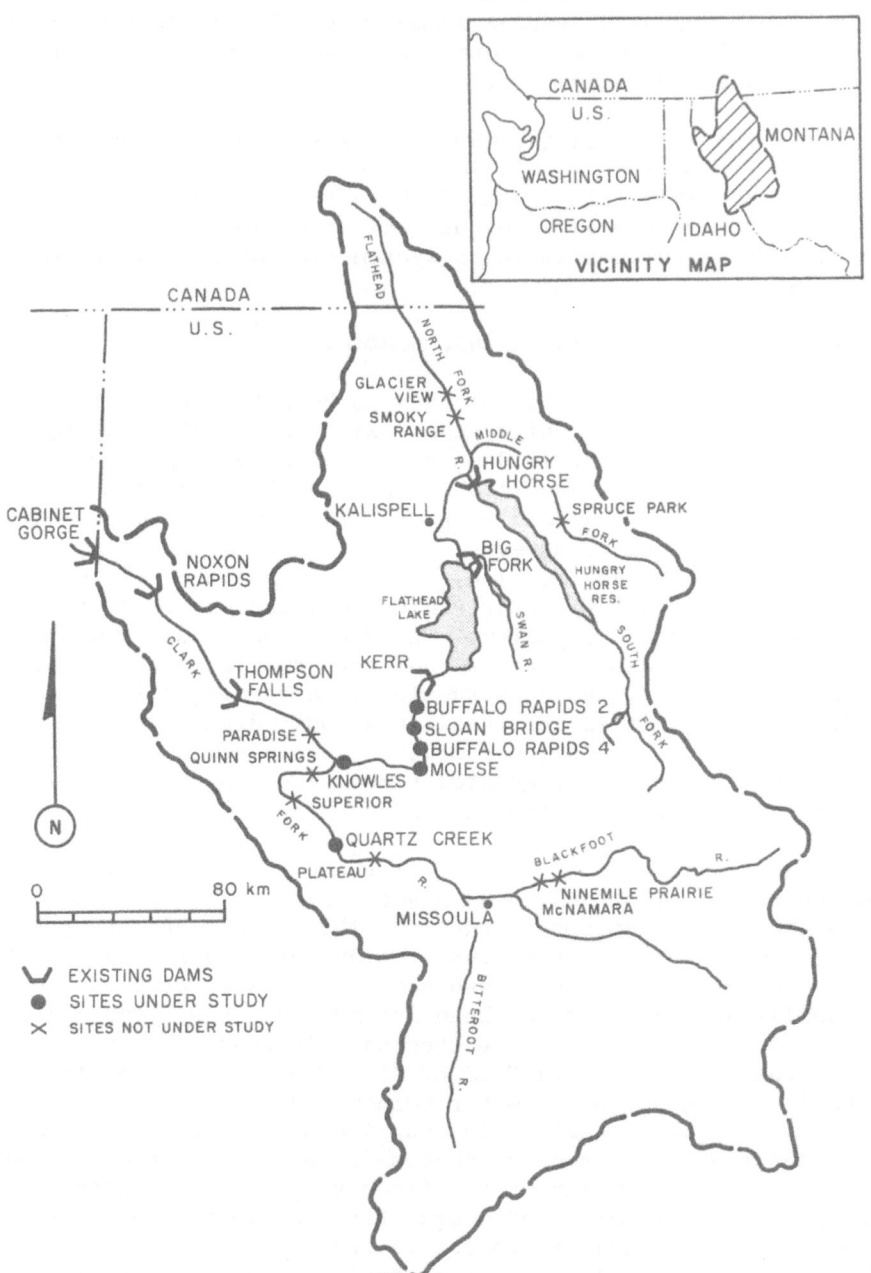

Fig. 5. Dam sites and regulated river segments in Flathead River/
         Clark Fork Basin, Montana.

A number of fish species in Flathead Lake make annual spawning runs into unregulated tributaries. The most notable of these, due to their popularity as sport fish, are the Dolly Varden (*Salvelinus malma*), cutthroat trout (*Salmo clarki clarki*) and kokanee salmon (*Oncorhynchus nerka nerka*). Forage species also migrate upstream to spawn. Kokanee salmon are an important food for bald eagles and grizzly bears during the fall spawning run.

The effects of regulation are both profound and subtle in the stream system. Classical stabilization and armoring of, the river bottom accompanied by profuse growths of periphytic algae characterize the minimum (4.3 m^3 sec^{-1}) flow channel in the South Fork below Hungry Horse Dam. Insect species diversity and biomass is significantly reduced in comparision to unaltered river segments (Stanford and Hauer, 1978). Stanford and Gaufin (in press) reported that 38 Plecoptera species probably resided in the South Fork prior to impoundment; these species presently exist in unaltered segments upstream of Hungry Horse Reservoir. Only five species, however, remain in the reservoir tailwaters. The annual thermal regime is relatively constant; weekly mean temperatures fluctuate only slightly around 7°C, compared to an annual thermal amplitude in unregulated segments of 0-18°C (see Fig. 2 in Ward and Stanford, this volume).

The more profound effects of hypolimnial discharges from Hungry Horse Reservoir are substantially reduced in the mainstream river by unregulated discharges from the North and Middle Forks. Annual spring floods redistribute substrata and scour periphyton. Benthic habitat in the main river is abundant and heterogeneous; Stanford and Gaufin (1974) provided evidence that hyporheic development is extensive. The annual thermal regime is only slightly modified; mean temperatures are decreased 1-5°C during summer and winter temperatures are elevated 1-3°C. Thus, the river does not completely freeze over during the winter, due to warming effects of hypolimnial waters from the South Fork. The benthic insect community is diverse; 42 Plecoptera and 38 Tricoptera species are known to have reproducing populations in the mainstream river, and individuals tend to be larger and more robust than in upstream areas. No deleterious effects of winter discharges from Hungry Horse Reservoir are evident in Flathead Lake, but effects of hypolimnial release on nutrient dynamics of the lake should be considered.

Subtle and possibly chronic effects of stream regulation have been documented in the mainstream Flathead, however, and relate to slight thermal modification and periodic de-watering of lateral margins of the stream channel. Stanford and Gaufin (in press) reported that late summer discharges during 1973 prevented emergence of a dominant stonefly species *Claassenia sabulosa*, due to depression of mean daily temperature below the 18°C emergence cue for the species.

Similar, but less well documented effects were observed during *Iso-capnia* spp. emergence in April, 1974. On several other occasions substantial numbers of benthic insects and larval fish were observed stranded on the de-watered, lateral margins of the river channel following sudden flow reductions in the mainstream. Several of the insects (e.g., *Diocosmoecus*) and fish (e.g., cutthroat fry) are species known to seek refuge in backwater pools. Upon reduction of water level, shallow portions of these pools were isolated and organisms were easy prey for herons (Stanford and Hauer, 1978). Major mortality of *Utacapnia* and *Capnia* nymphs was documented in 1973 when a sudden drawdown permitted icing conditions to prevail in shoreline areas where these species had congregated just prior to emergence. The quantitative effects of such mortality surely contribute to variance in population numbers from year to year and may constitute a chronic, deleterious effect. It is also suggested that small changes in temperature may delay movements of migratory fishes into the river (Schumacher and Hanzel, Montana Dept. Fish and Game, unpubl. report).

The effects of regulation in Flathead Lake and the lower river below Kerr Dam are even less understood. It is significant that prior to volume regulation, kokanee salmon spawned predominately on the lake shoreline, whereas adfluvial populations now dominate the fishery (Hanzel, 1976). Shoreline spawning activities have been sharply reduced. Apparently fall drawdown now occurs before the majority of lake-spawned salmon eggs hatch, leaving the redds high and dry. Ecological conditions in the lower river are virtually unknown, although recent surveys by the U.S. Fish and Wildlife Service indicate that a productive northern pike fishery exists along with numerous forage species.

South Platte River (Missouri Basin), Colorado

The Front Range of the Rocky Mountains in Colorado is drained by several medium-sized and high-gradient rivers that flow from high altitude headwaters through spectacular canyons on their course to the high plains (e.g., Cache la Poudre River, Big Thompson River, Saint Vrain River, South Platte River, Arkansas River). In the mountains the rhithron habitat of these rivers is quite similar and, except in tailwater segments, contain similar benthic biota (see Stanford and Reed, 1974; Ward, 1975a) and trout fisheries. These rivers are variously regulated.

The South Platte River has been extensively impounded and trans-mountain diversions now greatly augment natural flows to supply potable water to metropolitan Denver (Fig. 6). Ward (1974; 1975b, 1976b) quantified effects of hypolimnial discharges from Cheesman Reservoir

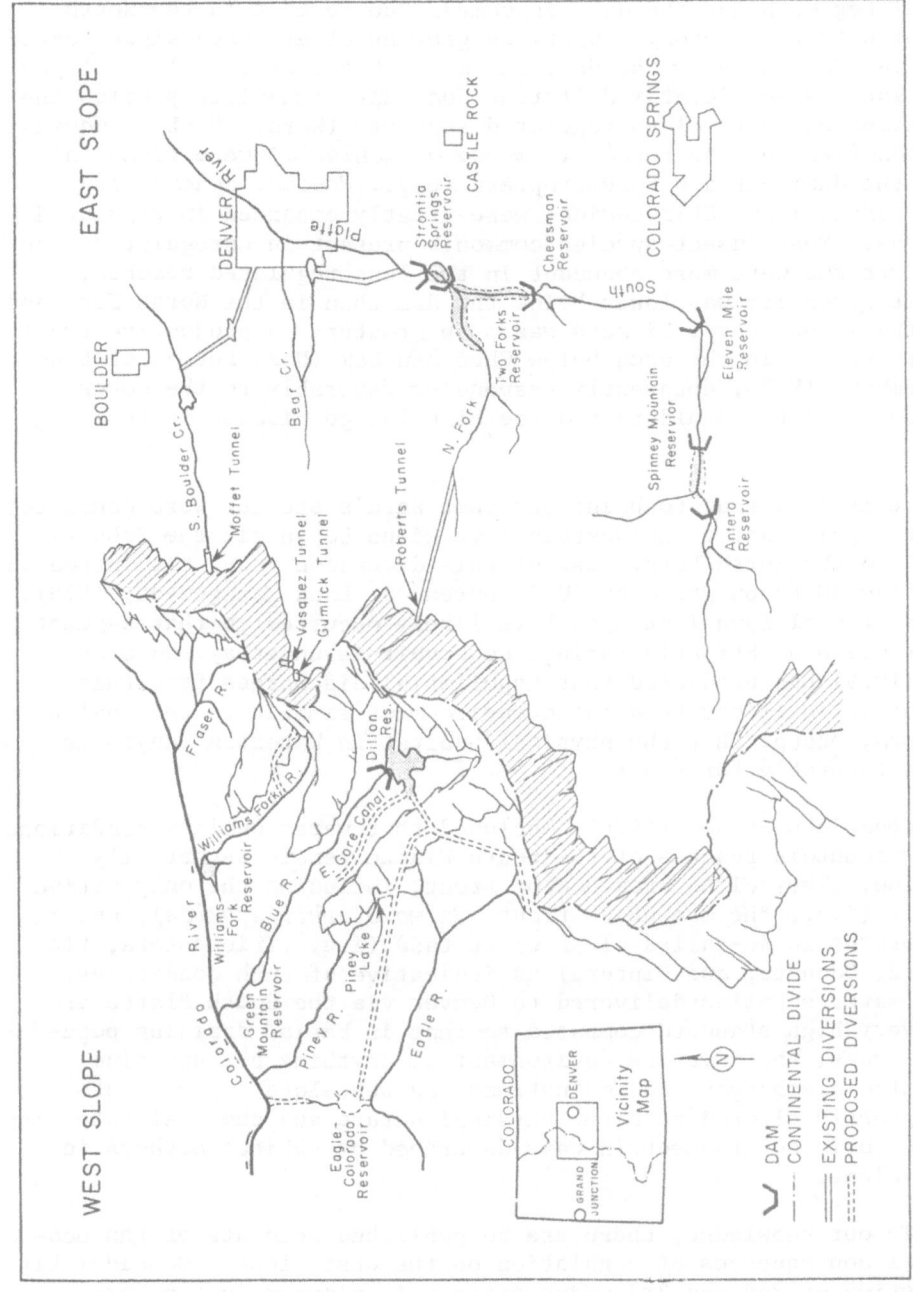

Fig. 6. Transmountain water diversions and mainstream impoundments regulating the South Platte River and tributaries to the Colorado River in Colorado.

on the downstream ecosystem (i.e., in Cheesman and Waterton Canyons).
Hypolimnial releases considerably dampened and delayed the annual
thermal regime below the dam, in comparison to that in the North
Fork or Waterton Canyon. Luxurious growths of attached algae coated
the river bottom below the dam, and macrophytes were well developed.
There was a significantly different community immediately below the
dam, compared to the less regulated segments (Ward, 1976b). *Gammarus lacustris* and gastropods (*Lymnaea auricularia*) were found only
below the dam; certain Ephemeroptera (e.g., *Baetis, Ephemerella,
Tricorythodes*) and Chironomidae were greatly enhanced in regulated
sections. Most insect species commonly present in unregulated Front
Range streams were more abundant in the less regulated reaches.
Species diversity was lower below the dam than in the North Fork and
downstream, but total biomass was much greater. A productive trout
fishery exists in the area below Cheesman Dam (U.S. Bureau of Land
Management, 1978), apparently responding favorably to the cover
provided by large boulders and the high forage biomass in that segment.

     It is important to point out that Ward's studies were conducted
before significant transmountain diversions began via the Roberts
Tunnel to the North Fork. Use of this diversion route has increased
mean flow 30 to 60 per cent (U.S. Bureau of Land Management, 1978),
and additional faunal changes have likely occurred in that segment.
Construction of Strontia Springs Reservoir is underway and Stanford (1979) has predicted that hypolimnial discharges from this
impoundment will create a lotic environment similar to that below
Cheesman, except that the physical habitat in Waterton Canyon is
not as favorable for trout.

     Hendricks et al. (1976) concluded that water quality conditions
in the mountain reaches of the South Platte remain essentially
pristine. They cited presence of trout, including the only native
species (i.e., the Greenback trout, *Salmo clarkii stomias*), and invertebrates in so-called clean water taxa (i.e., Trichoptera, Plecoptera, Odonata, and Diptera) as indicative of such conditions.
While water supplies delivered to Denver via the South Platte are
of a very high standard compared to that in basins draining populated areas, the riverine environment is anything but pristine.
Regulated discharges from mainstream and westslope impoundments
have grossly altered both the physical nature and chemical integrity
of the lotic environment in ways described by subject authors in
this volume.

     To our knowledge, there are no published accounts of the eco-
logical consequences of regulation on the west slope. Considerable
dewatering of 2nd and 3rd order streams is planned in the Gore
Creek drainage. Data have recently been collected below Dillon

Reservoir on the Blue River, below Ruedi Reservoir on the Frying-
pan River, and downstream from Granby Dam in the headwaters of the
Colorado River (Ward, unpubl.). Transmountain diversion may be
chronically impacting the mainstream Colorado River, as well.

Brazos River (Brazos Basin), Texas

The Brazos River, like other rivers in southwestern U.S.A., is
characterized by erratic, spate-related flow and extremely high sediment and dissolved solids loading. In most of its course the river
has an unstable, sandy bottom; salinity in headwater areas exceeds
15 ppt, due to erosion of salt laden formations (e.g., halite and
gypsum) in West Texas (U.S. Army Corps of Engineers, 1973). Several
large, mainstream impoundments not only trap much of the sediment
load, but also act as "sinks" for dissolved solids. For this reason,
the Brazos River is rather unique in that its water becomes less
saline as it moves toward the Gulf Coast.

Morris Shepard Dam, located in Palo Pinto County, Texas, was
built in 1941 to impound Possum Kingdom Reservoir. Prior to
impoundment, river discharge in this segment varied from more than
$1500 \text{ m}^3 \text{ sec}^{-1}$ during spates, to little or no flow during the dry
periods. Since construction of the 45 m-high dam, leaks around
the penstocks have maintained a base flow of $0.42-0.56 \text{ m}^3 \text{ sec}^{-1}$;
hypolimnial discharges for power generation increase flow 1-2 orders
of magnitude over base flow volume at unpredictable intervals. These
clear water discharges have created a rubble-bottom habitat by sluicing fine sediments from limestone substrata. Alternating pools and
rubble riffles persist nearly 150 km downstream before sediments
derived from side tributaries again predominate.

The annual thermal range varies from 9°C to 32°C, depending on
the distance from the dam. At base flow water discharged from the
dam is quickly modified by air temperature. Higher volume discharges
for power generation, however, moderate water temperature several
degrees, again depending on distance from the dam. Greatest temperature changes (i.e., 4-8°C) due to power generation occur during
warm weather within 30 km of the dam. Beyond 50 km, thermal modification is less than 2°C (Zimmerman and Richmond, in press; Coulter
and Stanford, unpubl.).

Water discharged from Possum Kingdom Reservoir contains 10 to
80 per cent less dissolved solids than river waters upstream. Although the hypolimnion is characterized by low concentrations of
dissolved oxygen and high concentrations of H_2S, air drafts in the
penstocks cause sufficient gas exchange to alleviate toxicity problems downstream. Other chemical parameters vary little within the
150 km river segment significantly altered by discharge effects
and generally reflect chemical conditions in the reservoir (Coulter
and Stanford, unpubl.).

The Possum Kingdom tailwaters are inhabited by an extraordinar-
ily diverse flora and fauna, at least by comparison to other lotic
systems in this locality. Periphyton (e.g., *Cladophora* sp.) is
abundant but does not coat the bottom. The macroinvertebrate community is dominated by 33-38 species of insects, compared to only
10-15 at sites above the reservoir (McClure and Stewart, 1976; Rhame
and Stewart, 1976; Coulter and Stanford, unpubl.). Poole and
Stewart (1976) documented vertical distribution to 40 cm within the
substrata at the Dark Valley site located 30 km downstream, although
most of the organisms resided in the upper 10 cm of the limestone
rubble.

Simuliidae and Chironomidae are abundant members of the riverine community with standing crops exceeding 4,000 individuals m^2
during some periods (Poole and Stewart, 1976). At least seven species
of caddisflies occur; biomass is dominated by *Cheumatopsyche campyla,*
C. lasia, and *Hydropsyche simulans* (Rhame and Stewart, 1976). Cloud
and Stewart (1974a,b) documented the typical dawn-dusk peaks of insect
drift in the river but did not assess the effects of fluctuating
discharges on drift response. The mayfly *Choroterpes mexicanus,* is
also very abundant (e.g., 130-2700 m^2) in the benthic community
(McClure and Stewart, 1976). All of the above insects are known to
be multivoline in the tailwaters and exhibit long, asynchronous emergence periods.

Only one stonefly, *Neoperla clymene* is present in the tailwaters, although *Perlesta placida* and *Hydroperla crosbyi* are
relatively abundant in other local streams. *N. clymene* produces one
generation per year in the tailwaters but also has a long emergence
period (Vaught and Stewart, 1974).

The megalopteran, *Corydalus cornutus,* is fairly common in benthic
samples. Stewart et al. (1973) demonstrated that larvae of this
large carnivorous species fed on the abundant filter feeders (e.g.,
Simuliidae and *Cheumatopsyche* spp.).

Although the insect community remains comparatively diverse
and sustains high biomass, Coulter and Stanford (unpubl.) found
significant differences in quantitative data from 1970-71 and 1978-79.
The mayfly, *Isonychia sicea manca,* was extremely abundant in 1970-
71 samples but was not found in 1978-79. Other species (e.g.,
Neoperla clymene; Tricorythodes sp.; *Chimarra obscura*) also declined
significantly in numbers, suggesting possible competitive exclusion
by *Cheumatopsyche* spp. and *Choroterpes mexicanus.*

Fish populations in the Possum Kingdom tailwaters also reflect
influence of substrate, temperature, and salinity changes. Although
species diversity is lower in the tailwaters than above the reser-

voir and data may be strongly influenced by sampling technique, biomass in tailwaters is two to three times greater. Also, certain species (e.g., *Cyprinodon rubrofluviatilis; Hybognathus placitus*) are found above the reservoir, but not in the tailwaters, and conversely for other fishes (e.g., *Percina sciera; Etheostoma spectabile; Campostoma anomalum*). The red shiner, *Notropis lutrensis*, one of the more commonly occurring minnows in Texas streams, is very abundant in Possum Kingdom tailwaters. Very few exist above the reservoir. Preliminary laboratory testing indicates these shiners and other indigenous minnows are intolerant of salinities greater than 11 ppt (Anderson and Beitinger, N. Tex. St. Univ., unpubl.).

Using electrophoresis, Richmond and Zimmerman (1978) isolated a "cool-water" isozyme of malate dehydrogenase (MDH) in populations of red shiners in tailwater areas significantly influenced by hypolimnial discharges (i.e., within 60 km of the dam). This isozyme was not found in populations from other segments of the Brazos River or other waters in Texas and Oklahoma. The presence of the "cool-water" isozyme was accompanied by a high degree of genic heterozygosity (heterosis). Zimmerman and Richmond (in press) demonstated that in Possum Kingdom tailwaters heterosis in *Notropis lutrensis* regressed significantly against location of maximum summer temperature change; fish in areas most thermally influenced by reservoir discharge exhibited a more adaptative or plastic genotype. They suggested such populations are better able to contend with the unpredictable discharge schedule in the tailwater segment of the river.

The Brazos River below Possum Kingdom Dam has received a greater diversity of hypothesis-oriented study than any other regulated river segment in North America. Although some of the studies are yet incomplete and others are needed, it is clear that this river segment is responding to regulation differently than other systems. Prior to impoundment this section of the Brazos was not very productive; however, no quantitative pre-impoundment records exist. The river itself is of little value as a potable or irrigation supply, due to the high salt content. The river segment should be recognized for its scientific value, however, as detailed, comparative data are badly needed to fully appreciate the myriad of ways organisms adapt or fail in response to stream regulation. Presently, plans for a pumped-storage facility, which would inundate most of the 150 km segment described here, are proceeding.

RESEARCH DIRECTION AND DISCUSSION

To date research in regulated streams of North America has been relatively limited in scope, although more detailed studies are in progress. The ability to predict more than obvious effects of an impoundment or diversion project have not been forthcoming in most cases, due primarily to a lack of knowledge on the functional processes in these altered environments. Alteration of community

structure is virtually site-specific and there are few studies
reporting enough detail to be of much comparative value. Cummins
(1974; this volume) summarized important structural and functional
relationships that characterize stream ecosystems. These relationships are relatively well understood in unregulated low order streams,
but the rigorous environment in some regulated streams presents
organisms with an unpredictable set of resource gradients. Other
regulated streams, however, exhibit increased predictability and
resource availability (Ward and Stanford, this volume).

Research should be undertaken to investigate functional responses (e.g., rates of productivity; nutrient limitation and cycling,
trophic relationships) in regulated streams. As pointed out above,
such functional responses must be evaluated on at least two temporal
scales (diel and seasonal), especially in streams regulated by hypolimnial release. Measurements of physico-chemical parameters over
annual periods usually demonstrate a trend toward constancy or loss
of heterogeneity in such environments (e.g., low amplitude thermal
regima, homogeneous benthic habitat due to armoring, and heavy uniform
growths of periphyton). On the other hand, regulated streams are
also characterized by sudden, extreme flow fluctuations, that very
significantly affect habitat availability, thermal constancy, and
possibly nutrient availability. Research discussed above and in the
subject reviews in this volume has shown that community structure
in regulated streams may be a function of either environmental constancy or shock phenomena, or both.

ACKNOWLEDGMENTS

We thank Dr. Andrew Sheldon and Mrs. Jan Peters for their help
in identifying free-flowing rivers. Thanks also to Drs. K. W. Stewart
and S. W. Szczytko and Mrs. S. A. Appert, Mr. F. R. Hauer, and two
anonymous reviewers for their constructive comments on the manuscript.

REFERENCES

Bentley, W.W., and Raymond, H.L., 1976, Delayed migrations of yearling
    chinook salmon since completion of lower monumental and Little
    Goose Dams on the Snake River, *Trans. Am. Fish. Soc.* 105:422-424.
Berkowitz, D. A., and Squires, A. M., 1969, "Power Generation and
    Enviromental Changes," MIT Press, Cambridge, Mass.
Boehmer, R.J., 1973, Ages, lengths, and weights of paddlefish caught
    in Gavins Point Dam tailwaters, Nebraska, *Proc. S.D. Acad. Sci.*,
    52:140-146.
Cloud, T. J. Jr. and Stewart K. W., 1974a, The drift of mayflies
    (Ephemeroptera) in the Brazos River, Texas, *Kansas Entomol.
    Soc.*, 47(3):379-396.

Cloud, T. J. Jr. and Stewart, K.W., 1974b, Seasonal fluctuations and periodicity in the drift of caddisfly larvae (Trichoptera) in the Brazos River, Texas *Entomol. Soc. Am.* 67:805-811.

Cummins, K. W., 1974, Structure and function of stream ecosystems, *BioScience* 24:631-641.

Dolan, R., Hayden, A., Howard, A., and Johnson, R., 1977, Environmental management of the Colorado River within the Grand Canyon, *Environmental Management* 1:391-400.

Dolan, R., Howard, A., and Gallenson, A., 1974, Man's impact on the Colorado River in the Grand Canyon, *Am. Sci.*, 62:392-401.

Dudley, R.G., and Golden, R. T., 1974, Effect of a hypolimnion discharge on growth of Bluegill (*Lepomis macrochirus*) in the Savannah River, Georgia, School of Forest Resources, University of Georgia, Athens, Georgia, and Environmental Resources Center, Georgia Institute of Technology, Atlanta, Georgia.

Edwards, R.J., 1978, The effect of hypolimnion releases on fish distribution and species diversity, *Trans. Am. Fish. Soc.* 107:71-77.

Efford, I. E., 1975, Assessment of the impact of hydro-dams, *J. Fish. Res. Board Can.*, 32:196-209.

Ehrlich, P. R., Ehrlich, A. H., and Holden, J. P., 1977, "Ecoscience: Population, Resources Environment," W. H. Freeman, San Francisco.

Fraser, J.C., 1972, Regulated stream discharge for fish and other aquatic resources: An annotated bibliography, FAO Fish. Tech. Pap., 112.

Friberg, D.V., 1972, Paddlefish abundance and harvest within a population lacking recruitment, Big Bend Dam Tailwaters, 1969071, Job Completion Rept., S.Dak. Dept. Game Fish Parks, Pierre, S.Dak.

Geen, G.H., 1974, Effects of hydroelectric development in Western Canada on aquatic ecosystems, *J. Fish. Res. Board Can.* 31:913-927.

Geen, G. H., 1975, Ecological consequence of the proposed Moran Dam on the Fraser River, *J. Fish. Res. Board. Can.*, 32:126-135.

Geraghty, J.H., Miller, D.W., Van Der Leeden, F., and Troise, F. L., 1973, "Water Atlas of the United States," Wat. Info. Centr. Publ. Port Washington, N.Y.

Gill, D., 1971, Damming the Mackenzie: A theoretical assessment of the long-term influence of river impoundment on the ecology of the Mackenzie River Delta, Proc. Peace-Athabaska Delta Symposium, Univ. Alberta, Edmonton.

Gore, J.A., 1978, A technique for predicting in-stream flow requirements of benthic macroinvertebrates, *Freshwater Biol.* 8:141-151.

Hagen, R. M., and Roberts, E.B., 1973, Ecological impacts of water storage and diversion projects, *in*: "Environmental Quality and Water Development," Goldman, C.R., McEvoy, J., and Richardson, P.J., eds, W.H. Freeman, San Francisco.

Hanzel, D.A., 1976, The seasonal area and depth distribution of fish populations in Flathead Lake, Job Performance Reports, Mt. Dept. Fish and Game, Kalispell, Mt.

Hayes, E.T., 1979, Energy resources available to the United States, 1985 to 2000, *Science* 203:233-239.

Hendricks, D. W., Asce, M., and Bluestein, M.H., 1976, Response of South Platte to effluent limitations, *Proc. Am. Soc. Civ. Eng.* 102:745-768.

Hoffman, C.E., and Kilambi, R. V., 1971, Environmental changes produced by cold-water outlets from three Arkansas reservoirs. Publ. No. 5, Wat. Res. Centr. Univ. Ark. Fayetteville, Ark.

Hubbs, C., and Pigg, J., 1976, The effects of impoundments on threatened fishes of Oklahoma, *Ann. Okla. Acad. Sci.* 5:133-177.

Krizek, R. J., Osallany, S.C., and Kazadi, G. M., eds, 1971, "Pumped storage development and its environmental effects," Am. Wat. Resour. Assoc. Series No. 15, Urbana, Ill.

McClure, R. G., and Stewart, K. W., 1976, Life cycle and production of the mayfly *Choroterpes* (*Neochoroterpes*) *mexicanus* Allen (Ephemeroptera:Leptophlebiidae). *Ann. Entomol. Soc. Am.* 69:134-144.

Netboy, A., 1974, "The Salmon: Their Fight for Survival," Houghton Mifflin Co., Boston, 613p.

Orsborn, J. F., and Allman, C. H., eds., 1976, "Instream Flow Needs", *Am. Fish. Soc.*, Bethesda Maryland. Vol. 2.

Pacific Northwest River Basin Commission, 1972, "Instream Flow Requirement Workshop", Vancouver, B.C.

Poole, W.C., and Stewart, K.W., 1976, The vertical distribution of macrobenthos within the substratum of the Brazos River, Texas. *Hydrobiologia* 50:151-160.

Rhame, R. E., and Stewart, K. W., 1976, Life cycles and food habits of three Hydropsychidae (Trichoptera) species in the Brazos River, Texas, *Am. Entomol. Soc.*, 102:65-99.

Richmond, M. C., and Zimmerman, E. G., 1978, Effect of temperature on activity of allozymic forms of supernatant malate dehydrogenase in the Red Shiner, *Notropis lutrensis*, *Comp. Biochem. Physiol.*, 61B:415-419.

Robinson, W. L., 1978, The Columbia: a river system under seige (part one of two) *Oregon Wildlife*, 33:3-12.

Stanford, J. A., 1979, The Foothills Project: comments on inadequacies of environmental impact analyses and evaluation of alternative actions, EPA Research, Series Publ. (in press).

Stanford, J. A., and Gaufin, A. R., 1974, Hyporheic communities of two Montana Rivers, *Science*, 185:700-702.

Stanford, J. A, and Gaufin, A. R., in press, Ecology and life histories of Plecoptera in the Flathead Rivers, Montana, *Arch. Hydrobiol. Suppl.*

Stanford, J. A, and Hauer, F. R., 1978, Preliminary observations on the ecological effect of flow regulation in the Flathead River, Montana, Rept. U.S. Bureau Reclamation, Boise, Idaho.

Stanford, J. A., and Potter, D. S., 1976, Limnology of the Flathead Lake-River ecosystem, Montana: a perspective, *in*: "Proc. of the Symposium on Terrestrial and Aquatic Ecological Studies of the Northwest", Soltero,R., ed., Eastern Washington State College Press, Cheney.

Stanford, J. A., and Reed, E. B., 1974, A basket sampling technique
    for quantifying riverine macrobenthos, *Wat. Res. Bull.*
    10:470-477.
Stewart, K. W., Friday, G. P., and Rhame, R. E., 1973, Food habits
    of hellgrammite larvae, *Corydalus cornutus* (Megaloptera:Coryda-
    lidae) in the Brazos River, Texas, *Annal. Entomol. Soc. Am.*
    66:959-963.
Townsend, G. H., 1975, Impact of the Bennett Dam on the Peace-
    Athabaska delta, *J. Fish. Res. Board. Can.*, 32:171-176.
U. S. Army Corps of Engineers, 1973, Brazos River Basin, Texas,
    Natural salt pollution study, Dept. of U.S. Army Engr. Dist.,
    Ft. Worth, Texas.
U. S. Bureau of Land Management, 1978, "The Proposed Foothill Pro-
    ject. Final Environmental Impact Statement". Department of
    Interior. Washington, D.C.
U. S. Federal Power Commission, 1976, "Hydroelectric Power Resources
    of the United States,: FPC-P43, Washington, D.C.
U. S. Fish and Wildlife Service, 1976, "U.S. Reservoir Inventory,
    National Reservoir Research Program," Fayetteville, Ark.
Vaught, G. L. and Stewart, K. W., 1974, The life History and ecol-
    ogy of the stonefly *Neoperla clymene* (Newman) (Plecoptera:
    Perlidae) *Annal. Entomol. Soc. Am.*, 67:167-178.
Wade, D. T., White, R. G., and Mate, S., 1978, A study of fish and
    aquatic macroinvertebrate fauna in the South Fork Boise River
    below Anderson Ranch Dam, Idaho Cooperative Fishery Research
    Unit, Univ. of Idaho; Idaho Dept.. Fish and Game.
Ward, J. V., 1974, A temperature-stressed stream ecosystem below a
    hypolimnial release mountain reservoir, *Arch. Hydrobiol.*,
    74:247-275.
Ward, J. V., 1975a, Bottom fauna-substrate relationships in a
    northern Colorado trout stream: 1945 and 1974, *Ecology*,
    56:1429-1434.
Ward, J. V., 1975b, Downstream fate of zooplankton from a hypo-
    limnial release mountain reservoir, *Verh. Internat. Verein.
    Limnol.*, 19:1798-1804.
Ward, J. V., 1976a, Effects of thermal constancy and seasonal temper-
    ature displacement on community structure of stream macroinvert-
    ebrates, in "Thermal Ecology II," Esch, G. W., and McFarlane,
    R. W., eds., ERDA Symposium Series (CONF-750425).
Ward, J. V., 1976b, Comparative limnology of differentially regulated
    sections of a Colorado mountain river, *Arch. Hydrobiol.*, 78:319-342.
Ward, J. V., 1976c, Effects of flow pattern below large dams on stream
    benthos: a review, "Instream Flow Needs Symposium, Vol. II,"
    Orsborn, J. F., and Allman, C. F., eds. *Am Fish. Soc.*
Ward, J. V., and Short, R. A., 1978, Macroinvertebrate community
    structure of four special lotic habitats in Colorado, U.S.A.,
    Verh. Internat. Verein. Limnol., 20:1382-1387.

Wesche, T. A., 1974, Relationship of discharge reductions to available trout habitat for recommending suitable streamflows, Water Resources Research Institute, University of Wyoming.

Wiltzius, W. J., 1978, Some factors historically affecting the distribution and abundance of fishes in the Gunnison River, Final report, Colorado Div. Wildlife, Fort Collins, Colo.

Zimmerman, E. G., and Richmond, M. C., in press, Heterosis at an enzyme locus in fish inhabiting a thermally coarse-grained environment, *Trans. Am. Fish Soc.*

Section III
Special Topics

ADAPTIVE STRATEGIES OF *AMPHIPSYCHE* LARVAE (TRICHOPTERA:

HYDROPSYCHIDAE) DOWNSTREAM OF A TROPICAL IMPOUNDMENT

P. J. Boon

Department of Zoology
The University
Newcastle-upon-Tyne, NE1 7RU, England

INTRODUCTION

The biological influence exerted by lakes on their outflows has
been recognized since the early days of stream research (e.g.,
Krawany, 1930). However, studies on regulated rivers are comparatively recent, since only in the last two decades have hydroelectric,
irrigation, and other flow-manipulative schemes assumed importance.
Much of this work involved overall river surveys and the effects of
regulation upon whole communities (e.g., Ward, 1976; Armitage,
1978), although occasional studies have concentrated on specific
aspects of lake outflow ecology, e.g., temperature (Lehmkuhl, 1972),
flow fluctuations (Trotzky and Gregory, 1974), and drift (Radford
and Hartland-Rowe, 1971).

The abundance of filter-feeders in lake outlets has frequently
been observed and correlated with a characteristically high level of
suspended food (e.g., Müller, 1954; Cushing, 1963). In a regulated
river, filter-feeders will be distributed as a function of the
balance between feeding advantages (i.e., increased suspended particulates) and the stresses imposed by flow regulation. It is therefore likely that certain outflow organisms will possess adaptations
(morphological, physiological, or behavioral) enabling them to
withstand such stresses rather better than their competitors and so
to exploit more effectively the available resources. However, in
the literature, specific examples of such adaptations are rare.
This paper seeks to describe and discuss the strategies adopted by
one particular species--*Amphipsyche meridiana*--in adapting to life
in a regulated river downstream of a tropical impoundment. The

paper further shows how this adaptability has also enabled *A. meridiana* to exploit a new and vacant niche provided by the construc- tion of a dam wall.

Apart from Ulmer's original descriptions (1951, 1957), I have been unable to discover any other reference to this species and very few to the genus as a whole. There are some morphological descriptions of *Amphipsyche* (e.g., Lepneva, 1964) but very little is known of its ecology [apart from studies by Chutter (1963, 1968) on the Vaal River in South Africa]. Although the results reported herein are based upon a restricted period of observation, weekly sampling provided ecological data previously unavailable for a species that employs effective adaptive strategies in coping with flow regulation in a tropical river.

THE STUDY AREA

Lake Rawapening is situated in Central Java, 45 km south of Semarang and 460 m above sea level (Fig. 1). It was formed in 1916 after construction of a dam on the River Tuntang and now plays an important role in irrigation, fish production, and hydroelectric power generation. The surface area of approximately 2,500 ha in the rainy season falls to about 650 ha during the dry season, a signif- icant proportion of which (about 120 ha) is now covered with water hyacinth (*Eichornia crassipes*). Because of the problems this plant creates, a number of policies have been suggested for its removal. One control measure has been merely the occasional flushing out of large quantities of *Eichornia* into the River Tuntang by opening the dam. This necessitates a high water level in Lake Rawapening, so that water supplies to the hydroelectric plant are not reduced. These conditions were met during the summer of 1978, when an abnor- mally long rainy season caused a considerable rise in lake level. Studies were carried out from late July to early September 1978, and the timing of water release during part of this period is shown in Fig. 2.

The Tuntang is a hard-water, neutral-pH river containing a wide range of substrate types varying from small stones to huge boulders. Two stony riffles were selected for sampling, one near the dam wall, the other further downstream (Table 1).

METHODS

Field

Suspended particles. Samples of river water were taken in small polyethylene bottles and 5-ml aliquots filtered through 0.45-µ

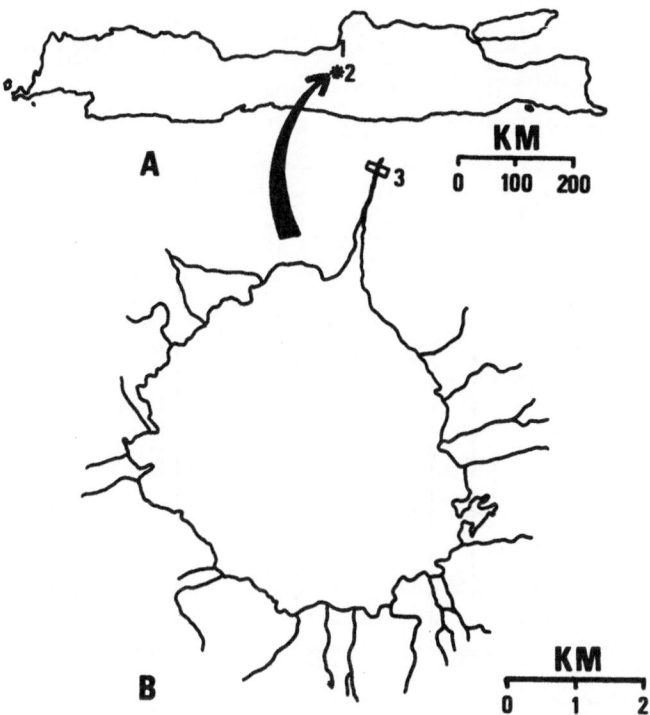

Fig. 1.   The study area (A = Java, B = Lake Rawapening, 1 = Semarang,
          2 = Lake Rawapening, 3 = dam on River Tuntang).

membrane filters.   These were cleared with immersion oil and mounted
in Euparal.

    Benthic sampling.   Six × 10-stone samples were taken from
Station 1 at weekly intervals from 17 August 1978 to 7 September
1978.   Animals and debris scrubbed free of stones were preserved in
the field, and approximate estimates made of stone surface area
by measuring the two longest diameters.

    Collection of feeding nets.   These were removed individually
from rock surfaces, together with the associated larva, and
immediately preserved in 70% alcohol and glycerine.

Laboratory

    (1)   Trichoptera larvae were sorted from debris and identified:
two of the three species to generic, and the third to specific,

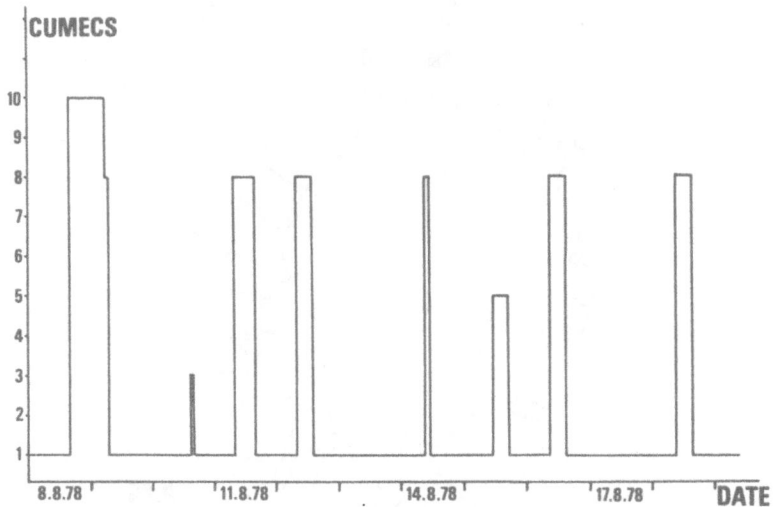

Fig. 2.   Flow volume in the River Tuntang downstream of the dam
          (baseline arbitrary--very low flow).

Table 1.   Description of Sampling Stations

Station	Distance from Dam (km)	Width (m)	Depth (cm)	Mean Flow (cm s^{-1})	Mean Flow (cm s^{-1}) (Dam Open)
1	0.2-0.3	4.0-5.5	7-20	68	--
2	4.5	11.0	15-37	113	171

level by reference to Ulmer (1951, 1957). Animals were separated
into instars as described elsewhere (Boon, in press).

(2) Nets were mounted in Euparal and examined under ×500
magnification. Measurements of mesh length and width were made
with a graduated eye piece.

RESULTS AND DISCUSSION

Suspended particles

Whether the dam was open or closed, 93-97% of suspended particles
measured at Station 1 were less than 5 μ. However, with the dam
open there was a marked increase (from 2.0 to 10.9 × 10^6/ℓ) in the
number of larger paticles (>10 μ), causing a high degree of turbidity.
Although most of these particles were not identified, there was a
noticeable increase (from 15 to 30% of all particles > 10 μ) in
phytoplankton density.

*Distribution of Net-Spinning Trichoptera--Adaptations
for Coexistence*

Samples from Station 1 consisted almost entirely of *A.
meridiana* and *Cheumatopsyche* sp., whereas those from Station 2 (see
later section) also contained *Hydropsyche* sp. (Table 2). Populations
of coexisting hydropsychids have been recorded by several workers
(e.g., Williams and Hynes, 1973; Wallace, 1975; Mackay, 1978; Boon,
1978, in press) and various segregation strategies postulated.
These may be broadly categorized as temporal, behavioral, or spatial.
Temporal differences in larval development (demonstrated for other
coexisting species by, e.g., Boon, in press; Hildrew and Edington,
in press) may reduce competition by separating periods of high
energy demand for each species (Oswood, 1976). In addition, it may
also ensure that the mesh sizes of two or more species at any given
time are different, thus assisting food partitioning. In a tropical
environment, where development is rapid, more than one cohort may be
present, leading to a wider spread of instars and thus mesh sizes.
At Station 2 on the River Tuntang, five instars for each species
were found coexisting (Fig. 3), favoring a potentially high
utilization efficiency of available seston.

The variation in net mesh sizes for those species and instars
examined is shown in Table 3 and Fig. 4. Marked differences exist
between species, although to what extent there is overlap in mesh
size for other unexamined instars is uncertain. However, there is
certainly the potential for food partitioning for at least a part of

Table 2. Percentage Species Composition of Net-
Spinning Trichoptera from Stations 1
and 2

Species	Station 1 (%)	Station 2 (%)
Amphipsyche meridiana	12	41
Cheumatopsyche sp.	87	15
Hydropsyche sp.	<1	44

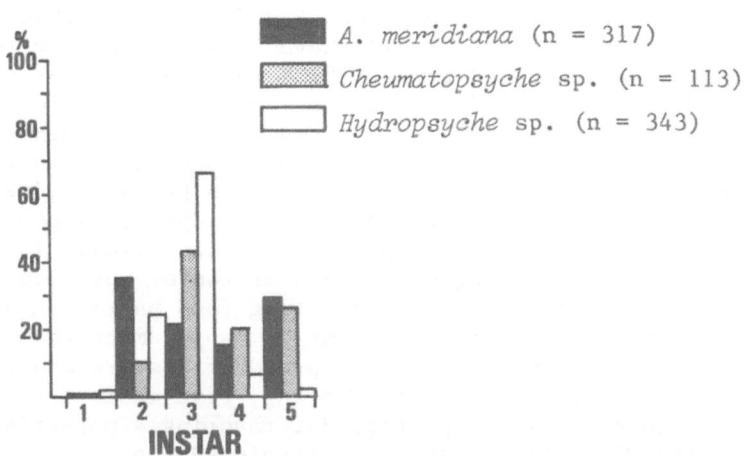

Fig. 3.   Instar distribution of *A. meridiana, Cheumatopsyche* sp.,
and *Hydropsyche* sp. at Station 2 on 3 September 1978.

Table 3. Feeding-Net Mesh Sizes of Species from River Tuntang

Instar	Species	Longer Diameter (μ) $\bar{x}$ (±95% CL)	Shorter Diameter (μ) $\bar{x}$ (±95% CL)	Mesh Area (μ^2) $\bar{x}$ (±95% CL)
4th	A. meridiana	34 (33 - 35)	27 (26 - 28)	925 (892 - 958)
5th	A. meridiana	37 (36 - 38)	31 (30 - 32)	1196 (1149 - 1243)
3rd	Hydropsyche sp.	59 (58 - 60)	45 (44 - 46)	2646 (2548 - 2744)
5th	Cheumatopsyche sp.	150 (145 - 155)	108 (105 - 111)	16375 (15566 - 17184)

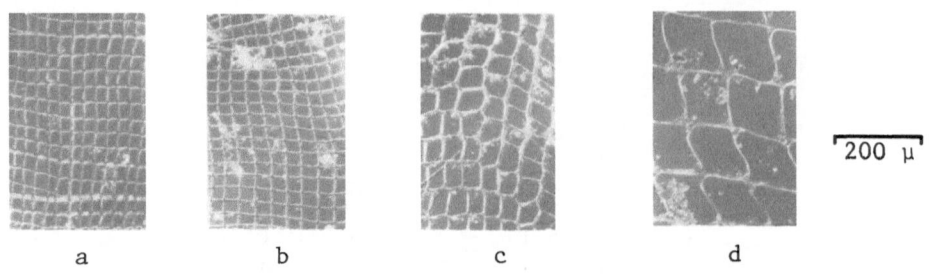

200 µ

a   b   c   d

Fig. 4.   Larval feeding nets: (a) *A. meridiana* (5th instar),
          (b) *A. meridiana* (4th instar), (c) *Hydropsyche* sp. (3rd
          instar), and (d) *Cheumatopsyche* sp. (5th instar).

each species' life cycle.   (The results of gut analysis will be
presented in a later paper.)

*The Colonization of Vesicular Rocks--An Adaptation to
a Fluctuating Environment*

     Spatial segregation of species may be achieved by differential
colonization of stone surfaces (e.g., Malas and Wallace, 1977).
Substrate diversity also promotes coexistence, provided the organisms
involved are sufficiently adaptable.   The River Tuntang provides an
excellent example of this and also illustrates one way in which the
stresses of a fluctuating flow regime may be minimized.

     In parts of the river (particularly Station 2), the substrate
includes small volcanic rocks formed from vesicular lava and pitted
with crevices (vesicles) (Fig. 5).   These vesicles occasionally
extend up to 3 cm into the rock (mean sizes: length, 7.1 mm ±
1.2 mm; width, 4.0 mm ± 0.8 mm; depth, 5.8 mm ± 1.3 mm).   Initial
sampling showed large numbers of caddis larvae living inside the
vesicles and building feeding nets on the rock surface.

     I removed 15 individual non-vesicular rocks from Station 2 and
removed all animals.   A smaller number of similar-sized vesicular
rocks (mean dimensions 20 × 15 cm) were compared and the results
presented in Table 4.   The mean number of animals per rock was three
times greater on the vesicular than on the non-vesicular substrate,
and this increase was almost entirely the result of the significant
increase ($P < 0.001$) in the numbers of *A. meridiana* (approximate
density of 7,300 m^{-2} compared to 700 m^{-2} for non-vesicular rock).
No significant differences in population levels on both substrate
types were found for the other two species.   The ability to colonize
small rock vesicles must require a certain degree of respiratory
adaptability, as many vesicles, especially deeper ones, are likely

3 cm

Fig. 5. Vesicular rock.

to have very low current flows through them. However, the exploitation of this habitat by *A. meridiana* appears to confer four major
advantages:

(1) *Enforcement of spacing.* The presence of vesicular rocks
not only assists in the spatial segregation of species between substrate types but also facilitates spacing of individual larvae on
the rock surface. Glass and Bovbjerg (1969) demonstrated that
Cheumatopsyche larvae dispersed in a density-related fashion and
that, when numbers were high, this dispersal was frequently enforced
by aggressive behavior. Rock vesicles are likely to reduce the
chances of such encounters with other individuals, either of the
same or of different species, assuming that one vesicle is inhabited
by only one larva. Of 46 vesicles examined, 40 had only one occupant.
When more than one larva was present, the vesicle often appeared
subdivided by larval secretions. (In one particularly large vesicle
I found 14 *A. meridiana* larvae of 4 different instars.) A second
feature of a vesicular rock is its variation in vesicle sizes which
should enable maximum space utilization for populations containing
several instars. Vesicle measurements support this suggestion,
showing a tendency for 4th instars of *A. meridiana* (mean vesicle
volume 41.4 mm^3) to occupy smaller vesicles than 5th instars (mean
vesicle volume 154.3 mm^3) ($P < 0.001$).

(2) *Protection against predation.* The extent to which this is
important in the River Tuntang is not known. However, the difficulty experienced in extracting larvae from vesicles, especially the
deeper ones, shows that they may function extremely efficiently as
refuges from predators.

(3) *Protection against dislodgement during high water levels.*
During the period of study, the River Tuntang was subjected to
severe water-level fluctuations, coupled with a marked increase in
suspended solids. Hynes (1970) has pointed out that it is a combination of these two factors that can be devastating to invertebrate
populations, and so the colonization of vesicular rock by *A. meridiana*

Table 4. Distribution of Trichoptera Larvae on Vesicular and Non-vesicular
Substrates

Substrate Type	No. of Larvae Per Rock ($\bar{X} \pm$ 95% CL)	No. of *A. meridiana* Per Rock ($\bar{X} \pm$ 95% CL)	No. of *Cheumatopsyche* sp. Per Rock ($\bar{X} \pm$ 95% CL)	No. of *Hydropsyche* sp. Per Rock ($\bar{X} \pm$ 95% CL)
Vesicular	149 (67 – 333)	105 (48 – 228)	5 (2 – 14)	39 (6 – 240)
Non-vesicular	53 (27 – 104)	21 (11 – 42)	8 (4 – 16)	23 (9 – 60)

is a strategy conferring obvious selective advantage. If such rocks
move (as they frequently will) during spate conditions, then the
animal can remain unharmed inside, merely spinning a new net when
the situation stabilizes.

(4) *Increased resistance to desiccation.* In a regulated
environment, such as the River Tuntang, and especially in a monsoon
climate, there will be times when the water level falls and exposes
rock surfaces. Some simple field and laboratory experiments were
carried out to investigate the role of vesicles during such periods
of exposure.

Several rocks were removed from a riffle at Station 2 and
observed (initially in the field, and later in the laboratory) to
determine whether a complete emigration of animals occurred or
whether some larvae remained within vesicles, despite total exposure.
After exposure, a proportion of the population leaves the rock,
presumably in search of more favorable conditions, while the
remainder stay within the vesicles (Fig. 6). Although numbers of
Cheumatopsyche sp. and *Hydropsyche* sp. are small, the results suggest
that these species are less inclined to remain *in situ* than is *A.
meridiana* (Table 5).

The ability of *A. meridiana* to resist desiccation was shown
remarkably in rock 4; even after 56 hours of exposure, 55 larvae
were found alive inside rock vesicles. It is not possible to make
absolute estimates of the proportion of any species dying within
vesicles as the drying and shrinking of dead larvae made them
particularly difficult to find. However, in rocks 5, 6, and 7
(searched after 78 hours), 14, 33, and 60 5th-instar *A. meridiana*
were found dead inside vesicles, suggesting that after a certain

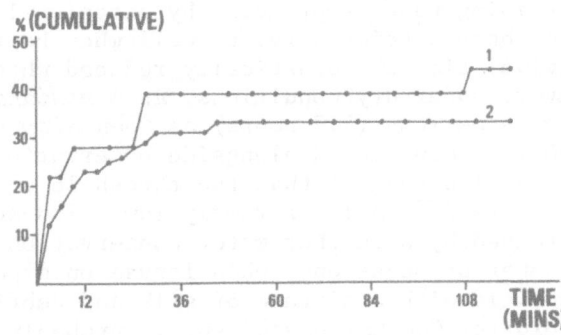

Fig. 6. The cumulative percentage of all Trichoptera larvae
         emigrating from vesicular rocks 1 and 2.

Table 5. Number of Larvae Migrating from, or Remaining Within,
         Vesicular Rocks (Periods: Rocks 1 and 2, 2 hours;
         Rock 3, 24 hours; Rock 4, 56 hours)

Species	Rock 1		Rock 2		Rock 3		Rock 4	
A. meridiana	5^a	11^b	40^a	81^b	50^a	25^b	108^a	55^b
Cheumatopsyche sp.	3	0	5	1	0	0	4	0
Hydropsyche sp.	0	0	0	3	44	4	17	2

[a]Number of larvae emigrating during given period.

[b]Number of larvae removed from vesicles after given period.

time animals remaining show no tendency to migrate and so will not
survive unless reflooded within 2-3 days.

     Fig. 7 shows that A. meridiana larvae appear to adopt two dif-
ferent strategies according to instar, smaller instars tending to
emigrate, larger ones tending to remain within vesicles. This may
well be related to the rate of desiccation, as the surface
area/volume ratio decreases with increasing instar (2nd, 0.80; 3rd,
0.52; 4th, 0.33; 5th, 0.25).

     Animals removed from vesicles in rock 4 (i.e., after 56 hours)
frequently appeared to be completely dry, so an experimental comparison of survival was made with larvae kept on dry filter paper.
Excess surface moisture was removed from individual larvae, which
were then put into covered petri dishes, some containing dry filter
paper, some containing paper kept moist by occasional drops of
water. Larvae of both species survived well when kept moist (>120
hours), but survival time was drastically reduced when animals were
kept dry. However, under dry conditions, A. meridiana survived for
approximately twice as long (3.7 hours) as Cheumatopsyche sp. (1.6
hours) (P < 0.001). When viewed alongside observations on rock
vesicles, these results suggest that the threshold moisture content
for survival in A. meridiana is extremely low. Its survival time is
undoubtedly increased by a further water conservation strategy,
observed on a number of occasions, when larvae on exposed rocks
covered their vesicle with a mixture of silk and debris. The
moisture threshold for Cheumatopsyche sp. is probably higher, which
may partly explain its more rapid emigration following rock exposure.
(This could be further studied with suitably controlled laboratory
experiments).

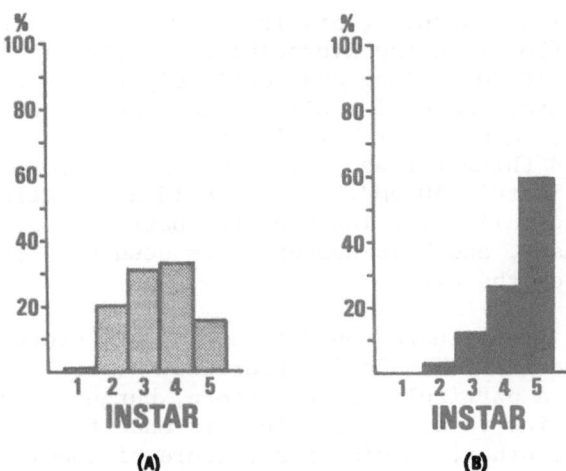

Fig. 7.   Instar analysis of *A. meridiana* in vesicular rocks;
(a) animals emigrating, and (b) animals removed from
vesicles.

The vesicular rock habitat provides an ideal way of coping with
a rigorous environment. Those animals colonizing such a rock
probably have the advantage of being able to channel a greater
proportion of energy into growth, as less will be required merely to
maintain their position in the current. Larval tubes spun from silk
and weighted with gravel (which were observed in other places) are
rarely used, as the vesicle now provides all the protection needed.
There are problems, however, mainly associated with the risks of
exposure; the risk of remaining inside the vesicle and possibly
dying is set against the risk of leaving and searching for another
suitable habitat.

*Colonization at the Dam Wall--An Adaptation
to a Vacant Niche*

The data so far presented, particularly concerning vesicular
rocks, categorize *A. meridiana* as an exploitative and opportunistic
species. Observations made at the dam wall itself further support
this view.

A few meters upstream from the dam, a number of barrier screens
are sited to prevent debris from entering the hydroelectric pipe.
These are metal structures, some 4–5 m long × 0.5 m wide, subdivided
along their length by struts 2–3 cm apart. On one occasion, I
noticed workers at the dam site removing solid aggregations of

Trichopteran larval tubes. (This is apparently done regularly; otherwise, the flow into the hydroelectric system is impeded.) Each aggregation was 10-20 cm long and completely filled the 2- to 3-cm gap between struts. Several small samples were taken for analysis and observation. A block of larval material is shown in Fig. 8a. Five hundred and three larvae, all *A. meridiana*, were removed from a block of approximately 50 cm^3. Specific identification was made of adults caught while emerging from the barrier screens. Emergence occurred at sunset, and huge numbers were seen crawling up the metal struts and out of the water.

The blocks are situated on the screens at intervals, giving the screens a horizontally "banded" appearance. (This may well be correlated with a particular flow pattern, but no information is available regarding this.) Each block is made from individual tubes fixed one to the other, forming a structure of considerable strength and rigidity. The blocks appear to be generally of two main layers, facing opposite directions, one into the current and one towards the leeward side. In the river below the dam, *A. meridiana* strengthens its silk tube with gravel. Larvae in vesicular rocks rarely use silken tubes because adequate protection is found within vesicles. In barrier screens, the adopted strategy, in the absence of suspended gravel or suitably sized protective crevices, is to obtain strength from each other's tubes (see Fig. 8b).

An obvious problem that arises is the way in which animals construct effective feeding nets in such a dense aggregation. Examination of a block shows traces of net material, apparently fastened by silk threads onto a number of larval tubes. When portions of net are examined, they frequently contain two components (Fig. 8c). One is an area of haphazard threads, apparently consisting of several layers one on top of another, and the other is a region of regular mesh. Mean measurements of regular meshes were 77 μ (±2) × 49 μ (±2), with a mean area of 3779 μ^2 (±213). From these results, two aspects of net-spinning behavior are particularly interesting. First, it seems likely that larvae are feeding out of what must technically be described as the same net. Even if each larva has an individual net portion solely for its own use, the fact that net material is attached to several tubes implies at least a certain degree of passive cooperation. It may well be that at a certain density an individual net per larva is no longer a functional possibility. Second, there were differences in mesh size between animals living in barrier screens and those living in the river. Barrier screen nets have mesh areas three times greater than nets from the river. Most literature on this subject implies that, within a given range, a particular species and instar will spin a net of constant dimensions, and Kaiser (1965) has shown how mesh size may be determined by morphological features, such as the distance between maxillary palps. One might ask, therefore, whether

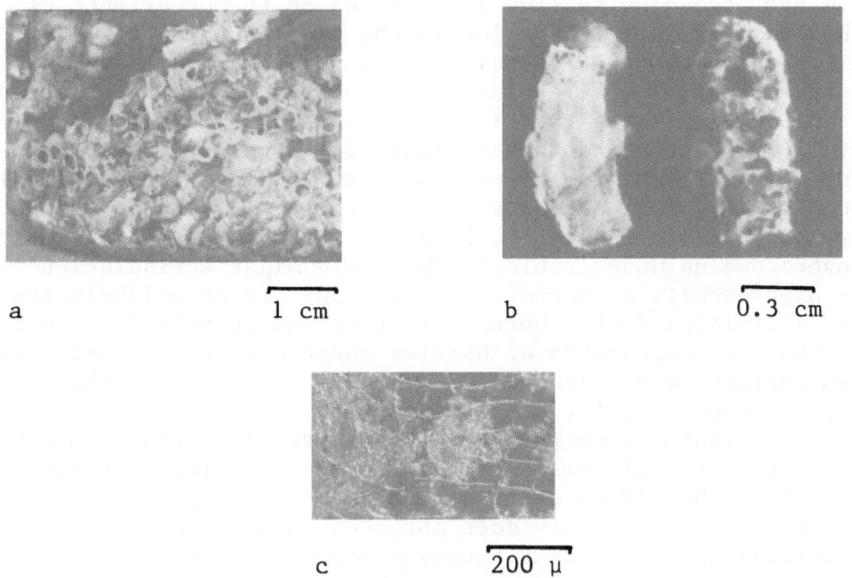

a   1 cm   b   0.3 cm

c   200 μ

Fig. 8.   (a) Larval aggregation from barrier screen; (b) larval tube
          from barrier screen cf tube from river; and (c) net from
          larval aggregation.

animals living in barrier screens are a different but closely
related species of the *Amphipsyche* in the river. Larvae from screens
were positively identified as *A. meridiana*; and, although only two
specimens from the river were reared to adults, both of these were
also identified as *A. meridiana*. In addition, no consistent morphological differences can be seen between larvae from both sites, and
so the most reasonable conclusion must be that they are the same
species.

     If this is so, how might such a large difference in mesh size
be explained? Some recent work by Fey and Schumacher (1978) has
shown that mesh sizes of *Hydropsyche pellucidula* may be regulated by
a temperature-dependent rate of net production, rather than by any
morphological control. It would seem quite reasonable to suggest
that *A. meridiana* may also be modifying its net-spinning behavior in
response to certain ecological stimuli. It may be that a high
density of fine-meshed nets would reduce their filtering efficiency,
especially if some regions of the net were built from layers contributed by different larvae. Further information is needed, both on
physical parameters, such as flow and temperature, and on behavioral
responses to high density situations, before firm conclusions can be
made.

In the preceding section I commented on the occurrence of spacing and territorial behavior in the Hydropsychidae. In relation to this, Johnstone (1964) obtained evidence that hydropsychid larvae stridulate by using a file-like series of ridges on the underside of the head, and suggested that it might be used in territorial behavior. Edington (1965) proposed that stridulation might be interpreted as "a conventional mechanism to prevent overcrowding," conventional inasmuch as "it would not be related directly (except in its evolutionary history) to those factors like food shortage which would make overcrowding undesirable." What role might stridulation play in the high-density *A. meridiana* situation? It is unlikely that food is a limiting factor here, but space undoubtedly is, especially as the "banded" appearance of barrier screens suggests that only certain portions are suitable for colonization. Within these limits, overcrowding, far from being undesirable, in a sense becomes a necessity, inasmuch as the group gains stability from its density. *A. meridiana* certainly possesses large, well-developed stridulatory ridges but can hardly be using stridulation as a spacing mechanism. Whether stridulation simply does not occur under these conditions, or is fulfilling a different function when an individual is part of an aggregation remains unanswered.

GENERAL DISCUSSION

Net spinning in fluctuating flow conditions is bound to create problems. Radford and Hartland-Rowe (1971) found comparatively fewer Trichoptera in a regulated than in an unregulated river and suggested that this may have been because of the inability of some net spinners to maintain operative nets over a wide range of flow conditions. However, nets of *A. meridiana* (which are both resistant to collapse upon exposure and resilient to handling) may well be able to function in both low and high flows. Their structure is in sharp contrast to the closely related genus *Macronema*, which has a mesh size of 5 × 40 µ (Wallace and Sherberger, 1974).

The problem of having a net functional over a wide range of conditions is probably less important in an area of high food availability than that of maintaining position on the substrate. It is here that *A. meridiana* has an advantage in its ability to exploit the vesicular rock habitat. Fisher and LaVoy's (1972) freshwater "intertidal" zone makes an interesting comparison. Much lower diversities and densities of invertebrates were recorded on areas that were periodically exposed than on those that were continuously flooded. Fisher and LaVoy suggested that sufficient time has not been available for the evolution of complex communities suited to this recent and man-made biotope. Although this interpretation may be broadly true, faunal distribution must also depend upon the previous history of the river and the type of flow regime before

regulation. Thus, in the River Tuntang, *A. meridiana* may have
developed adaptive strategies to deal with naturally occurring
seasonal fluctuations, which now also assist in resisting additional
man-made stresses.

The colonization of artificial and stable structures, i.e., at
the dam wall itself, provides another means of avoiding an unpredictable environment. The dam on the River Tuntang is certainly not
a unique example of such a colonization. Fremling (1960) found high
densities of *Hydropsyche* in similar situations on the upper Mississippi, where swarms of emerging adults reached pest proportions.
In the River Tuntang, the difference between larvae in the river and
those in barrier screens poses an interesting question. Is the
diversity apparent within the species an expression of the individual's
capability for behavioral variation or is the population behaviorally
polymorphic? Gillett (1979), in a recent article on the vector
concept, comments on the range of behavior patterns within mosquito
populations and suggests that "the vector concept must be accepted
as a statistical one concerning gene frequencies in populations."
The type of meticulous experimental analysis required to reach this
conclusion may also be the way of clarifying the *A. meridiana*
situation.

ACKNOWLEDGMENTS

I would like to thank Dr. F. Göltenboth (Universitas Kristen
Satya Wacana, Central Java) for arranging research facilities,
Agus Kristyanto for his assistance in field work, Dr. P. C. Barnard
[British Museum (Natural History)] for identification of specimens,
and Miss L. Tarn for typing the manuscript.

I would also like to acknowledge for their financial support:
The Royal Society (Scientific Investigations Grant); The Scientific
Research Society of North America; and the Department of Zoology,
University of Newcastle-upon-Tyne.

REFERENCES

Armitage, P. D., 1978, The impact of Cow Green Reservoir on inverte-
    brate populations in the River Tees, *F.B.A. Annu. Rep.*,
    46:47-56.
Boon, P. J., 1978, The pre-impoundment distribution of certain
    Trichoptera larvae in the North Tyne river system (northern
    England), with particular reference to current speed,
    Hydrobiologia, 57:167-174.

Boon, P. J., In press, Studies on the spatial and temporal distri-
    bution of larval Hydropsychidae in the North Tyne river system
    (northern England), *Arch. Hydrobiol.*

Chutter, F. M., 1963, Hydrobiological studies on the Vaal River in
    the Vereenigung area. Part 1. Introduction, water chemistry,
    and biological studies on the fauna of habitats other than
    muddy bottom sediments, *Hydrobiologia*, 21:1-65.

Chutter, F. M., 1968, On the ecology of the fauna of stones in the
    current in a South African river supporting a very large *Sim
    ulium* (Diptera) population, *J. Appl. Ecol.*, 5:531-561.

Cushing, C. E., 1963, Filter-feeding insect distribution and
    planktonic food in the Montreal River, *Trans. Am. Fish. Soc.*,
    92:216-219.

Edington, J. M., 1965, Effect of water flow on populations of net-
    spinning Trichoptera, *Mitt. Int. Verein. Theor. Angew. Limnol.*,
    13:40-48.

Fey, J. M., and Schumacher, H., 1978, Zum Einfluss wechselnder
    Temperatur auf den Netzbau von Larven der Köcherfliegen-Art
    Hydropsyche pellucidula (Trichoptera:Hydropsychidae), *Entomol.
    Germ.*, 4:1-11.

Fisher, S. G., and LaVoy, A., 1972, Differences in littoral fauna
    due to fluctuating water levels below a hydroelectric dam, *J.
    Fish. Res. Board Can.*, 29:1472-1476.

Fremling, C. R., 1960, Biology and possible control of nuisance
    caddisflies of the upper Mississippi River, *Res. Bull. Iowa
    Agric. Exp. Stn.*, 483:856-879.

Gillett, J. D., 1979, The vector concept, *Bull. R. Entomol. Soc.
    Lond.*, 3:17-22.

Glass, L. W., and Bovbjerg, R. V., 1969, Density and dispersion in
    laboratory populations of caddisfly larvae (*Cheumatopsyche*,
    Hydropsychidae), *Ecology*, 50:1082-1084.

Hildrew, A. G., and Edington, J. M., In press, Factors facilitating
    the coexistence of hydropsychid caddis larvae (Trichoptera) in
    the same river system, *J. Anim. Ecol.*

Hynes, H. B. N., 1970, "The Ecology of Running Waters," Liverpool
    Univ. Press, Liverpool.

Johnstone, G. W., 1964, Stridulation by larval Hydropsychidae
    (Trichoptera), *Proc. R. Entomol. Soc. Lond. (A)*, 39:146-150.

Kaiser, P., 1965, Über Netzbau und Strömungssinn bei den Larven der
    Gattung *Hydropsyche* Pict. (Ins., Trichoptera), *Int. Rev. Ges
    Hydrobiol.*, 50:169-224.

Krawany, H., 1930, Trichopterenstudien im Gebiet der Lunzer Seen, II
    and III, *Int. Rev. Ges. Hydrobiol.*, 23:417-427.

Lehmkuhl, D. M., 1972, Change in thermal regime as a cause of reduc-
    tion of benthic fauna downstream of a reservoir, *J. Fish. Res.
    Board Can.*, 29:1329-1332.

Lepneva, S. G., 1964, Fauna of the U.S.S.R., Trichoptera. Larvae
    and pupae of Annulipalpia, *Zool. Inst. Acad. Sci., USSR*, 88:

Mackay, R. J., 1978, Larval identification and instar association in some species of *Hydropsyche* and *Cheumatopsyche* (Trichoptera:Hydropsychidae), *Ann. Entomol. Soc. Am.*, 71:499-509.

Malas, D., and Wallace, J. B., 1977, Strategies for coexistence in three species of net-spinning caddisflies (Trichoptera) in second-order southern Appalachian streams, *Can. J. Zool.*, 55:1829-1840.

Müller, K., 1954, Faunistisch-ökologische Untersuchungen in nordschwedischen Waldbächen, *Oikos*, 5:77-93.

Oswood, M. W., 1976, Comparative life histories of the Hydropsychidae (Trichoptera) in a Montana lake outlet, *Am. Midl. Nat.*, 96:493-497.

Radford, D. S., and Hartland-Rowe, R., 1971, A preliminary investigation of bottom fauna and invertebrate drift in an unregulated and a regulated stream in Alberta, *J. Appl. Ecol.*, 8:883-903.

Trotzky, H. M., and Gregory, R. W., 1974, The effects of water flow manipulation below a hydroelectric power dam on the bottom fauna of the upper Kennebec River, Maine, *Trans. Am. Fish. Soc.*, 103:318-324.

Ulmer, G., 1951, Köcherfliegen von den Sunda-Inseln, Pt. 1, *Arch. Hydrobiol.*, Suppl. 19:1-528.

Ulmer, G., 1957, Köcherfliegen von den Sunda-Inseln, Pt. 3, *Arch. Hydrobiol.*, Suppl. 23:109-470.

Wallace, J. B., 1975, Food partitioning in net-spinning Trichoptera larvae: *Hydropsyche venularis, Cheumatopsyche etrona*, and *Macronema zebratum* (Hydropsychidae), *Ann. Entomol. Soc. Am.*, 68:463-472.

Wallace, J. B., and Sherberger, F. F., 1974, The larval retreat and feeding net of *Macronema carolina* Banks (Trichoptera:Hydropsychidae), *Hydrobiologia*, 45:177-184.

Ward, J. V., 1976, Comparative limnology of differentially regulated sections of a Colorado mountain river, *Arch. Hydrobiol.*, 78:319-342.

Williams, N. E., and Hynes, H. B. N., 1973, Microdistribution and feeding of the net-spinning caddisflies (Trichoptera) of a Canadian stream, *Oikos*, 24:73-84.

EFFECTS OF ELEVATED STREAM TEMPERATURES BELOW A SHALLOW RESERVOIR ON A COLD WATER MACROINVERTEBRATE FAUNA

John J. Fraley

Montana Cooperative Fishery Research Unit
Montana State University
Bozeman, Montana 59715

INTRODUCTION

The Madison River is one of Montana's finest cold-water streams, supporting a highly productive fishery that yields nearly 100,000 trout to anglers each year (Vincent, 1969). It comprises 116 km of the 727 km of blue ribbon trout water in the state.

Shortly after 1900 the Madison River was impounded to form Ennis Reservoir. Water from the reservoir is used to produce power at the Madison Power Plant located immediately below Ennis Dam. The reservoir is wide and shallow and acts as a heat trap, raising summer temperatures of downstream waters (Heaton, 1961; Vincent, 1977). These increased summer water temperatures appear to be threatening the blue ribbon trout fishery downstream from the reservoir. Recent studies (Vincent, 1977) have shown that rainbow and brown trout greater than 28 cm in length had slower growth rates in the river downstream from the reservoir (lower river) than upstream (upper river). Conversely, trout less than 28 cm grew faster downstream from the reservoir. The reasons for these differences in growth rates were not clearly understood, but the effects of high mean summer water temperatures (approaching 19°C) on trout metabolism or food supply have been suggested as possible causes.

The purpose of this study was to evaluate the effects of Ennis Reservoir on the thermal regime and the aquatic macroinvertebrate communities of the lower Madison River. Field research was conducted from July 1976 through October 1977.

DESCRIPTION OF THE STUDY AREA

The Madison River arises in northwestern Yellowstone National Park, Wyoming, and flows through Madison and Gallatin counties in southwestern Montana. It is formed by the confluence of the Gibbon and Firehole rivers at an elevation of 2074 m. The river then flows approximately 220 km northward through broad valleys and joins the Jefferson and Gallatin rivers to form the Missouri River near Three Forks, Montana. The elevation is 1235 m at the river's mouth.

The study area was located on the lower 90 km of the river (Fig. 1). Stations 1 and 2 were located 22 and 7 km above Ennis Reservoir in the upper Madison Valley, Station 3 was located 3 km below Ennis Reservoir at the head of the Beartrap Canyon, and Stations 4 and 5 were located 20 and 56 km below the reservoir in the lower Madison Valley. Ennis Reservoir had a surface area of 1530 ha and an average depth of less than 3.5 m. It did not stratify, and winds mixed its waters continuously throughout the ice-free period.

The Madison River is a moderately hard (mean total alkalinity 100 mg l^{-1}), calcium bicarbonate stream with similar anion and cation concentrations above and below Ennis Reservoir (Matney and Garvin, 1978; Fraley, 1978). The flow of the river is regulated by discharges from Hebgen and Ennis reservoirs and is relatively stable compared to unregulated mountain streams of similar size. Mean flow during the study period at Ennis Reservoir was 46.9 m^3 sec^{-1}, compared to the 38-year mean of 51 m^3 sec^{-1} (U.S.G.S., 1976).

METHODS

Water temperatures were monitored at each of the five sampling stations with either a Taylor, Wecksler, or Foxboro continuous-recording thermograph. Temperatures were recorded throughout the study period except during January, February, and March 1977. Calibration of the thermographs was checked weekly from late May to September. Temperatures from the thermograph charts were digitized and analyzed with computer programs. Water temperatures were measured at sites 1, 1.5, and 3 km below Station 3 from mid-June to mid-July 1977, with maximum-minimum thermometers.

Macroinvertebrates were sampled monthly with 0.20 m^2 artificial substrate samplers similar to those of Hester and Dendy (1962). Four samplers were placed in riffle areas at each of the five sampling stations. All samplers were placed in areas of similar depths and current velocities, with the plates positioned parallel to the flow. After approximately 30 days, the samplers were sur-rounded by a net (mesh size 500 μm) to prevent loss of organisms

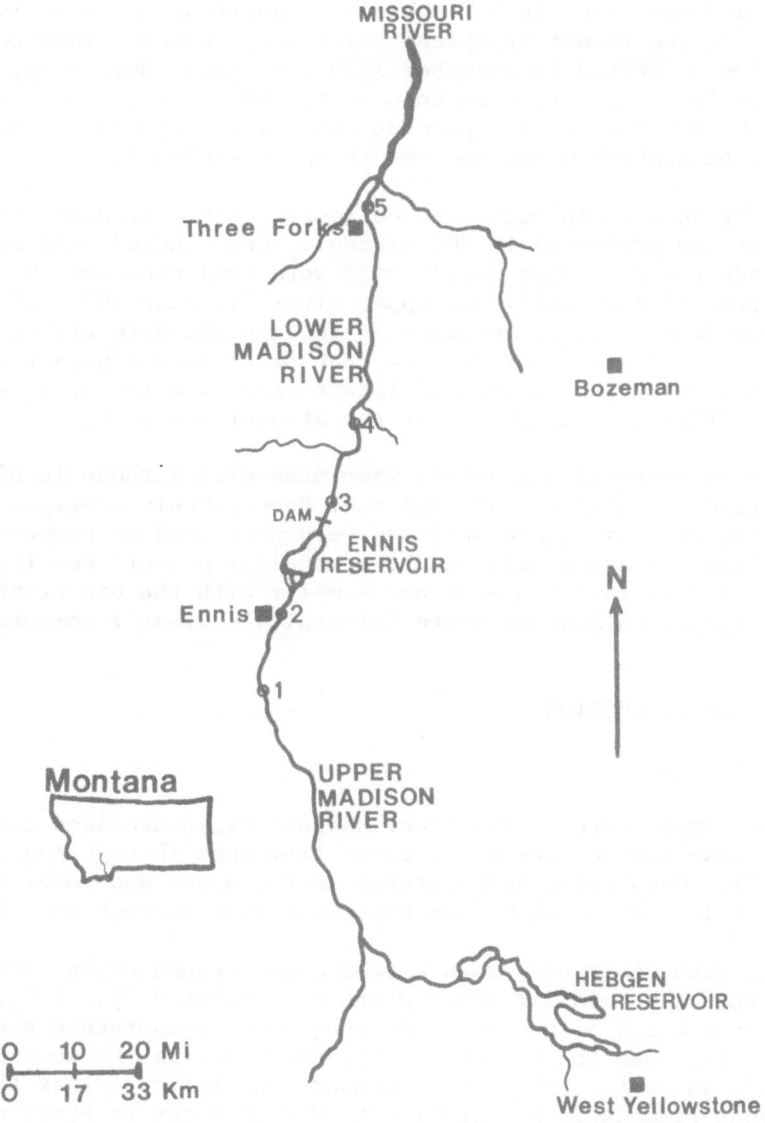

Fig. 1. Map of the study area.   Numbers indicate the location of the sampling stations.

and removed from the stream. The material on each sampler was removed and preserved in 10% formalin. Samples were taken to the laboratory, where each was washed on a U.S. Series Number 30 screen (mesh size 600 μm) and the aquatic macroinvertebrates were separated, washed, and preserved in 70% ethanol. Invertebrates were identified to genus or the lowest practical taxon and counted. Invertebrates in samples collected in December 1976 and March, May, July, and September 1977 were dried in an oven at 100°C and weighed on a Mettler H-16 balance. The percent composition by dry weight of each order at each station for each month was calculated.

Adult aquatic insects were collected with a standard fine-mesh sweep net and preserved in 70% ethanol. One hundred male and 100 female adult *Pteronarcys californica* were collected from both the lower river (7 June 1977) and upper river (18 June 1977) within the first two days of their emergence. The insects were killed with chloroform fumes in the field and oven dried to a constant weight in the laboratory. The weights of insects from the two areas were compared statistically by use of the Student's T test.

Linear regression analyses were made with methods in Dixon and Massey (1969). Analysis of variance, Newman-Keuls multiple comparison tests, and Chi-Square analyses were performed on numbers of invertebrate taxa and total numbers, ordinal numbers and total weights of macroinvertebrates per sampler with the pre-programmed MSUSTAT system on Montana State University's Sigma 7 computer.

RESULTS AND DISCUSSION

Water Temperature

The temperature of the lower Madison River averaged 3.4°C greater than that of the upper river from June through August 1977 (Table 1). Mean water temperatures in the upper and lower river varied by less than 1.5°C from September 1976 through May 1977.

The mean water temperature from June through August was 1.4°C cooler than the mean air temperature at Station 2 (U.S.D.C., 1977). At Stations 4 and 5, however, the mean water temperature was 2.0 and 0.6°C warmer than the mean air temperature during the same period. These air and water temperature comparisons indicate that thermal enrichment from Ennis Reservoir extended at least to Station 4, 22 km downstream from the dam. Irrigation return may have accounted for the slight increase in mean water temperature at Station 5.

The average diurnal temperature fluctuation was similar at Stations 1, 2, 4, and 5 but reduced by 80% at Station 3, 3 km below

Table 1. Mean Water Temperatures and Average Diurnal Fluctuations (°C) at Stations on the Madison River, June
Through August 1977

Station	1	2	3	4	5
Mean temperature	15.3	16.2	19.1	18.9	19.3
Average diurnal fluctuation	5.5	5.0	1.1	6.4	5.4

Ennis Dam. Maximum and minimum thermometers showed this pattern of
thermal constancy broke between 4.5 and 6 km below the dam.

Temperatures greater than 17°C have been considered above
optimum for cold-water organisms (Gaufin, 1962; Nebeker, 1971a;
Nebeker and Lemke, 1968; Gaufin and Hern, 1971; Brett et al., 1969).
The percent of time water temperatures exceeded 17°C at each station
for each month from May through September 1977 is presented in
Fig. 2. In May, water temperatures did not exceed 17°C more than
10% of the time at any station. From June through August, 17°C was
exceeded 77% of the time in the lower river but only 30% of the time
in the upper river. During the average day from June through
August, cold-water organisms in the lower river were subjected to
temperatures above 17°C for about 18 hours. In the upper river,
temperatures were above 17°C for only seven hours of the average
day. In September, temperatures were above 17°C about 20% of the
time in the lower river and 5% of the time in the upper river.

Water temperatures exceeding 21°C severely stress cold water
organisms (Gaufin, 1962; Gaufin and Hern, 1971; Nebeker and Lemke,
1968, 1971a). From June through August 1977, water temperatures
of 21°C were exceeded 1, 5, 23, 25, and 29% of the time at Stations
1 through 5.

Thermal enrichment of the lower Madison river has probably
increased over time. Mean water temperatures for the period August
15-31, 1961, were 16.6 and 17.6°C at locations near Stations 2 and
3. However, for this period in 1977, mean water temperatures at the
same locations were 16.2 and 18.4°C. Water temperatures were higher
at Station 3 in 1977, even though the mean air temperature in 1977
was 3°C cooler than in 1961. Furthermore, Vincent (1978) correlated
the available air and water temperatures from 1961 through 1977 and
found a significant warming trend in water temperatures near Station
3. This increased warming is probably the result of the reservoir
becoming shallower from the continuing deposition of sediment.

STATION

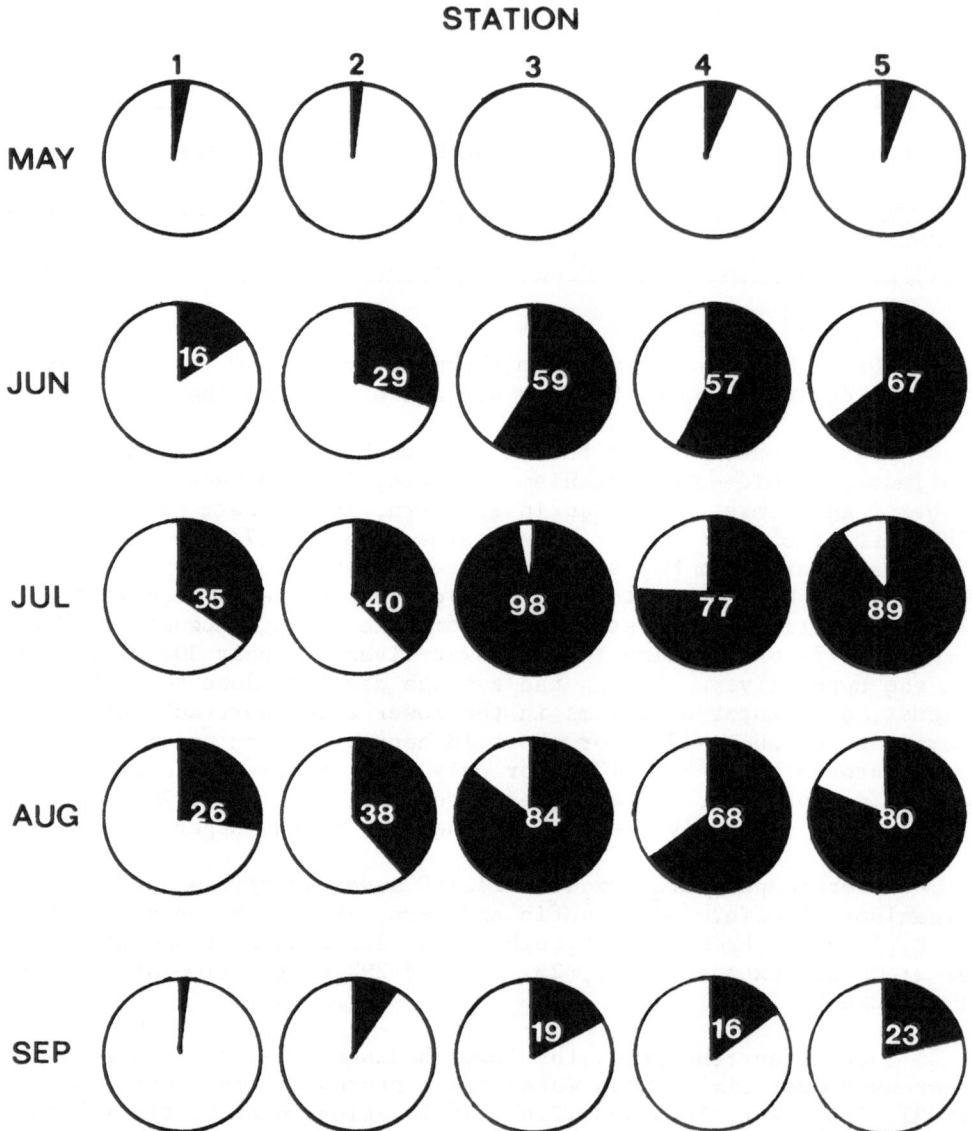

Fig. 2.   The black portion represents the percent of time (numbers) that water temperatures exceeded 17°C from May through September, 1977 at stations on the Madison River.   Circles without numbers indicate values of less than 10%.

MACROINVERTEBRATES

Thirty-six, 33, 35, 33, and 26 artificial substrate samplers were recovered at Stations 1 through 5 from September 1976 through September 1977. Some samplers were lost because of water-level fluctuation, ice formation, vandalism, or fouling by heavy algae growths. Fifty-six taxa belonging to 11 orders were collected from the 163 samplers. Four taxa were taken only in the upper river, 14 were collected only in the lower river, and 38 were found in both sections. Jaccard and Czekanowski similarity coefficients (Clifford and Stephenson, 1975) of the macroinvertebrate communities at each station indicated that there were distinct community types at Stations 1 and 2, at Station 3, and at Stations 4 and 5.

The total number of taxa collected at Stations 1 through 5 were 36, 41, 30, 42, and 39. Analysis of variance and the Newman-Keuls multiple comparison tests showed that Station 3 had a significantly lower mean number of taxa per sampler than all other stations ($P < 0.01$). Mean number of taxa per sampler at Stations 1, 2, 4, and 5 were not statistically different.

The frequencies of occurrence of taxa collected on artificial substrates were calculated by dividing the number of samples in which a taxon occurred at a station by the total number of samples collected at that station. *Ephemerella grandis, Arctopsyche inermis, Glossosoma* spp., and *Physa* spp. had frequencies of occurrence at least four times greater in the upper river than in the lower river. *Choropterpes albiannulata, Tricorythodes minutus, Hydroptila* sp., *Leucotrichia pictipes, Zumatrichia notosa, Parargyractis confusalis, Hyalella azteca*, and *Asellus* sp. were collected only in the lower river or had frequencies of occurrence at least four times greater in the lower river than in the upper river. *Ephemerella inermis, Baetis* sp., and *Amiocentrus aspilus* had similar frequencies of occurrence in both sections.

Because water temperature appeared to be the major physical-chemical factor that differed between sections of the river, the frequencies of occurrence of taxa were correlated to the percent of time over 17°C at each of the five stations and tested statistically with methods of Dixon and Massey (1969). Taxa with negative correlation coefficients significant to the 80% level or greater were classed as cold water preference organisms. Taxa with positive correlation coefficients significant to the 80% level or greater were classed as warm water preference organisms. Taxa with nonsignificant correlation coefficients appeared to exhibit no thermal preference and were classed as eurythermal organisms. Taxa occurring at frequencies of less than 10% at all stations were not classified.

The results of the classifications by thermal preference are shown in Table 2. *Physa* spp., *A. inermis*, *Antocha* sp., *Helicopsyche borealis*, *Glossosoma* spp., *Lepidostoma veleda*, *Ephemerella grandis*, and *Psychomyia flavida* demonstrated the strongest negative responses to water temperatures above 17°C with correlation coefficients significant to the 95% level or greater.

Hydroptila sp., *Heptagenia elegatula*, and *Cheumatopsyche* spp. showed the strongest positive responses to temperatures over 17°C with correlation coefficients significant to the 95% level or greater. *Tricorythodes minutus* and *Hydropsyche* spp. showed statistically significant positive responses to temperatures over 17°C at the 90% level, and *Asellus* sp. had a correlation with temperatures over 17°C significant to the 80% level. *Amiocentrus aspilus*, *Claassenia sabulosa*, *Ephemerella inermis*, *Oecetis avara*, and *Baetis intermedius* had the lowest correlations to 17°C or greater temperatures, thus showing a eurythermal response.

An average of 622 invertebrates per sampler was collected. Numbers were highest in samples collected in June or July and lowest in December or March. The average total number of macroinvertebrates per sampler was 218, 228, 1386, 840, and 323 at Stations 1 through 5. Analysis of variance was performed and station means were grouped according to the Newman-Keuls multiple comparison test. Stations 3, 4, and 5 had significantly greater numbers of invertebrates than Stations 1 and 2, and Stations 3 and 4 had significantly greater numbers than Station 5 (P < 0.01). The lower numbers at Station 5 resulted from smaller sample sizes because of sampler loss. Brenda and Proffitt (1974) and Coutant (1962) also found increased numbers of invertebrates in warmed sections of streams. Ward (1976) reported increased numbers of invertebrates at a regulated site below a hypolimnial discharge dam, although other workers (e.g., Lehmkuhl, 1972) have found severe reductions in benthos below deep release dams.

The percent composition of macroinvertebrate numbers at each station by major taxa is shown in Fig. 3. The percent composition of Trichoptera, Diptera, Ephemeroptera, and Plecoptera was similar at Stations 1, 2, 4, and 5. At Station 3, however, the composition of Trichoptera was significantly greater (P < 0.01) and the composition of Plecoptera and Diptera less than at all other stations. Ward (1976) also reported uneven ordinal composition at regulated sites, with Plecoptera being the most severely reduced group.

Hydropsyche spp., *Brachycentrus occidentalis*, *Cheumatopsyche* spp., *Baetis intermedius*, *Simulium* sp., *Ephemerella inermis*, and Chironomidae were the numerically abundant subordinal taxa. *B. occidentalis* was the numerically dominant taxon collected in the upper river. The macroinvertebrate community in the upper river was

Table 2. Thermal Preferences of Taxa from the Madison River Based on the Correlation of Frequency of Occurrence with the Percent of Time Above 17°C at Each Station (Correlation Coefficients are in Parentheses)

Cold Water Preference Forms		Warm Water Preference Forms		Eurythermal Forms	
Physa spp.	(-0.97)[a]	Hydroptila sp.	(0.94)[b]	Pteronarcys californica	(-0.66)
Arctopsyche inermis	(-0.96)[a]	Heptagenia elegantula	(0.94)[b]	Atherix variegata	(-0.62)
Antocha sp.	(-0.96)[a]	Cheumatopsyche spp.	(0.88)[b]	Hesperoperla pacifica	(-0.46)
Lepidostoma veleda	(-0.94)[b]	Hydropsyche spp.	(0.84)[c]	Epeorus sp.	(-0.39)
Helicopsyche borealis	(-0.94)[b]	Tricorythodes minutus	(0.81)[c]	Amiocentrus aspilus	(-0.26)
Glossosoma spp.	(-0.93)[b]	Asellus sp.	(0.80)[d]	Claassenia sabulosa	(-0.01)
Ephemerella grandis	(-0.91)[b]	Simulium spp.	(0.77)[d]	Ephemerella inermis	(0.12)
Psychomyia flavida	(-0.89)[b]	Parargyractis confusalis	(0.73)[d]	Oscecetis avara	(0.17)
Brachycentrus occidentalis	(-0.69)[d]	Leucotrichia pictipes	(0.72)[d]	Baetis intermedius	(0.30)
		Zumatrichia notosa	(0.70)[d]	Skwala paralella	(0.31)
				Ferrissia sp.	(0.31)
				Rhithrogena undulata	(0.34)
				Paraleptophlebia heteronea	(0.38)
				Microcyloeppus sp.	(0.53)
				Choroterpes albiannulata	(0.59)

[a] Significant to the 99% level.
[b] Significant to the 95% level.
[c] Significant to the 90% level.
[d] Significant to the 80% level.

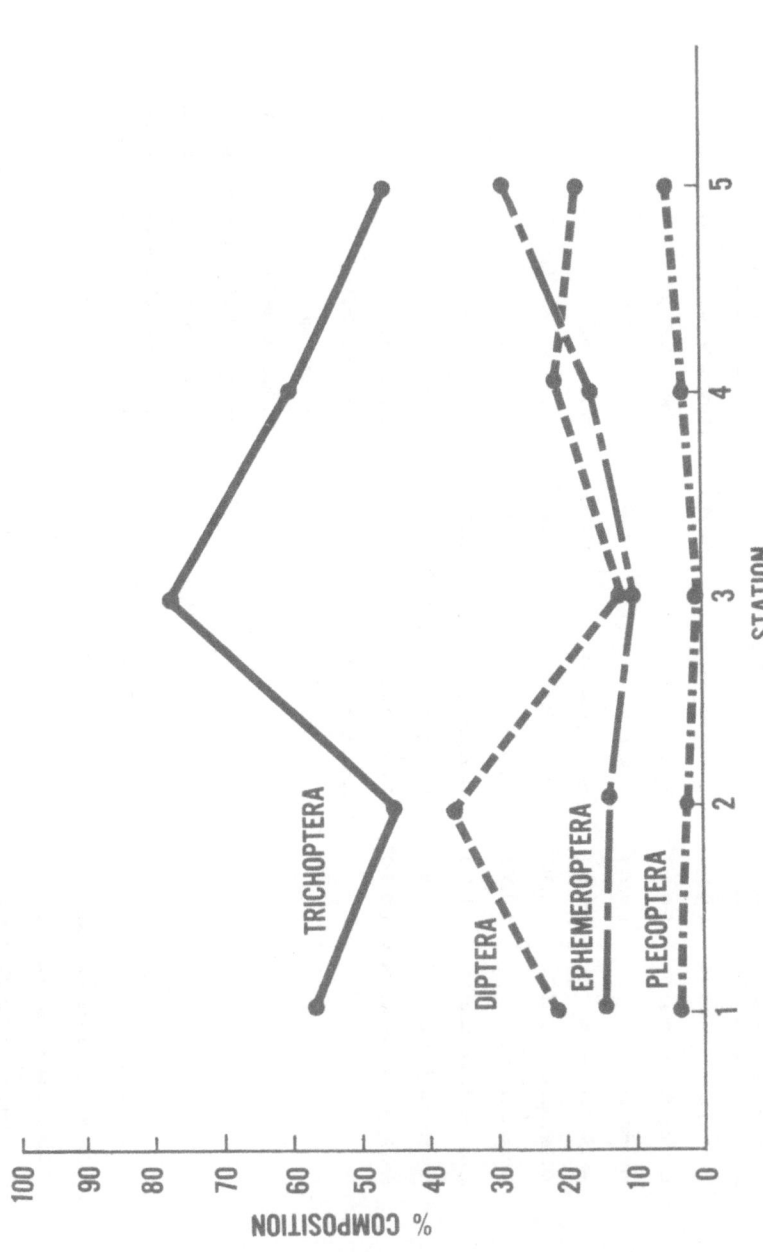

Fig. 3. Percent ordinal composition of numbers of macroinvertebrates collected on artificial substrates in the Madison River, September 1976 through September 1977.

dominated by forms previously classed as cold water and eurythermal, while the community in the lower river was dominated by forms classed as warm water and eurythermal (Table 2). At Station 3, 67% of all invertebrates collected were *Hydropsyche* spp. Ward (1976) and Hynes (1970) reported that *Hydropsyche* numerically dominated the fauna at some regulated sites. Ward (1976) reported that the presence of high numbers of collectors such as *Hydropsyche* and *Cheumatopsyche* below dams may be the result of the exclusion of invertebrate predators through thermal modifications or increased planktonic food supply from the reservoir. Coutant (1962) found large numbers of *Hydropsyche* in a heated riffle not located below a dam.

The average dry weight of macroinvertebrates per sampler at Stations 1 through 5 was 211, 368, 745, 419, and 607 mg. The mean dry weight at Station 3 was significantly greater than at Stations 1 and 2. The results of analyses of dry weights (Fig. 4) showed similar ordinal compositions at Stations 1, 2, 4, and 5. Ordinal composition at Station 3 was shifted toward higher composition of Trichoptera and lower composition of Plecoptera (P < 0.01).

Adult Aquatic Insects

Fifty species of adult aquatic insects were collected, of which 14, 11, 24, and 1 were species of Plecoptera, Ephemeroptera, Trichoptera, and Lepidoptera. Forms collected commonly in the upper river but not taken on the lower river were the plecopterans *Suwallia lineosa* and *S. pallidula*, the ephemeropterans *Rhithrogena undulata* and *Epeorus* sp., and the trichopterans *Rhyacophila coloradensis, Glossosoma montana, Arctopsyche inermis*, and *Hydropsyche jewetti*.

Forms collected commonly on the lower river but not found on the upper river were the plecopteran *Isoperla patricia*, the ephemeropteran *Tricorythodes minutus*, the trichopterans *Nectopsyche* sp., *Leucotrichia pictipes*, and *Zumatrichia notosa*, and the lepidopteran *Parargyractis confusalis*. Twenty-one species were collected on both the upper and lower river.

The adults of some species present in both sections of the river appeared earlier on the lower river than on the upper river. The plecopterans *Pteronarcys californica, Claassenia sabulosa*, and *Hesperoperla pacifica* were collected about two weeks earlier and *Isoperla fulva* appeared about one month earlier in the lower river. The trichopteran *Helicopsyche borealis* was collected about one month earlier and the ephemeropterans *Ephemerella grandis* and *E. inermis* several weeks earlier in the lower river. The earlier emergences in the lower river were probably caused by the higher temperatures that

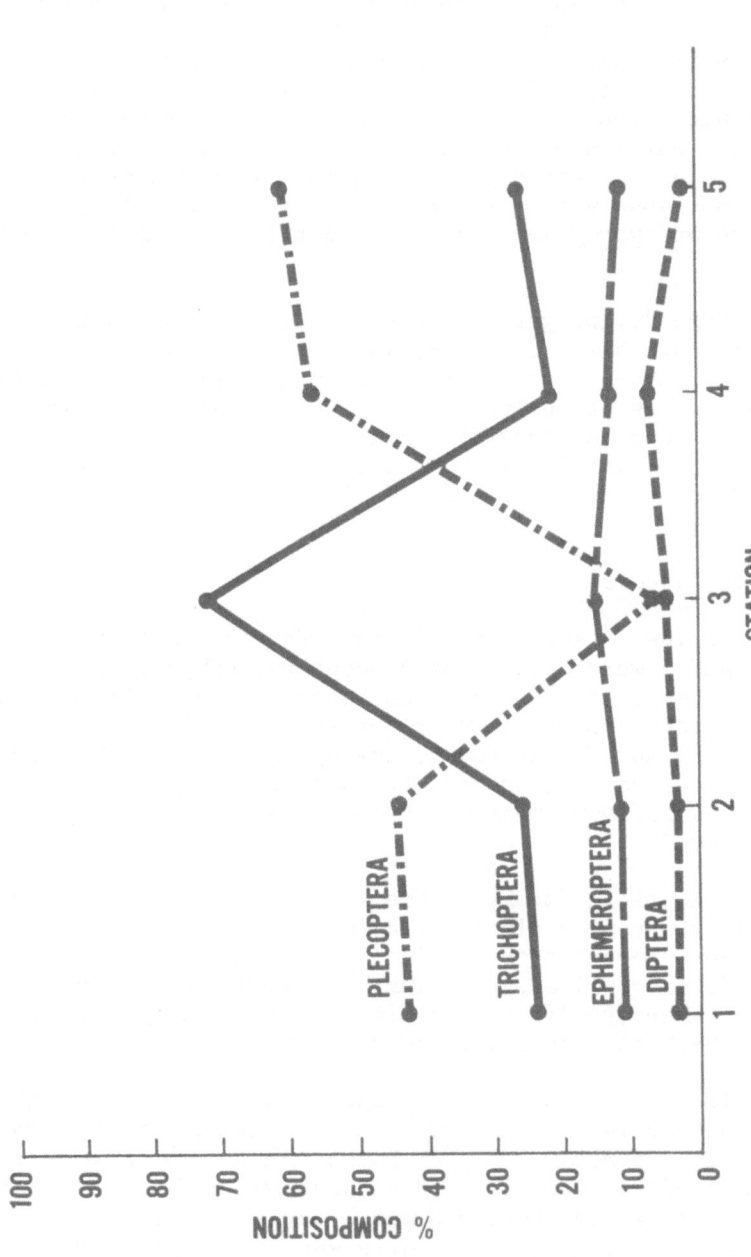

Fig. 4. Percent ordinal composition of dry weights of macroinvertebrates collected on artificial substrates in the Madison River, December 1976 and March, May, July, and September 1977.

occurred there.   Nebeker (1971a) also found that some species of
aquatic insects emerged earlier in warmer sections of a stream.
Hynes (1976) suggested that water temperature was a major factor
controlling the timing of emergence of many species of Plecoptera.
Langford (1971), however, did not find earlier emergences of
Plecoptera and Ephemeroptera in a stretch of a river warmed by
cooling water from a power plant.

   Pteronarcys californica emerged in large numbers upstream and
downstream from Ennis Reservoir, where there was no pattern of
thermal constancy; however, no emergence of this species was observed
for approximately 4 km below the reservoir, where a pattern of
thermal constancy existed.   This emergence failure probably resulted
from the lack of proper temperature cues caused by the dampening of
maximum temperatures by the reservoir during the emergence period.
Comparisons of emergence dates and temperature data at other sites
on the river from 1972 through 1978 indicate that the emergence
timing of *P. californica* is controlled by maximum temperatures
rather than by mean temperatures or temperature summation (Fraley
and Vincent, unpublished data).   Stanford (1975) reported that
although temperature summation controlled the development of nymphal
Pteronarcella badia, temperature cues appeared to time their
emergence.

   The weights of *P. californica* adults collected in 1977 from the
lower river were 13% less (P < 0.01) than those in the upper river.
Preliminary analysis of weights obtained in 1978 also showed that *P.
californica* adults weighed less in the lower river.   The lighter
weights of *P. californica* in the lower river may have resulted from
lower fat reserves caused by the higher metabolic costs required by
temperatures above optimum levels.   Fat reserves are the important
source of energy in plecopteran nymphs just before emergence
(Oberndorfer and Stewart, 1977).   Nebeker (1971b) has suggested that
food storage in *P. dorsata* was probably less in nymphs raised at
20°C than at colder temperatures, because of increased metabolic
costs.   During the seven days before the emergence of *P. californica*
in the upper river, the average maximum temperature and percent of
time over 17°C were 14.5 and zero.   During the seven days before the
emergence in the lower river, the average maximum temperature and
percent of time over 17°C was 21.0 and 39.   Sweeney and Vannote
(1978) found decreased adult weights of some aquatic insects raised
in thermal regimes above or below optimum levels.   They attributed
this phenomenon to altered nymphal growth rates and timing and rate
of adult tissue development.

CONCLUSIONS

Increased stream temperatures caused by Ennis Reservoir at
Stations 3, 4, and 5 appeared to (1) stimulate increases in invertebrate numbers, (2) shift invertebrate community dominance to warm-
water forms, and (3) cause earlier emergence of some insects.
Thermal constancy at Station 3 appeared to (1) stimulate great
increases in invertebrate numbers, (2) decrease diversity, and
(3) cause emergence failure of *Pteronarcys californica*. Thermal
modifications may have operated through direct or indirect routes to
exert these effects on the macroinvertebrates (see Ward and Stanford,
this volume). Increased planktonic food supply may have contributed
to the increase in invertebrate numbers and decrease in diversity at
Station 3.

ACKNOWLEDGMENTS

I would like to express my appreciation to those who assisted
me during this study. Dr. W. Gould directed my master's research
from which this paper was written. Drs. G. Roemhild, D. Denning,
W. Lange, and R. Newell provided identifications of selected aquatic
invertebrates. Drs. D. Burkhalter and D. Reichmuth wrote the
computer programs used to reduce and analyze the temperature data.
R. Luedtke, R. Oswald, and Dr. J. Stanford reviewed the manuscript.
This research was supported by the Montana Cooperative Fisheries
Research Unit with cooperation and equipment provided by Region 3 of
the Montana Fish and Game Department.

REFERENCES

Aagaard, F. C., 1969, "Temperature of Surface Waters in Montana,"
      U.S. Dep. Interior Geol. Surv. and Montana Fish and Game
      Commission, 613 p.
Brenda, R. S., and Proffitt, M. A., 1974, The effects of thermal
      effluents on fish and invertebrates, p. 438-447, *in*: "Thermal
      Ecology," J. W. Gibbons and R. R. Sharitz, eds., AEC
      publication.
Brett, J. R., Shelborn, J. E., and Shoop, C. T., 1969, Growth rate
      and body composition of fingerling sockeye salmon, *Oncorhynchus
      nerka*, in relation to temperature and ration size, *J. Fish.
      Res. Board Can.*, 26:2363-2394.
Brown, C. J. D., Chairman, Stream Classification Committee, 1965, "A
      Classification of Montana Fishing Streams," U.S. Fish and
      Wildl. Serv., Billings, Montana, 2 p.
Clifford, H. T., and Stephenson, W., 1975, "An Introduction to
      Numerical Classification," Acad. Press, Inc., New York,
      229 p.

Coutant, C. C., 1962, The effect of a heated water effluent upon the macroinvertebrate riffle fauna of the Delaware River, *Proc. Penn. Acad. Sci.*, 36:58-71.

Dixon, W. F., and Massey, F. J., Jr., 1969, "Introduction to Statistical Analysis," McGraw-Hill Book Co., New York, 683 p.

Fraley, J. J., 1978, "Effects of Elevated Summer Water Temperatures Below Ennis Reservoir on the Macroinvertebrates of the Madison River, Montana," M.S. Thesis, Montana State Univ., Bozeman, 120 p.

Gaufin, A. R., 1962, Environmental requirements of Plecoptera, p. 105-110, *in*: "Third Seminar on Biological Problems in Water Pollution," C. M. Tarzwell, ed.

Gaufin, A. R., and Hern, S., 1971, Laboratory studies on the tolerance of aquatic insects to heated waters, *J. Kansas Entomol. Soc.*, 44(2):240-245.

Heaton, J. R., 1961, "Temperature Study of the Madison River Drainage," Job Completion Report, Fed. Aid Proj. F-9-R-9., Job. No. 11 B., Montana Dep. Fish and Game, 10 p.

Hester, F. E., and Dendy, J. S., 1962, A multiple plate sampler for aquatic macroinvertebrates, *Trans. Am. Fish. Soc.*, 91(4):420-421.

Hynes, H. B. N., 1970, "The Ecology of Running Waters," Univ. Toronto Press, 555 p.

Hynes, H. B. N., 1976, Biology of Plecoptera, *Annu. Rev. Entomol.*, 21:135-153.

Langford, T. E., 1971, The distribution, abundance and life history of stoneflies and mayflies in a British river warmed by cooling water from a power station, *Hydrobiologia*, 38(2):339-377.

Lehmkuhl, D. M., 1972, Change in thermal regime as a cause of reduction of benthic fauna downstream of a reservoir, *J. Fish. Res. Board Can.*, 29:1329-1332.

Matney, C. E., and Garvin, W. H., 1978, "Agricultural Water Quality in the Gallatin and Madison Drainages," Rep. to the EPA by Blue Ribbons of the Big Sky Country Areawide Planning Organization, 139 p.

Nebeker, A. V., 1971a, Effect of water temperatures on nymphal feeding rate, emergence and adult longevity of the stonefly, *Pteronarcys dorsata*, *J. Kansas Entomol. Soc.*, 44:21-26.

Nebeker, A. V., 1971b, Effect of temperature at different altitudes on the emergence of aquatic insects from a single stream, *J. Kansas Entomol. Soc.*, 44:26-35.

Nebeker, A. V., and Lemke, A. E., 1968, Preliminary studies on the tolerance of aquatic insects to heated waters, *J. Kansas Entomol. Soc.*, 41(3):413-418.

Oberndorfer, R. Y., and Stewart, K. W., 1977, The life cycle of *Hydroperla crosbyi*: (Plecoptera:Perlodidae), *Great Basin Nat.*, 37(2):260-273.

Stanford, J. A., 1975, "Ecological Studies of Plecoptera in the Upper Flathead and Tobacco Rivers, Montana," Ph.D. Diss., Univ. Utah, Salt Lake City, 241 p.

Sweeney, B. W., and Vannote, R. L., 1978, Size variation and the distribution of hemimetabolous aquatic insects: Two thermal equilibrium hypotheses, *Science*, 200:444-446.

U.S. Department of Commerce, Weather Bureau, 1977, "Climatological Data," Vol. 80, 300 p.

Vincent, E. R., 1969, "Madison River Creel Census," Res. Proj. Rep., Fed. Aid Proj. F-9-R-16, Job No. I(a), Montana Dep. Fish and Game, 12 p.

Vincent, E. R., 1977, "Madison River Temperature Study," Job Prog. Rep., Fed. Aid Proj. F-9-R-25, Job No. II(a), Montana Dep. Fish and Game, 10 p.

Vincent, E. R., 1978, "Madison River Temperature Study," Job Prog. Rep., Fed. Aid Proj. F-9-R-25, Job No. II(a), Montana Dep. Fish and Game, 9 p.

Ward, J. V., 1976, Comparative limnology of differentially regulated sections of a Colorado mountain river, *Arch. Hydrobiol.*, 78(3):319-342.

HYDROBIOLOGY OF SOME REGULATED RIVERS

IN THE SOUTHWEST OF FRANCE

H. Décamps, J. Capblancq, H. Casanova, et J. N. Tourenq

Laboratoire d'Hydrobiologie et
Service de la Carte de la Végétation du C.N.R.S.
29, rue Jeanne-Marvig, 31055 Toulouse Cedex, France

INTRODUCTION

Hydroelectric development has brought about changes in many streams below dams, adding further to the disturbance caused earlier by construction of canals and weirs for navigation (Tourenq et al., 1978). The structure and function of these ecosystems stem directly from man's modifications of the water flow. The consequences to hydrobiology are being studied on several rivers in southwest France.

RIVERS STUDIED AND THE WATER FLOW

The Lot and the Truyère

The river Lot is 491 km long and has a drainage area of about 11,800 km^2, its principal tributary, the Truyère, accounting for 63% of the mean annual flow. The Lot is characterized by extremely low water levels in summer; one year in ten the August monthly flow is less than 9 m^3/sec, whereas the annual flow averages around 150 m^3/sec. The area of land upstream from the junction of these two rivers accounts for 46% of the drainage area. Between 1933 and 1962, nine hydroelectric dams were built in this area. Together, these dams represent a stocking capacity of 22 × 10^6 m^3 on the Lot (2 dams) and 538 × 10^6 m^3 on the Truyère (7 dams). Downstream from these reservoirs, in the middle and lower reaches of the Lot, is a succession of weirs, built between the 14th and 19th centuries for navigation. About one-third of these are equipped with small hydroelectric power stations. This part of the river appears,

therefore, as a series of water bodies or levels of differing size
(Fig. 1). As a result of this modification, the volume of water in
the river has been increased and the speed of flow during the
summer months reduced.

The purpose of the upstream reservoirs is to provide electricity during peak consumption hours. The flow downstream is
modified according to this use, especially when the level of the
river is low. During the low summer water level, the dams release
a base flow of 6 m^3/sec at the junction of the two rivers. During
4-6 hours each day, the Lot additionally carries a maximum of
40 m^3/sec and the Truyère 60 m^3/sec. As a result, the instantaneous
flow undergoes major fluctuations in the middle and lower reaches
of the river. Downstream the flow is further chaotically modulated
by 17 small factories located on the river's middle and lower
reaches. Fluctuations of the level caused by the release of water
from the dams are therefore made worse by further sudden changes
through regulation by the factories. Moreover, extraction of water
for various domestic, industrial, and agricultural purposes further
aggravates the severity of the low summer level.

The Lot, therefore, appears as a river with a very irregular
flow, which tends to be increased as a result of the hydroelectric
installations. This situation largely contributes to the quality

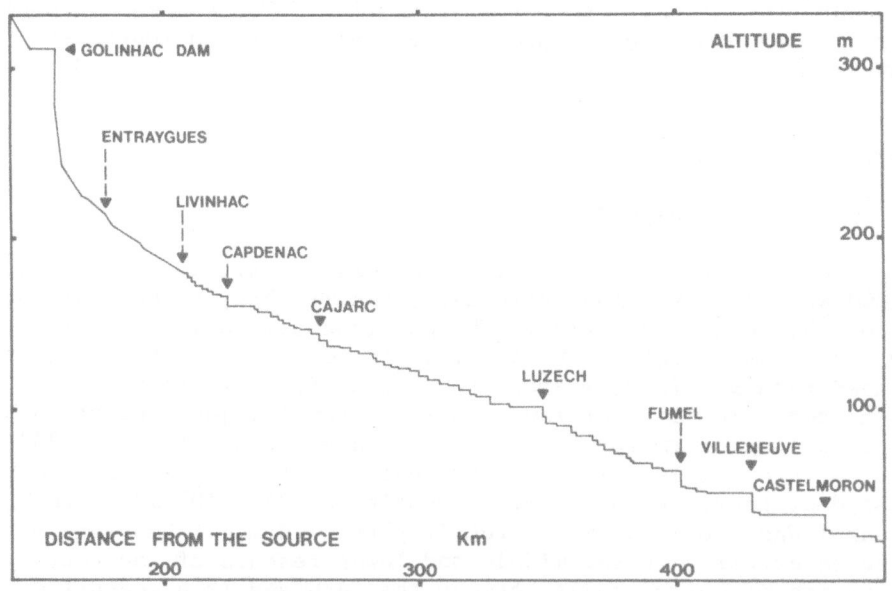

Fig. 1. Profile of the river Lot downstream from the hydroelectric
        dams (from Dauta, 1975).

of the water in the river. The chemical load is low in the upper reaches but increases downstream, where agricultural activities are more intense (Capblancq and Tourenq, 1978).

The Dordogne

The river Dordogne is almost 500 km long and has a drainage area of about 24,000 km^2. The mean flow during the year is 340 m^3/sec (reconstituted natural flow). The first dams were built on the lower reaches of the river in 1842 and 1843 and were followed by a third in 1904; these constructions stopped the migration of salmon up the Dordogne. Between 1935 and 1958, five hydroelectric dams were constructed on the upper reaches of the river and together have a stocking capacity in the region of 1×10^9 m^3. As with the first two rivers, the aim of these installations is to provide electricity according to energy requirements. As a result, downstream from this series of five dams, (1) a basic flow (reserved flow) of 5 m^3/sec remains during the low summer level, and (2) periodic use of the turbines rapidly changes the water level.

Furthermore, the dams are used to store water in winter and spring in order to fill the reservoirs during the summer. The consequences to the flow downstream can be seen clearly in Fig. 2. High spring water levels were characteristic of the period 1920–1930, before the dams were built. These high water levels cleansed the bed of the river at least some years. Since the construction of the dams upstream, the maximum spring flow never exceeds 450 m^3/sec; as a result, the sediment that accumulates during the year is no longer removed.

SUSPENDED MATTER

The river Lot was still clear a few years ago, but its turbidity has increased considerably since the dams were built. Various mineral and organic particles (algae and detritus) are carried by the river in the canalized section, the average amount varying considerably during the year (Table 1).

The problem of suspended matter in the river Lot is essentially linked to the problem of the rate of flow of the river. (1) Under natural conditions the Lot has extremely low summer water levels. (2) The construction of weirs between the 14th and 19th centuries led to the formation of a succession of artificial levels in the canalized part of the river. The 62 levels thus created have enclosed areas of sediment deposits and phytoplankton development. (3) During the 20th century the hydroelectric dams have suppressed the high spring water levels, as shown in Fig. 2, for the river

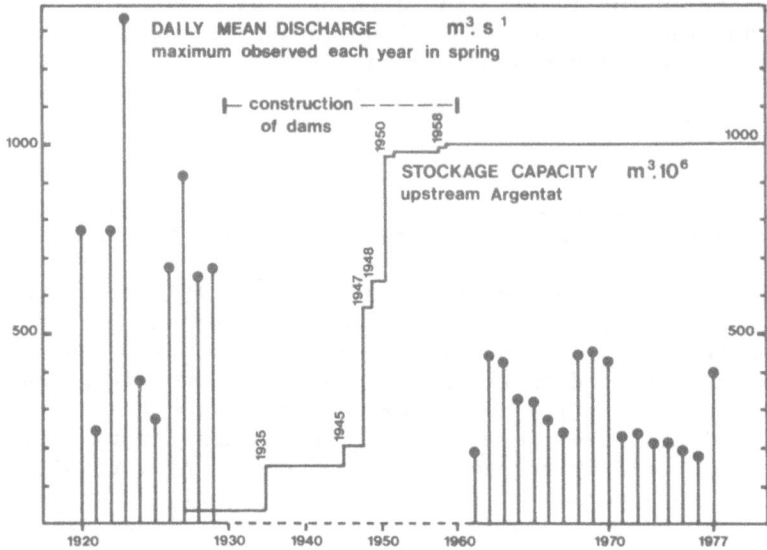

Fig. 2.   Effect of the construction of dams on the maximum daily
          mean discharge observed in spring from 1920 to 1977 at the
          town of Argentat, downstream from the series of dams built
          on the upper reaches of the river Dordogne (modified from
          Anon., 1974).

Table 1.   Average Amounts Observed at Different Dates in the
           River Lot: Mean and 95% CL, from 15 Sampling Sites
           Situated at Regular Intervals Along the Canalized
           Section of the Lot (from Décamps and Casanova-Batut,
           1978)

Date	Mineral Matter (mg/l)	Organic Matter of the Detritus (mg/l)	Algal Organic Matter (mg/l)
13 September 1974	3.23 ± 0.99	1.83 ± 0.75	0.49 ± 0.39
3 December 1974	1.40 ± 0.65	1.25 ± 0.62	0.15 ± 0.21
10 March 1975	16.74 $\frac{x}{\div}$ 1.59	3.58 ± 1.05	0.19 ± 0.24
22 July 1975	4.22 ± 1.14	2.74 ± 0.91	1.06 ± 0.57

Dordogne. In summer the dams cause an additional reduction of the flow and a discontinuous release of water. As a consequence, the finest sediments are alternately deposited and suspended (Angelier et al., 1978).

The chronic turbidity of the waters of the Lot is caused by suspended mineral and organic particles ranging in size from 1 to 20 μm. This size class of particles is composed mainly of phytoplankton and has the greatest influence on water turbidity (Table 2).

Finally, it can be seen that the flow of solid particles smaller than 20 μm increases between the upper and lower reaches of the canalized section of the Lot in the proportion of 1 to 4 for 10 m^3/sec and 1 to 2 for 16–19 m^3/sec and for 27 m^3/sec (Table 3). No increase was noted in a flow greater than 50 m^3/sec. Table 3 shows that most of this increase is in the group 2–16 μm (more particularly, 5–8 μm). These increases are caused by the proliferation of phytoplankton in the slowest-flowing waters.

PHYTOPLANKTON

Several authors have observed that water storage enhances the growth and persistence of planktonic forms (see Hynes, 1970; Oglesby et al., 1972; Whitton, 1975). The large dams on the upper

Table 2. Suspended Matter (Mean and 95% CL) in the Canalized Section of the Lot in August 1975: Distribution of Mineral and Organic Matter in the Various Size Classes and their Influence on the Turbidity (from Décamps and Casanova-Batut, 1978)

Size classes	Suspended Matter (mg/l)		Turbidity (F.T.U.)
	Mineral	Organic	
∅ > 32 μm	0.77 ± 0.18	0.58 ± 0.10	0.46 ± 0.12
32 > ∅ > 20	1.10 ± 0.17	0.73 ± 0.11	0.38 ± 0.07
20 > ∅ > 1	2.55 ± 0.62	4.21 ± 0.62	2.92 ± 0.33
1 μm > ∅	--	--	0.62 ± 0.22
Total	4.42 ± 0.83	5.52 ± 0.70	4.37 ± 0.41

Table 3. Evolution of the Flow of Solid Particles Smaller than
          20 μm (in 10^3 mm^3/sec) Between the Upper (210 km) and
          the Lower (372 km) Stretches of the Canalized Section
          of the Lot

Diameter of Particles (μm)	Flow					
	10 m^3/sec		16–19 m^3/sec		27 m^3/sec	
0–2	12[a]	8[b]	41[a]	41[b]	20[a]	24[b]
2–16	21	128	46	101	48	124
16–20	2	4	3	2	2	2
0–20	35	140	90	144	70	150

[a]Upper stretch

[b]Lower stretch

reaches of the Lot, which favor phytoplankton development, constitute
a source of inoculation for the middle and lower reaches, where, as
a result of the slower current, planktonic populations resume their
growth. The development of algae in the middle reaches of the Lot
has been studied by experimentally marking a mass of water (Capblancq
and Décamps, 1979). In a steady flow of 16 m^3/sec the volume of
phytoplankton increased about sixfold during fine weather, whereas
the volume of other matter carried by the river remained more or
less constant (Fig. 3).

Indeed, phytoplankton develops in the spring in the reservoirs
of the upper reaches of the river, some of which will be transported to the canalized section, which is composed of a series of
basins separated by weirs. This section of the river often carries
an abundance of nutrients but does not support zooplankton. Algae
can then proliferate during the summer, when the flow is reduced.
These also are essentially planktonic, originating from the reservoirs upstream, and find conditions favorable for their growth. In
all, almost 190 species and varieties have been found. More than
90% of the biomass is composed of diatoms (106 species) and
Euchlorophyceae (49 species). The ever-present diatoms account for
more than 98% of the biomass in winter and spring and just under
50% in summer. Their growth cycle is characterized by a first
pulse in spring, followed by a less substantial pulse in autumn.

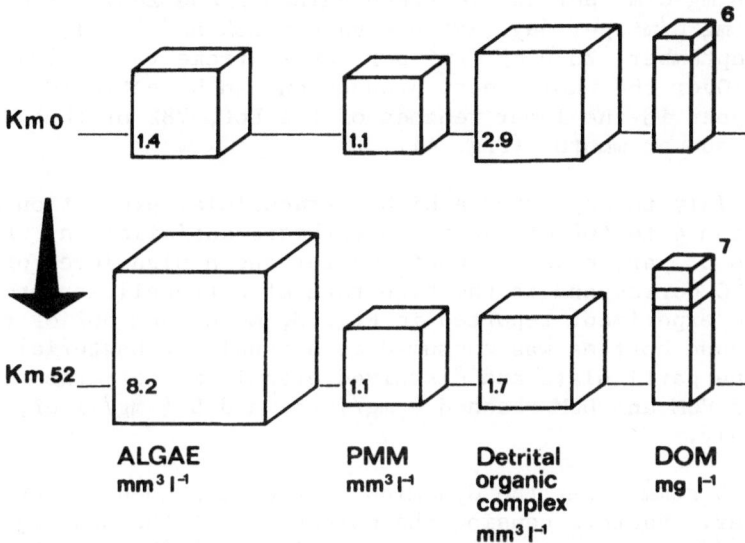

Fig. 3. Evolution of the suspended matter in a marked mass of
         water monitored over a 10-day period along 52 km of the
         canalized section of the river Lot in August 1975 (PMM:
         particulate mineral matter; DOM: dissolved organic
         matter). The increase of DOM is visualized by horizontal
         lines indicating levels of 5, 6, and 7 mg liter^{-1}.

Euchlorophyceae are abundant in August and September: The
Chlorococcales and Volvocales in the nannoplankton proliferate when
the flow is reduced. At the same time, blooms of blue-green algae
can develop, originating from the reservoirs upstream (*Microcystis,
Anabaena,* and *Aphanizomenon*).

     In spring the biomass of phytoplankton reaches 2-9 g wet
weight per cubic meter. The greatest values occur in summer in the
reservoirs upstream and in the canalized section of the river Lot,
where biomass can reach 15 g wet weight per cubic meter. Considerable variations of the mean summer biomass have been noted at the
same sampling site: 2.5, 12.6, and 5 g wet weight per cubic meter
in three successive years (Capblancq and Dauta, 1978). When the
flow is less than 10 m^3/sec, a stratification of the phytoplankton
develops, the diatoms sediment, the amount of phaeopigments increases
near the bottom, and the mobile forms--Volvocales, Chrysophyceae,
Cryptophyceae--concentrate near the surface.

     The seasonal cycle is closely linked to the rate of flow of
the water. It is characterized by low growth in winter and maximum

growth in summer. Between May and October the mean daily production is 68 mg C/m^2 per day (extreme values 2 and 245); from May to
June, 745 mg C/m^2 per day (extreme values 235 and 1377); and from
July to September, 2320 mg C/m^2 per day (extreme values 504 and
10,650). Over the whole year, production can be estimated at 275 g
C/m^2 per year in the lower reaches of the Lot, 78% of this during
the three summer months (Fig. 4).

From July to September a high extracellular production appears,
varying from 4 to 40% of the photosynthetic assimilation. Large
quantities of particulate organic matter and a high level of heterotrophic ^{14}C correspond to the high rate of extracellular production.
During the experiment reported in Fig. 3, more than 50% of the oxygen
in the opaque bottles was consumed as a result of bacterial degradation of the particulate and dissolved organic matter; the mean
amounts of POM and DOM reached 9 mg/liter and 5.4 mg/liter,
respectively.

Fig. 5 summarizes the dynamics of the suspended matter in the
river water. Factors causing the suspension of the non-algal
mineral and organic matter are in opposition to those factors
causing the decanting of the suspended matter and the increase in
algal biomass. In the water, interactions between transported
materials are numerous. Organic particles (algae and the detritus
complex) develop particularly in summer and increase in importance
as the flow is reduced. This algal development is attenuated by
the presence of suspended mineral particles. However, the products
of excretion and decomposition favor bacterial growth and contribute
to an increase in the turbidity during the low summer level. In
the river Lot, summer levels are too low, which leads to a proliferation of algae at a critical time for water quality. Indeed, it is
during the summer that problems are most acute: Stagnation occurs
when temperatures are high and sunlight at a maximum.

MACROPHYTES

Extensive beds of water crowfoot (*Ranunculus*) have developed
downstream from the dams in recent years. On the Dordogne they
occur along about 30 km of the river, starting just downstream from
the series of five dams. The first of these (Fig. 6), at the town
of Argentat, occupies an area of about 10 ha. In addition to the
dominant species *Ranunculus fluitans*, plant beds also contain
such species as *Potamogeton fluitans, Potamogeton crispus,
Myriophyllum spicatum, Callitriche hamulata, Fontinalis antipyretica,
Hygrohypnum luridum*, and *Leptodyctium riparium*. A similar situation
has developed under identical conditions in the Truyère, downstream
from a series of dams at the town of Entraygues. Here the dominant
species is a hybrid *Ranunculus fluitans* × *Ranunculus peltatus*

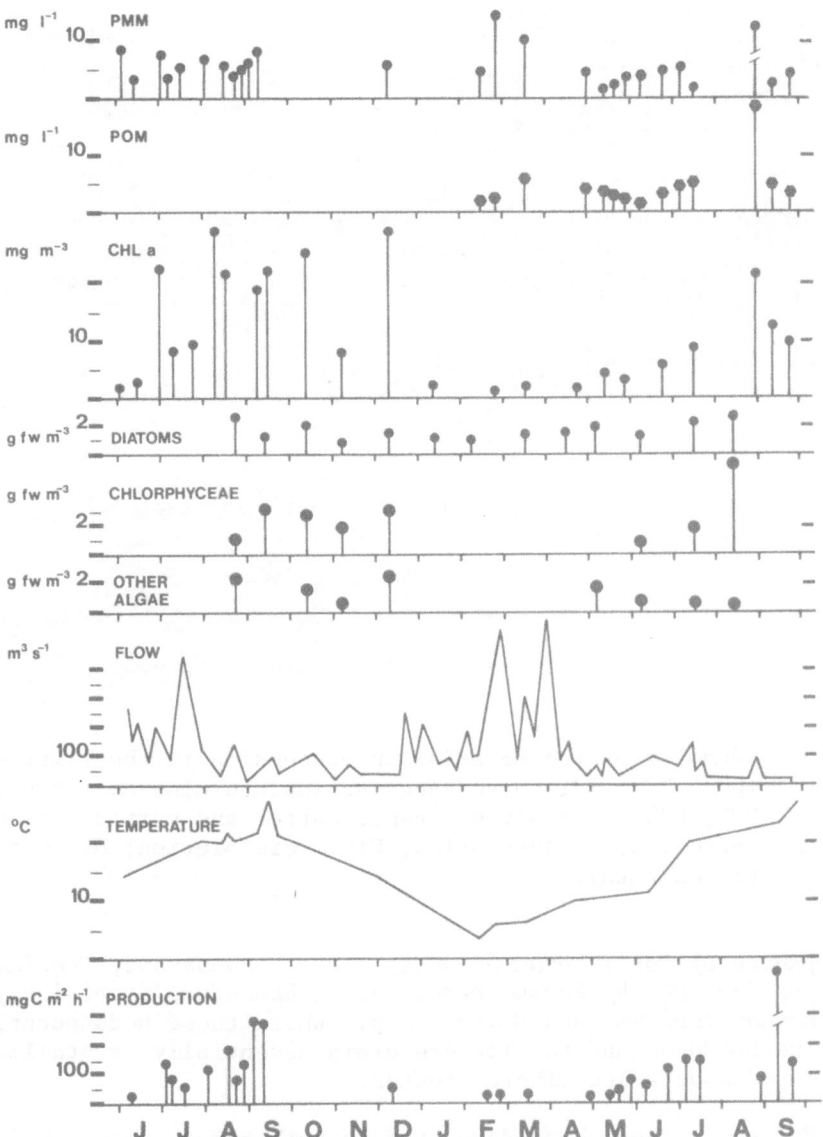

Fig. 4. Evolution of some biological and physical parameters in the river Lot during the year (modified from Capblancq and Dauta, 1978).

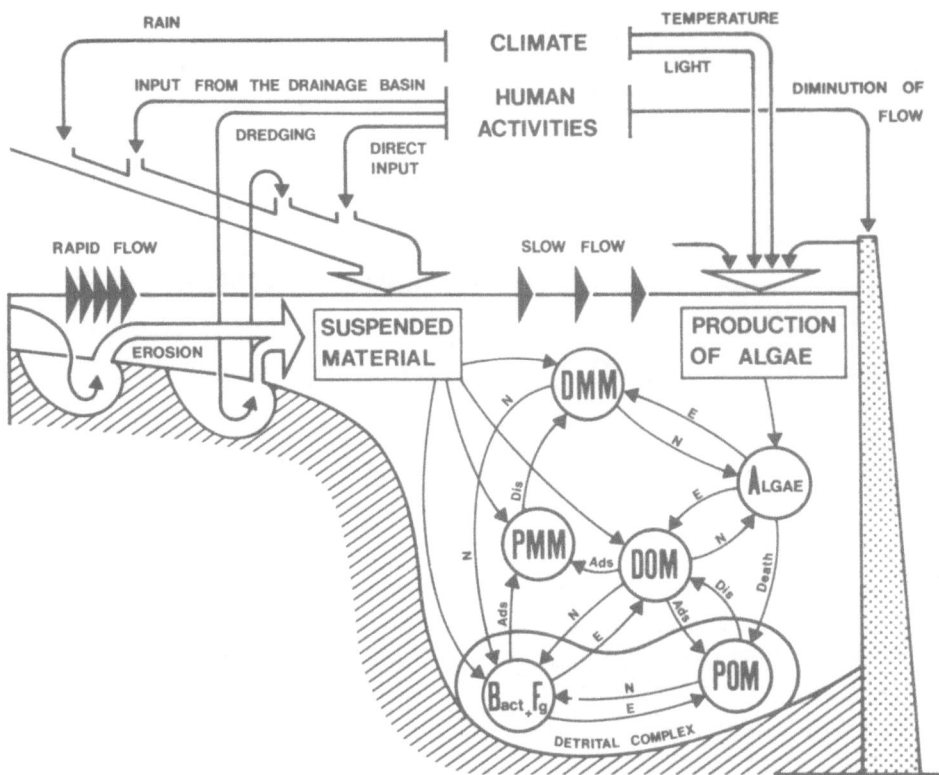

Fig. 5. Dynamics of the material in suspension in the river Lot
         (DMM, PMM: dissolved and particulate mineral matter;
         DOM, POM: dissolved organic matter and particulate organic
         matter; Ads: adsorption; Dis: dissolution; N: nutrients;
         E: excreta).

accompanied by *Potamogetum crispus, Elodea canadensis, Fontinalis
antipyretica, Brachythecium rivulare, Eurhynchium riparioides,
Hygrohypnum luridum,* and *Scapania* sp. Where these beds occur,
both the Dordogne and the Truyère drain essentially crystalline
basins and have a low mineral content.

     Ranunculus beds have developed in both rivers since the dams
were built upstream and result from a variety of factors:
(1) Terracing has led to an accumulation of sediments on the stones
of the river bed, (2) the use of the dams has entailed a reduction
of the flow during high spring water levels (Fig. 2) and, more
generally, during the low summer levels, and (3) downstream from
the dams the river is enriched in nutritive materials originating
from use of the water by the small towns through which it flows.

Fig. 6. Aerial view of a bed of *Ranunculus fluitans* in the river
Dordogne (Cl. Guillemyn).

As a result of the presence of these beds of aquatic plants
downstream from the dams, important changes have taken place in
(1) the chemical composition of the water, and (2) the invertebrate
fauna. Indeed, since these are deep-release dams, the water at the
outflow has a low oxygen content. During the day this is reoxy-
genated by the plants. Furthermore, they accumulate certain
nutrients during their growth, notably phosphorus-rich compounds
released during the decline of the plants at the end of the summer.
On an areal basis, macroinvertebrate density values may be up to 50
times greater in plant beds than in areas lacking macrophytes.
Within the beds of aquatic plants, the invertebrate populations
show a fairly homogeneous distribution according to biomass.

INVERTEBRATES

Downstream from the dams in the rivers Lot and Truyère, only a
few of those species more sensitive to water-level changes are
absent. Communities are of the lotic type, with several species
present just below the dams originating from the reservoirs upstream. The number of species of chironomids collected while
drifting reflects the influence of the dams just upstream; at the
junction of the rivers Lot and Truyère the species found drifting
belong to the lotic species in the Lot and to a mixture of the

lotic and lenitic species, with many Chironomini in the Truyère (Table 4). Downstream from the dams, even if those species more sensitive to water-level changes have been eliminated, for example, certain Plecoptera (Berthélemy and Laur, 1975), a fauna of running water is maintained, occasionally in high densities, notably in the beds of water crowfoot. The main change in the river Lot takes place in the canalized section, where the rheophilous forms give way to limnephilous species. Frequently, the bed of the canalized section shelters communities composed of more than 90% of oligochaetes and chironomid larvae.

FISH

Construction of dams on the Dordogne and the Lot have changed the structure of the fish communities.

In the Dordogne, anadromous species—*Lampetra planeri, Petromyzon marinus, Alosa alosa,* and *Salmo salar*—disappeared after the first dams were built on the lower reaches between 1842 and 1904. Downstream from the dams on the upper reaches, three species

Table 4. The Numbers of Species of Chironomids Collected Drifting in the River Lot (15 km Downstream from a Dam) and in the Truyère (2 km Downstream from a Dam) at the Junction of the Two Rivers (from Laville, in press)

	River Lot (Total Drift)	River Truyère	
Tanypodinae	4	6[a]	5[b]
Orthocladiinae	42	42	39
Chironomini	7	21	9
Tanytarsini	17	18	18
Total	70	87	71

[a]With species drifting from the reservoir upstream.

[b]Minus species drifting from the reservoir upstream.

are abundant, *Salmo trutta fario*, *Barbus barbus*, and *Anguilla anguilla*. Species such as *Scardinius erythrophtalmus*, *Chondrostoma nasus*, *Abramis brama*, and *Cobitis barbatula* occur only in small numbers and are steadily decreasing. Others, formerly abundant, have now almost completely disappeared: *Gobio gobio*, *Alburnus alburnus*, and *Ameiurus nebulosus*. The populations of *Rutilus rutilus*, *Leuciscus leuciscus*, and *Leuciscus cephalus* have also declined since the dams were built on the upper reaches of the river. Finally, species such as *Esox lucius*, *Cyprinus carpio*, and *Tinca tinca* are maintaining their numbers in the deepest parts of the stream. In all, apart from a few recent introductions (*Lucioperca lucioperca*, *Micropterus salmoides*, and *Acerina cernua*), populations of all species have declined since the dams were built on the upper reaches. This situation has been caused by (1) rapid changes of the water level in the spawning areas, (2) a lowering of the water temperature in spring and summer (because of the depth at which the water is taken from the dams), and (3) the buildup of mud on the river bed and the choking up of the interstices following the resuspension of solid materials by gravel workings.

A similar evolution can be noted in the river Lot since the period of dam construction (Fig. 7). A lowering of the water temperature caused a rapid decline in the numbers of white fish immediately below the reservoirs, the sole exception being *Leuciscus cephalus*. Here, the trout *Salmo trutta fario* has now become the dominant species, replacing *Leuciscus leuciscus*.

CONCLUSIONS

The rivers under study constitute completely transformed ecosystems, characterized by their instability and fragility. Their functioning results from the modification by man of their rate of flow and morphology. For instance, because of the operation of upstream dams, water levels are very low and therefore water may be retained in the middle and lower reaches of the Lot for a long time. In these reaches, solar radiation may produce thermal stratification (Caussade et al., 1978). It is essential, therefore, to combine ecology and hydrodynamics in order to predict the evolution of these systems and to provide the inhabitants of the valley with water resources of sufficient quantity and quality. Several objectives have been stipulated in the management plan of the Lot valley (Décamps, 1978): (1) drinking water, (2) industrial water, (3) irrigation, (4) allowances for consumption, (5) fish life, and (6) the needs of tourism and the environment. Various actions have been proposed to achieve these objectives: (1) control of domestic pollution, (2) control of industrial pollution, (3) increased water flows, (4) strict regulation of the low flows, (5) strict regulation of gravel works, (6) elimination of floating wastes, (7) management

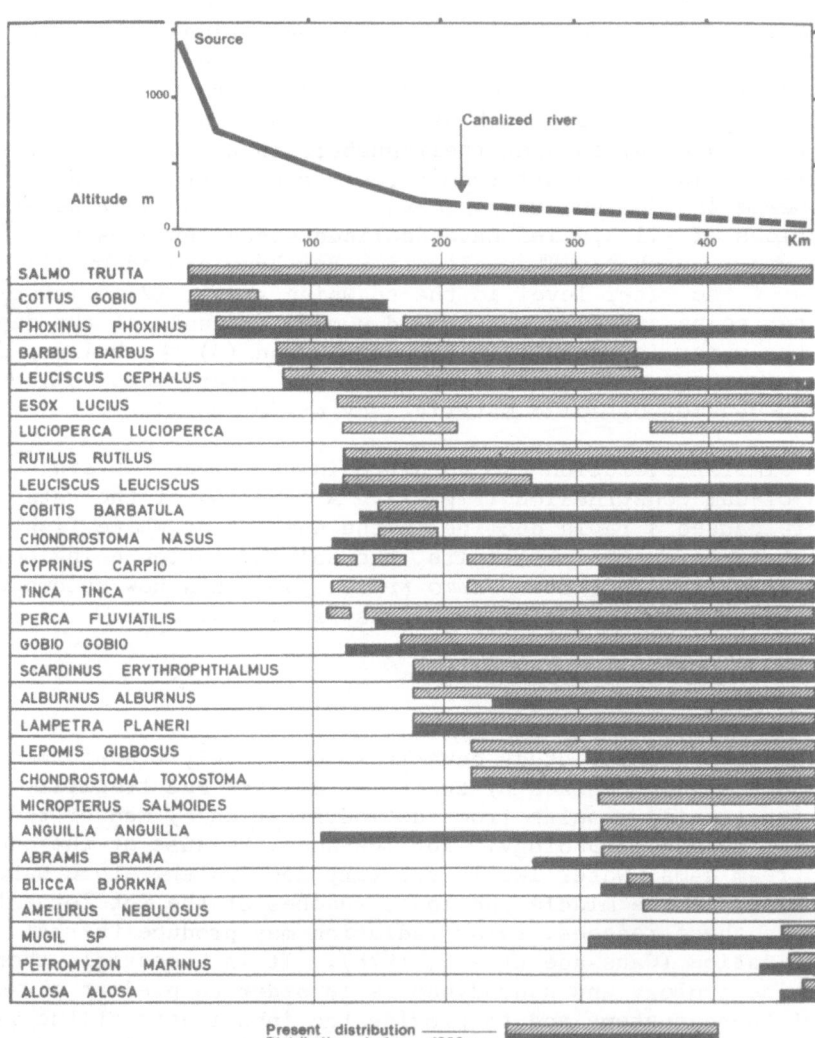

Fig. 7. Distribution of the fish fauna in the river Lot (modified
from Tourenq and Dauba, 1978).

of the banks of the Lot, and (8) general control of the management plan in collaboration with the local communities. Clearly, man's activities along the entire river basins, notably agricultural and forestry use of the land and urbanization, must be taken into account. Both the direct actions on the river itself and the indirect action on the drainage basin should be considered. We need more work linking these two aspects of the impact of man on regulated rivers.

REFERENCES

Angelier, E., Bordes, J. M., Lucchetta, J. C., and Rochard, M., 1978, Analyse statistique des paramètres physico-chimiques de la rivière Lot, *Ann. Limnol.*, 14:39-57.

Anon., 1974, Qualité des eaux de la Dordogne. "Approche de l'influence des réservoirs," Rapport Agence Bassin Adour-Garonne, p. 1-30.

Berthélemy, C., and Laur, C., 1975, Plécoptères et coléoptères aquatiques du Lot (Massif central francais), *Ann. Limnol.*, 11:263-286.

Capblancq, J., and Tourenq, J. N., 1978, Hydrochimie de la rivière Lot, *Ann. Limnol.*, 14:25-37.

Capblancq, J., and Dauta, A., 1978, Phytoplancton et production primaire de la rivière Lot, *Ann. Limnol.*, 14:85-112.

Capblancq, J., and Décamps, H., 1979, Dynamics of the phytoplankton in the river Lot, *Verh. Int. Verein. Theor. Angew. Limnol.*, 20:(in press).

Caussade, B., Chaussavoine, C., Dalmayrac, S., and Masbernat, L., 1978, Modélisation d'écosystèmes rivières: Application à un bief du Lot, *Ann. Limnol.*, 14:139-162.

Dauta, A., Décamps, H., Denat, D., Galharague, J., Giot, D., Massio, J. C., and Trolet, J. L., 1975, Evolution des matières en suspension et de la turbidité dans une masse d'eau marquée, *Rapport B.R.G.M.*, 239:1-31.

Dauta, A., 1975, Etude du phytoplancton du Lot, *Ann. Limnol.*, 11:219-238.

Décamps, H., 1978, Qualité des eaux et développement de la vallée du Lot, *Ann. Limnol.*, 14:163-179.

Décamps, H., and Casanova-Batut, T., 1978, Les matières en suspension et la turbidité de l'eau dans la rivière Lot, *Ann. Limnol.*, 14:59-84.

Hynes, H. B. N., 1970, "The Ecology of Running Waters," Liverpool Univ. Press.

Laville, H., 1979, Etude de la dérive des exuvies nymphales de Chironomides au niveau du confluent Lot-Truyère, *Ann. Limnol.*, (in press).

Oglesby, R. T., Carlson, C. A., and McCann, J. A., eds., 1972, "River Ecology and Man," Acad. Press, London.

Tourenq, J. N., Capblancq, J., and Casanova, H., 1978, Bassin
    versant et hydrologie de la rivière Lot, *Ann. Limnol.*,
    14:9-24.
Tourenq, J. N., and Dauba, F., 1978, Transformation de la faune des
    poissons dans la rivière Lot, *Ann. Limnol.*, 14:133-138.
Whitton, B. A., ed., 1975, "River Ecology," Blackwell, London.

EFFECTS OF TVA IMPOUNDMENTS ON DOWNSTREAM WATER QUALITY AND BIOTA

Peter A. Krenkel*, G. Fred Lee**, and R. Anne Jones**

*Water Resources Center, Desert Research Institute
Reno, Nevada; **Department of Civil Engineering, Colorado
State University, Fort Collins, Colorado 80523

INTRODUCTION

The Tennessee Valley Authority (TVA), a corporate agency of the
United States Government, was established in 1933 to develop the
resources of the Tennessee Valley. The Tennessee River is an integral part of the interconnected Inland Waterway System of the
United States, the final portion being the Tennessee-Tombigbee
Waterway, presently under construction. The 106,000-square-kilometer
watershed, shown in Fig. 1, encompasses portions of seven southeastern states with a population near 4 million people.

In the 1930s, the TVA area was subject to floods that claimed
many lives and inflicted severe property damage. In addition, navigation and commerce were restricted during periods of low flow.
The TVA system of nine large, multiple-purpose dams on the main
stem of the Tennessee River brought flood control and guaranteed
depths for navigation. Supplemented by the operation of multi-
purpose dams on the major Tennessee River tributaries, the TVA
reservoir system made the shoreline safe for industries and jobs
and has helped to restore thousands of hectares of agricultural
lands to productivity. The growth of the TVA system has not been
without adverse ecological effects, however.

A prime objective of the agency was to produce power at the
lowest possible cost. Beginning with low-cost hydroelectric power,
TVA is now the largest producer of electric power in the U.S.,
using a mixed system of 29 hydro plants, 12 coal-fired plants,
7 nuclear plants (operating, planned, or under construction), 48
gas turbines, 12 industrial dams, and 8 U.S. Army Corps of Engineers
dams.

Although the three principal purposes of the TVA reservoir
system (flood control, navigation, and power production) remain
unchanged, new demands for water have complicated operation of the
system. Recreation, municipal and industrial water supplies,
wastewater treatment, vector control, aquatic life, and other factors
must all be considered in managing the system. Another factor adding
to the complexity and cost of managing the TVA system is the growing
concern for protecting the quality of the environment that is reflected in increasingly stringent environmental legislation.

EFFECTS OF IMPOUNDMENTS ON DOWNSTREAM WATER QUALITY

Of prime interest in this discussion are the water quality
changes that result from impounding a free-flowing river, in particular as they affect downstream biota. Inasmuch as the mainstem
dams on the Tennessee River mostly discharge into backwaters, the
emphasis in this discussion will be on the tributary dams in eastern
Tennessee, shown in Fig. 1. As will be demonstrated subsequently,
the impoundment of water may have significant effects on water
quality parameters that are vital to aquatic life, the most drastic
changes being in temperature and dissolved oxygen.

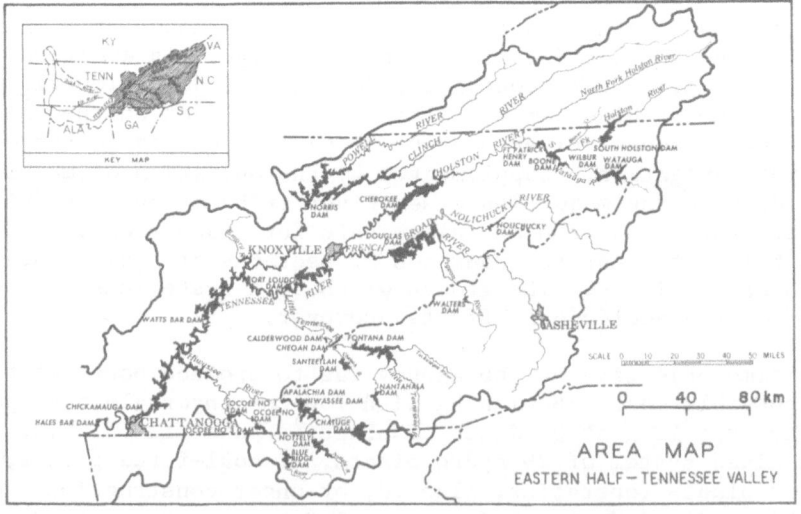

Fig. 1. Area map of TVA drainage basin (Churchill, 1964).

Thermal Structure in Reservoirs

The impoundment of a river usually markedly affects the thermal characteristics of the impounded water and of the water downstream from the dam. The resulting thermal structure of the impounded water can markedly affect the significance to water quality of certain contaminants, such as aquatic plant nutrients, within the reservoir as well as downstream from it. At the end of a winter season, the impounded water is usually of rather uniform quality and has a relatively low temperature. As summer approaches, the surface water and the incoming water temperatures are increased. The less dense water tends to "float" on the colder, higher density water already in the reservoirs. As a result of this phenomena, two strata are formed: the epilimnion, or surface stratum, and the hypolimnion, or lower stratum, which are separated by the metalimnion, a transition zone of rapid temperature-density change with depth. These conditions may exist until autumn, when the reservoir begins to exhibit a net heat loss. As the surface water becomes cooler and more dense, the metalimnion sinks, unstable conditions occur, and the reservoir mixes (overturns), returning to the conditions found at the beginning of the season.

These strata may possess different water quality characteristics. The water quality of the discharge from the dam during stratification will obviously depend on the depth of the discharge. A primary factor governing the water quality characteristics of the epilimnion and hypolimnion is the areal plant nutrient load. These loads (usually nitrogen and phosphorus) typically control the phytoplankton production in a reservoir. Vollenweider (1976), Lee et al. (1978), and Rast and Lee (1978) have demonstrated that for a wide variety of waterbodies in various parts of the world, phytoplankton production is usually correlated with the phosphorus load, normalized by mean depth and hydraulic residence time (filling time). Newbry et al. (1979) have recently demonstrated that the nutrient load-eutrophication response relationships developed by Rast and Lee (1978) for approximately forty U.S. waterbodies, based on Vollenweider's load-response model, is applicable to the fourteen TVA impoundments evaluated by Newbry et al. (1979). All TVA impoundments evaluated were found to have P load-hydrological-morphological characteristics typical of eutrophic waterbodies and would therefore likely have one or more of their beneficial uses impaired.

Probably the most important impact of the thermal stratification that occurs in many reservoirs is the inhibition of mass transfer between the hypolimnion and epilimnion. Oxygen depleted in the hypolimnion by oxygen-demanding materials, such as dead phytoplankton, is not replaced from the oxygen-rich epilimnion, and the water quality of the hypolimnion deteriorates. Rast and Lee (1978) and Lee et al. (1978) have shown that oxygen depletion in the hypolimnion can be correlated with the areal P load to the waterbody as normalized by the mean depth and hydraulic residence time. Therefore, the excessive P load to such a waterbody as a TVA impoundment may have a significant adverse effect on the water quality within the impoundment and, most important for this review, in the water downstream from it.

The thermal structure of an impoundment and the temperature of the inflowing waters may cause density currents within the impoundment. This appears to be prevalent in certain TVA impoundments. As shown in Fig. 2, the density currents may appear in three ways: the overflow, where the incoming water is lighter than the receiving water; the underflow, where the incoming water is heavier than the receiving water; and the interflow, where the incoming water has a density intermediate to that of a stratified-flow regime. Each of these density flows occurs in the TVA system and, as discussed in subsequent sections, may have different effects on downstream water quality.

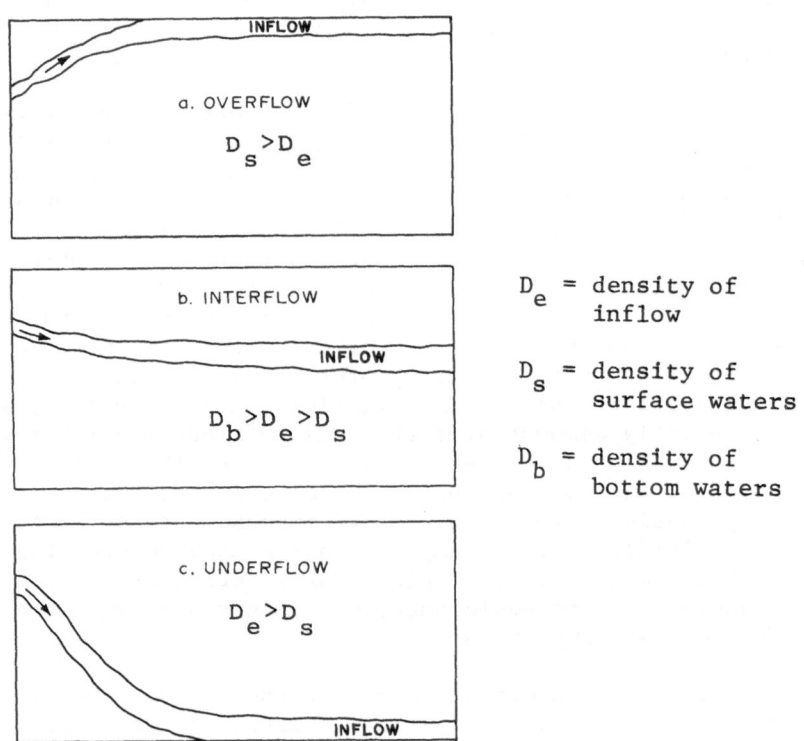

D_e = density of inflow

D_s = density of surface waters

D_b = density of bottom waters

Fig. 2. The forms of density currents (Parker and Krenkel, 1969).

Generally, all reservoirs in the TVA system discharge water from the hypolimnion rather than from the epilimnion in order to increase hydro-power efficiency. The mainstream reservoirs are sufficiently closely spaced that the discharge of one reservoir is into the backwaters of the next. This creates a density flow, similar to that depicted in Fig. 3, in which there is essentially no free-flowing river between reservoirs. Another, more typical, situation is when there are a number of kilometers of free-flowing river down-

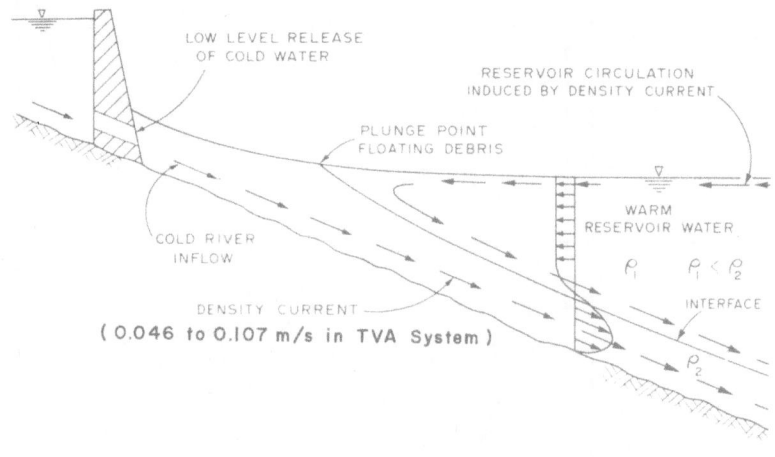

ρ_1 = density of surface water

ρ_2 = density of inflowing water

Fig. 3. Density-current underflow in a reservoir (Parker and Krenkel, 1969).

stream. In this situation an opportunity exists for establishing a riverine ecosystem and also for heating the reservoir discharge waters in summer. This heating could affect the formation of density currents in the downstream impoundment.

Effects of Thermal Structure on Water Quality Characteristics

A typical water characteristic profile for a eutrophic reservoir under stratified conditions is shown in Fig. 4. In addition to increased concentrations of iron and manganese in anoxic hypolimnia, many reservoirs show increased concentrations of sulfides in deep waters. The concentration of sulfides can be sufficiently high in

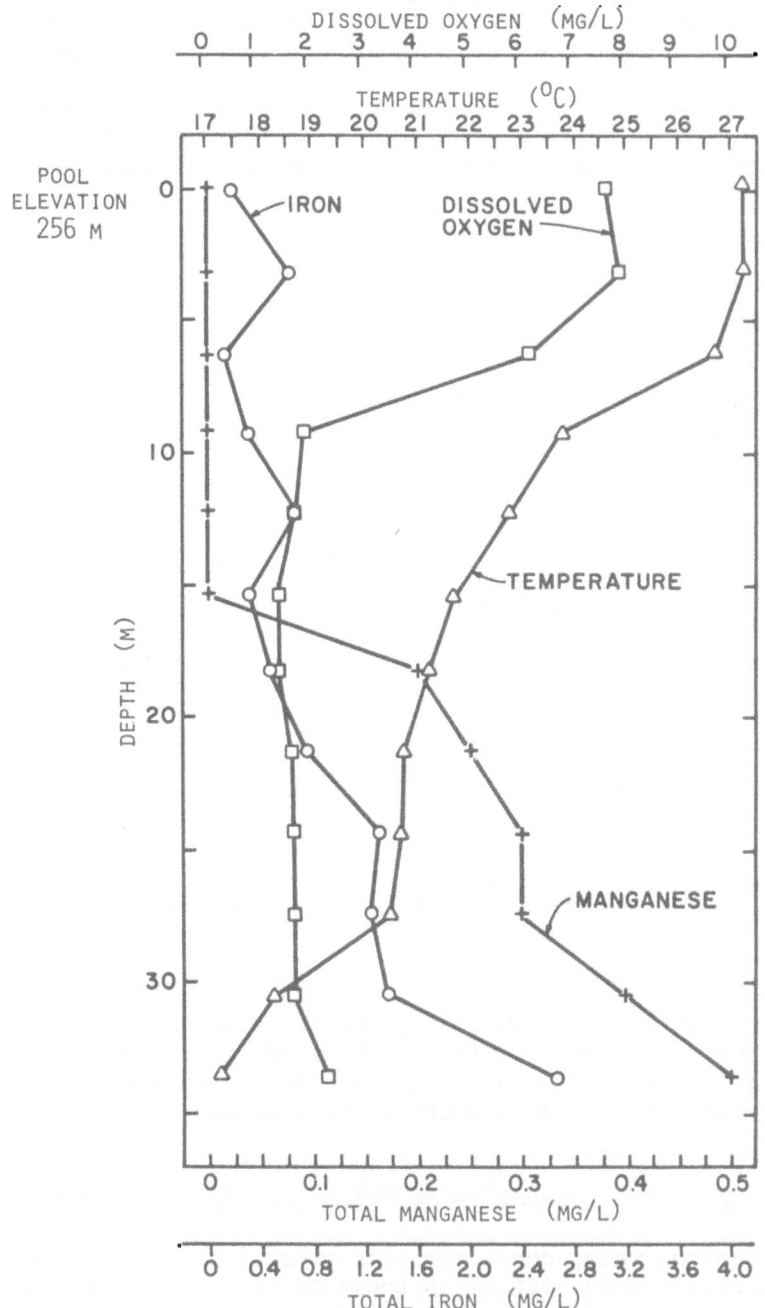

Fig. 4. Water quality characteristics profiles under stratified
conditions (adapted from Parker and Krenkel, 1969).

some reservoirs to cause large releases of sulfides to the atmosphere
downstream from the dam. This does not, however, appear to be a
problem associated with TVA reservoirs.

In hardwater systems, thermal stratification can promote a significant increase in the calcium concentration in hypolimnetic waters.
Precipitation of $CaCO_3$ in the surface waters (caused by increased pH
arising from photosynthesis and the loss of CO_2 resulting from elevated temperatures) tend to soften surface waters. The lower-pH,
colder, hypolimnetic waters tend to dissolve $CaCO_3$ particles that
rain down from the epilimnion as well as particles in the sediment.
There is evidence for this phenomenon occuring in Cherokee Reservoir
of the TVA system. For further discussion of the dynamics of $CaCO_3$
in freshwater systems, consult Morton and Lee (1968) and Lee and
Delfino (1969).

The importance of the location of the intake for hydro-power
generation (reservoir discharge) is demonstrated by comparing Fig.
4 with Fig. 5, which shows the process of selective withdrawal.

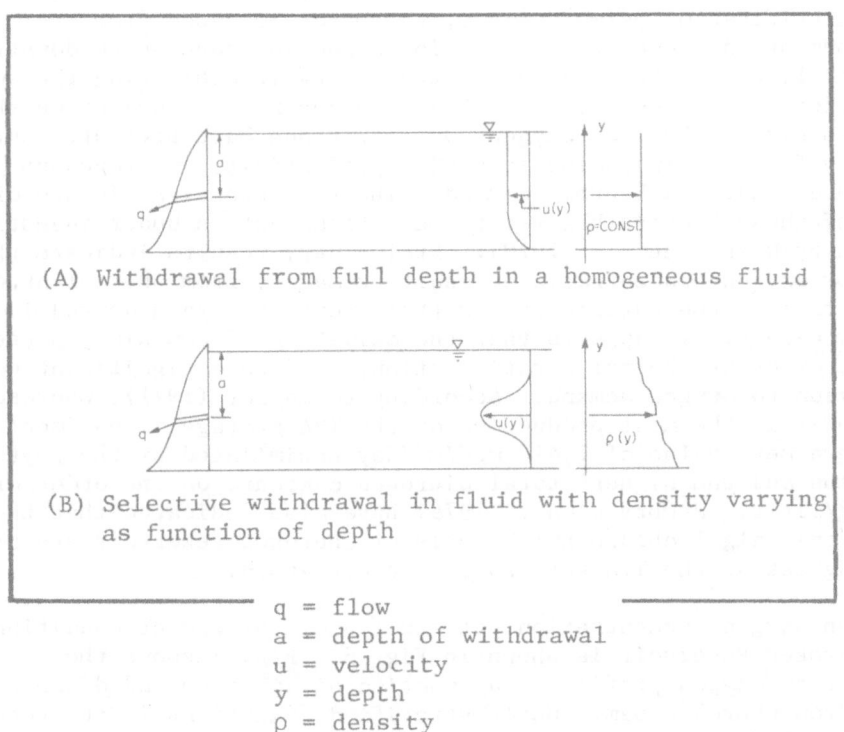

(A) Withdrawal from full depth in a homogeneous fluid

(B) Selective withdrawal in fluid with density varying
     as function of depth

q = flow
a = depth of withdrawal
u = velocity
y = depth
ρ = density

Fig. 5. The process of selective withdrawal (Elder and Garrison,
         1964).

Under stratified conditions, waters tend to be withdrawn from a
relatively narrow layer of approximately constant density. It is
possible that as a result of density currents within the reservoir,
waters may pass through the reservoir in relatively short periods
of time without ever completely mixing with the epilimnetic or
hypolimnetic waters of the waterbody. Under these conditions, the
downstream thermal and chemical characteristics would be similar to
those of water entering the reservoir. It is important to note
that density currents within waterbodies may occur for relatively
short periods of time. In Cherokee Reservoir in the TVA system,
they appear to be a significant feature in the reservoir only
during late summer and early fall. Investigators of water quality
downstream from a reservoir must be fully cognizant of the water
quality, physical-hydrodynamic characteristics, and the operation
of the reservoir.

Dissolved Oxygen

Because of the previously mentioned processes, the dissolved
oxygen concentration in the hypolimnion may become exhausted during
the stratification period. Thus, the water released from the
impoundment may be extremely low in oxygen and adverse to downstream
aquatic life. In the TVA system the reservoir exhibiting the worst
conditions regarding oxygen is Cherokee Reservoir, located on the
Holston River. The low oxygen concentrations have been attributed
to many factors, including excessive productivity, nitrogenous
oxygen demand, and benthic demand. The relative significance of
each of these factors has yet to be defined but is under investigation by Newbry et al. (1979). Preliminary results indicate that
the low oxygen concentrations result primarily from phytoplankton
production in the surface waters that exert an oxygen demand in the
bottom waters. It appears that the oxidation of ammonium, present
at the onset of thermal stratification, is also a significant contribution to oxygen demand. According to Taylor (1971), Cherokee
Reservoir is the most productive of the TVA storage impoundments,
having a mean value of 1,416 $mgC/m^2/day$ assimilated by the phytoplankton and the highest total nitrogen content, on the order of
2.7 mg/liter. Newbry et al. (1979) have also indicated that the
planktonic algal chlorophyll levels of Cherokee Reservoir are among
the highest of the TVA waterbodies investigated.

An oxygen concentration pattern typical of summer conditions
at Cherokee Reservoir is shown in Fig. 6. Fig. 7 shows the
downstream oxygen profile as a function of distance and discharge
rate from Cherokee Dam. Under stratified conditions in the reservoir,
the dissolved oxygen content may be below the US EPA-recommended
criterion of 5 mg/liter for a warm-water fishery for as much as 50 km
downstream.

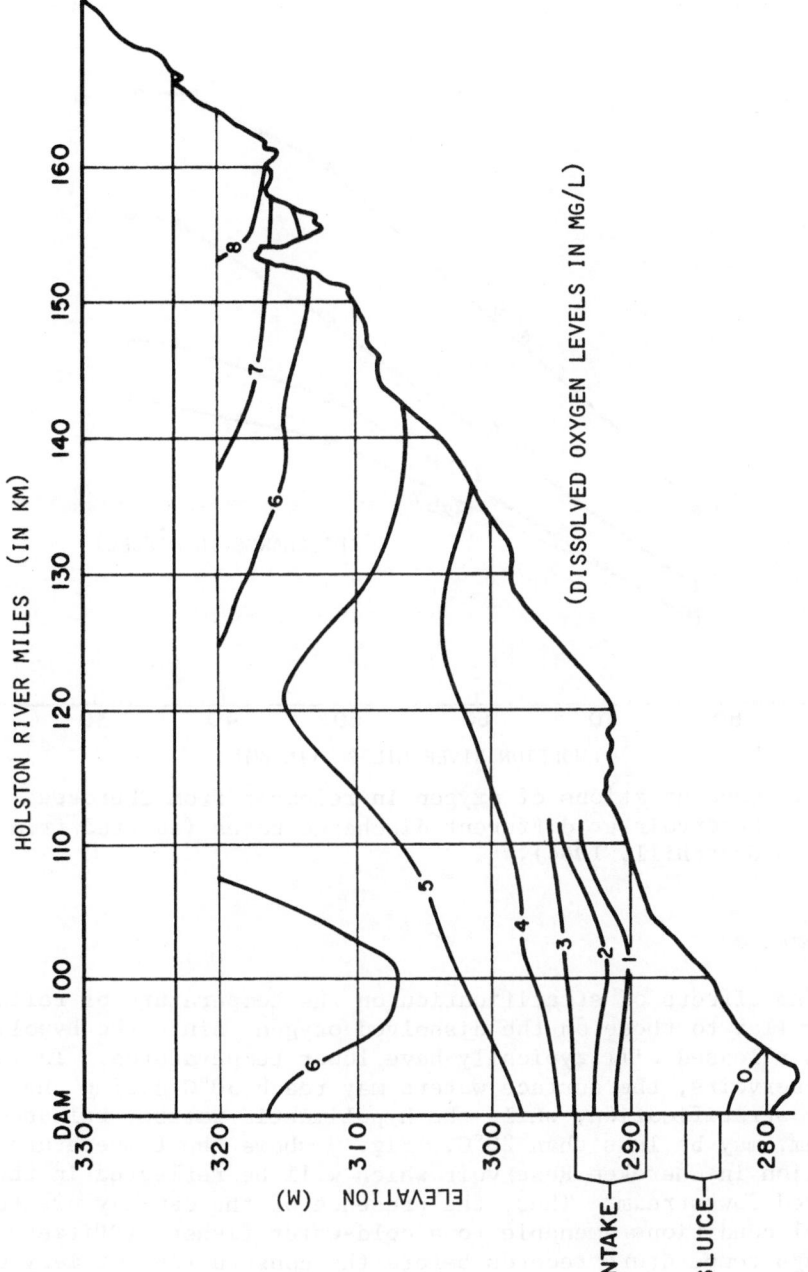

Fig. 6. Oxygen levels in Cherokee Reservoir (adapted from Churchill, 1964).

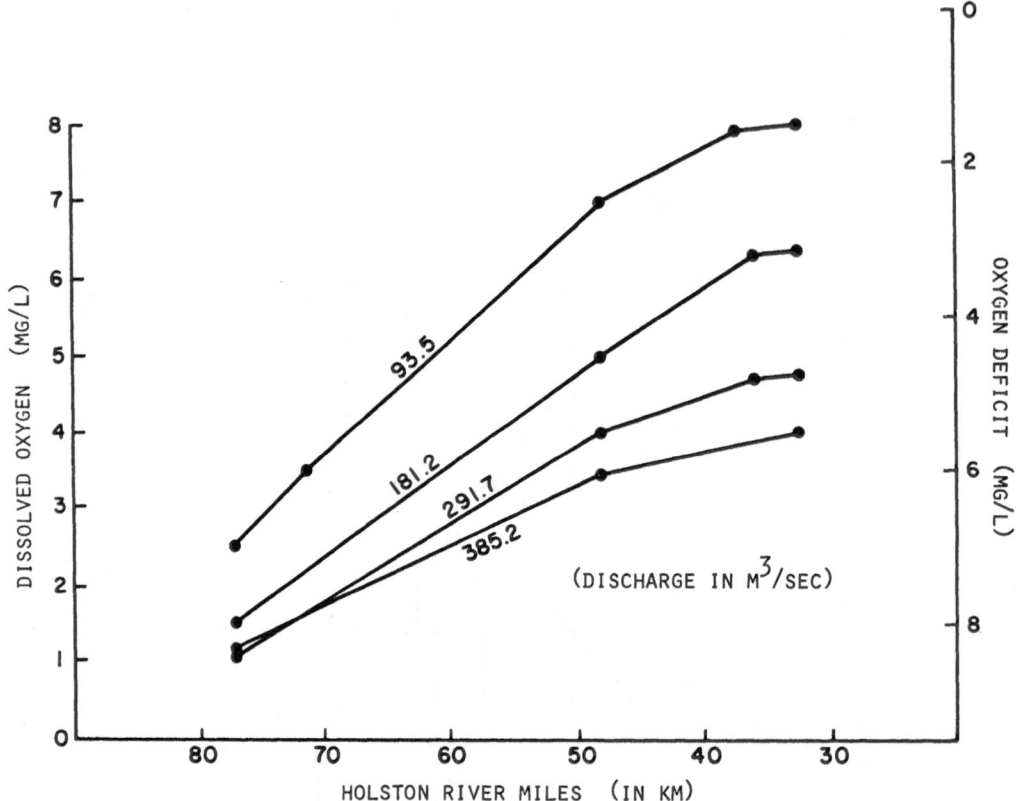

Fig. 7. Concentrations of oxygen in releases from Cherokee
         Reservoir at different discharge rates (adapted from
         Churchill, 1964).

Temperature

     The effects of stratification on the temperature of releases
are similar to those on the dissolved oxygen, since the hypolimnetic
waters released will typically have lower temperatures. In many
TVA reservoirs, the surface waters may reach 30°C during the
summer stratification, while the hypolimnetic waters, released from
the dam, may be less than 20°C. Fig. 8 shows the temperature distribution in Cherokee Reservoir which will be reflected in the water
released downstream. Thus, the presence of the dam may create
thermal conditions amenable to a cold-water fishery (Pfitzer, 1962).
Although temperature records before the construction of many of the
TVA dams are sparse, data reported by Krenkel and Novotny (1973)

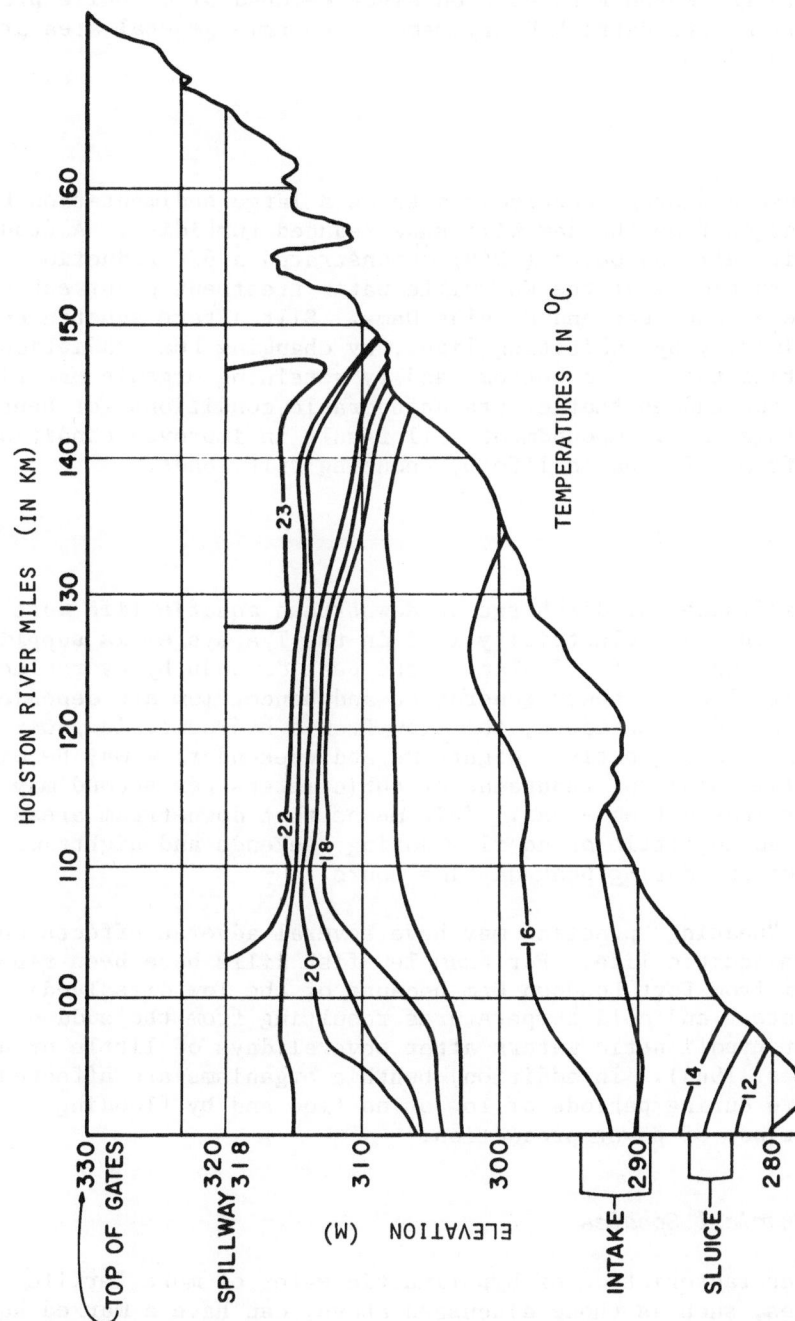

Fig. 8. Water temperatures in Cherokee Reservoir, 6 June 1945 (adapted from Churchill, 1964).

indicate that before the closure of upstream dams, the natural temperatures in the North Fork Holston River reached 31°C, while present
releases from Fort Patrick Henry Dam in the same general area are on
the order of 20°C.

Turbidity

Because a storage reservoir acts as a large sedimentation basin,
the discharges from the dam will show reduced turbidity. A study
done by Kittrell and Quinn (1949) demonstrated a 61% reduction
in annual turbidity at the Knoxville water treatment plant subsequent
to closure of Cherokee and Douglas Dams. Silt alters aquatic environments, primarily by inhibiting light, by changing heat radiation,
by blanketing the stream bottom, and by retaining organic material
and other substances that create unfavorable conditions for benthos.
Thus, in this case, impoundment will result in improved conditions
for some forms of aquatic life by reducing silt loads.

Flow

The influence of discharge on downstream aquatic life must
be noted. The base electricity load in the TVA system is supplied
by coal- and nuclear-fired plants, the peak demands by hydroelectric
power; hydroelectric power generation and hence flow are dependent
on power demand. Therefore, the flow from a hydroeletric power
facility may vary greatly. Nighttime and weekend flow may be minimal,
while daytime peaks of thousands of cubic meters per second may be
suddenly released from a dam. This means that downstream areas may
be subjected to little or no flow during weekends and nighttime hours
and to flooding during peak daytime hours.

This "peaking" practice may have several adverse effects on
downstream aquatic life. For example, fish kills have been reported
downstream from Fort Loudoun Dam because of the low dissolved-
oxygen content and cold temperatures resulting from the sudden
release of hypolimnetic waters after several days of little or no
flow (Jones, 1964). In addition, benthic organisms are affected
by exposure during periods of low or no flow and by flooding
during periods of power production.

Reduced Chemical Species

The characteristics of hypolimnetic water of more fertile
waterbodies, such as those discussed above, can have a marked adverse
effect on downstream water quality. The buildup of large concentrations of reduced chemical species (e.g., iron, manganese, and

sulfur) increases the total oxygen demand of hypolimnetic waters, which, for impoundments that discharge water from below the thermocline in the summer, increases the time-distance over which a balanced aquatic-organism population is not possible. Furthermore, oxygenation of iron and manganese species leads to the formation of insoluble hydrous metal oxides, which may form benthic deposits and coatings within the river downstream from the reservoir. Although not adequately studied, such deposits could significantly affect the numbers and types of organisms present in the river. In addition, because hydrous metal oxides tend to have high sorptive capacity for trace metals and many organics (Lee, 1975), the numbers and types of organisms could be changed as a result of changes in the trace element-compound composition of the water. In general, insufficient attention has been given to the role of aquatic plant nutrients in influencing biological and chemical characteristics of a river downstream from an impoundment. The OECD eutrophication modeling approach (Rast and Lee, 1978) has made significant advances in predicting changes in eutrophication-related water quality. Work needs to be done, however, to relate the nutrient load to a waterbody to the downstream water quality for those streams fed by hypolimnetic waters.

Other Water-Quality Parameters

The free CO_2 concentration in hypolimnetic waters is usually high because of bacterial activity, colder temperatures, and the barrier to gas exchange with the atmosphere. Although the pH of the hypolimnion is frequently lower than that of the epilimnion, the alkalinity is usually the same as that of the inflowing water. However, in hardwater systems, dissolution of calcium carbonate takes place in the hypolimnion, because of the presence of free CO_2, causing an increase in alkalinity.

Coliform bacteria, an indication of potential bacterial contamination, is usually reduced by impoundment. Other water quality parameters, if not changed by impoundments, are modified by storage and dilution, so that their concentrations in the discharge tend to be more uniform than in unimpounded waters.

EFFECTS OF IMPOUNDMENTS ON AQUATIC BIOTA

The changes in water quality caused by impoundment may have significant effects on downstream ecological systems. Although Wiebe (1958) stated that the TVA fishery is now at least 50 times that of the unimpounded river, Pfitzer (1962) noted that the construction of Cherokee Dam resulted in the loss of 52 miles (84 km) of very attractive and valuable stream as a fishery. Unfortunately,

little data are available from TVA to document the conditions of
fisheries downstream from impoundments.

Aquatic Macrophytes

The importance of aquatic plants can be demonstrated by
noting that the dissolved oxygen concentration in the Holston River
would fall below 3 mg/liter, even if no wastewater discharges were
present, because of the presence of aquatic macrophytes (Anon., 1973).
It should be noted that these plants, primarily *Potamogeton*, were
not prolific before construction of several upstream impoundments.
The presence of increased sediment because of lower flow velocities
and smaller depths apparently account for the increased density of
the plants.

The effect of aquatic plants on dissolved oxygen is a function
of the plant density and distribution, plant species, light intensity, water depth, turbidity, temperature, and oxygen concentration.
Meyer et al. (1943) showed that an increase in the depth of immersion
of aquatic plants reduced their photosynthetic activity; however, the
rate of apparent photosynthesis decreases less rapidly than light
intensity with depth of immersion. Peltier (1970) summarized
work at four locations and could not find relationships between
nitrogen and phosphorus in the water and relative macrophyte abundance.

There is no doubt that the construction and operation of the
TVA impoundments has exacerbated the proliferation of aquatic plants
in the system. Inasmuch as the oxygen demand of these plants may
be significant, it is imperative that the factors controlling the
growth of aquatic macrophytes be delineated. For example, will
limiting the phosphorus and nitrogen inputs to the reservoirs inhibit their growth?

Benthic Organisms

Tarzwell (1939) investigated the benthos of the Clinch River
below Norris Dam, the first high dam with low-level discharges in
the TVA system, shortly after water was impounded. Despite this
early beginning, relatively few detailed data on benthos of TVA
tailwaters have been published.

Pfitzer (1954) described the bottom fauna of the South Holston
tailwaters, which he considered typical of other established tailwaters in the Tennessee Valley. Compared to the benthos in the
unaltered river, tailwater benthos exhibited greater numbers and
volume. Numerical increases were attributed to large numbers of

simuliids, chironomids, *Gammarus* and *Hydropsyche*; increases in
volume, primarily to large numbers of snails. Benthic forms that
were reduced or eliminated included stoneflies *(Acroneuria internata,
Phasganophora,* and *Taeniopteryx nivalis)*, ephemerid mayflies, and
the hellgrammite *Corydalus cornutus.*

More recently, Isom (1971) reviewed benthic studies of the
reservoirs and tailwaters of the TVA system (with an emphasis on
Mollusca). He concluded that benthic fauna in the region "may
be limited by siltation, rheotactile deprivation, water level
fluctuation, increased hydrostatic pressure, light, and most pert-
inently by hypolimnetic oxygen deficiency in the storage impoundments." He noted that benthic fauna have been almost eliminated
from storage impounds in the valley. He further stated that
"benthic fauna below mainstream impoundments is typically rheophilic,
and includes mussels and residual populations of Pleuroceridae..."
A recent decline in mussels was attributed to both impoundment and
overharvest. In addition, a decline of snails and Pleuroceridae
was associated with alteration of habitats by impoundment.

Obviously, the reduction in benthic fauna will have an effect
on the fisheries, and, as noted by Eschmeyer (1950), the food chain
in impoundments relies on planktonic organisms. A significant adverse effect of TVA impoundments has been on mussels, which have
been reduced from 100 species to fewer than half that number.
Furthermore, the commercial take of mussels has been reduced from
10,000 tons (9,078 metric tons) annually in the 1950s to approximately 2,000 tons (1,816 metric tons) annually (Isom, 1969).

Pfitzer (1962) examined the food habits of trout in several
tailwaters below TVA impoundments. It was demonstrated that the
stomach contents reflected the composition of the benthic population. A paucity of aquatic insects was found, and fish stomachs
contained excessive amounts of algae. In the Cherokee tailwater,
the lack of insects and the absence of trout was attributed to
the unusual temperature pattern and long periods of no flow
during the summer months.

Plankton

Pfitzer (1962) abandoned efforts to determine numbers and types
of plankton in tailwaters because of the highly variable nature of
the system. This would be expected when the distribution of plankton in the reservoir is considered along with the intake of configu-
ration and selective withdrawal aspects of the sampled reservoirs.

It should be noted that the plankton composition of tailwaters
will depend on the elevation of the intake and the hydrodynamics of

the reservoir, which are site specific. The intakes in the TVA
system are near the reservoir bottom or, in a few deeper reservoirs,
near mid-depth. Thus, releases would be primarily from the hypolimnion during periods of stratification.

Fish

       As previously indicated, a paucity of data exist on the tail-
water fisheries below TVA impoundments. In an effort to understand
the factors responsible for the decline in fishing that occurred
and to apply appropriate management techniques, the Tennessee Game
and Fish Commission instigated the studies referred to by Pfitzer
(1962). The tailwaters studied were below South Holston, Watauga,
Wilbur, Norris, Cherokee, Douglas, Calderwood and Apalachia Dams
(all TVA dams), which create a total of some 500 km of tributary-
stream tailwaters. One tailwater was producing a promising trout
fishery, while others failed to support either trout or warm-water
species.

       Results of the pre- and post-impoundment studies showed drastic
changes in the fisheries. For example, pre-impoundment studies of
the Watauga River showed 32 species, of which only 11 remained in
the post-impoundment studies, with 6 species not present in the
pre-impoundment survey. Similarly, the pre-impoundment survey
of the South Fork Holston River identified 43 species, of which
only 16 remained in the post-impoundment studies, with 7 species
not found in the pre-impoundment surveys.

       The quality of water discharged from the dams affects the type
of tailwater fishery that can be supported. Probably the most
important water-quality parameter to be considered is temperature,
its magnitude, and its rate of change. Temperature in turn depends
upon the operation of the dam and the reservoir hydrodynamics.

       There is also no doubt that dissolved oxygen is a prime factor
in the viability of tailwater aquatic life. Even though Pfitzer
(1962) found no direct evidence of adverse effects of low dissolved
oxygen concentrations on fish in the tailwaters, low dissolved
oxygen concentrations have been shown to be adverse to aquatic
life. As with temperature, the tailwater dissolved oxygen concentrations depend on dam operation and reservoir hydrodynamics.

       The adverse effects of stratified flow discharges on the
available food supply must also be considered as affecting the
tailwater fishery. Again, this is a result of the reservoir
stratification and operation.

SUMMARY

It has been shown that reservoir construction and operation have significant effects on aquatic life, which vary depending upon the project objectives and the specific requirements of tailwater biota. The large variations in flow and associated water quality parameters may adversely affect downstream fisheries, influence the benthos, and, in general, lessen recreational opportunities.

Water quality parameters, such as temperature and dissolved oxygen, can greatly influence downstream aquatic life, particularly when adverse water quality conditions are combined with peaking power operations to yield periods of low flow, then sudden increases in cold water flow.

Additional information is needed to more clearly define the role of reservoir operation in maintaining downstream fisheries. Also, it is apparent that before meaningful biological studies can be pursued in reservoirs and tailwaters, the hydrodynamics of the systems must be delineated.

REFERENCES

Anon., 1973, Tennessee Valley Authority, Div. of Env. Planning, Chattanooga, Tennessee, unpublished data.

Churchill, M. A., 1964, Effects of density currents in reservoirs on water quality, in: "Proc. 3rd Annual Sanitary and Water Resources Engineering Conf.," P. A. Krenkel, ed., Vanderbilt Univ., Nashville, Tennessee.

Elder, R. A., and Garrison, J. M., 1964, The causes and persistence of density currents, in: "Proc. 3rd Annual Sanitary and Water Resources Engineering Conf.," P. A. Krenkel, ed., Vanderbilt Univ., Nashville, Tennessee.

Eschmeyer, R. W., 1950, "Fish and Fishing in TVA Impoundments," Tennessee Dep. Conservation.

Isom, B. G., 1969, The mussel resource of the Tennessee River, Malacologia, 7(2-3):397-425.

Isom, B. G., 1971, Effects of storage and mainstream reservoirs on benthic macroinvertebrates in the Tennessee Valley, in: "Reservoir Fisheries and Limnology," G. E. Hall, ed., Am. Fish. Soc., Special Publ. No. 8, Washington D.C.

Kittrell, F. W., and Quinn, J. J., 1949, Multi-purpose reservoirs aid downstream water supplies, Eng. News Record, May 26.

Krenkel, P. A., and Novotny, V., 1973, "The Assimulative Capacity of the South Fork Holston River and Holston River near Kingsport, Tennessee," Rep. to Tennessee Eastman Co., September.

Lee, G. F., 1975, Role of hydrous metal oxides in the transport of heavy metals in the environment, p. 137-147, in: "Progress in Water Technology," Vol. 17, Proc. Symposium of Transport of Heavy Metals in the Environment.

Lee, G. F., and Delfino, J. J., 1969, Discussion, "Equilibrium and Kinetic Aspects of Inorganic Chemical Quality in Reservoirs," Proc. Specialty Conf. on Current Research into the Effects of Reservoirs on Water Quality, January, 1968, ASCE Tech. Rep. No. 17, p. 51-72.

Lee, G. F., Rast, W., and Jones, R. A., 1978, Eutrophication of waterbodies: Insights for an age-old problem, *Environ. Sci. Tech.*, 12:900-908.

Meyer, B. S., Bell, F. H., Thompson, L. C., and Clay, E. I., 1943, Effect of depth of immersion on apparent photosynthesis in submerged vascular aquatics, *Ecology*, 24:393-399.

Morton, S. D., and Lee, G. F., 1968, Calcium carbonate equilibria in lakes, *J. Chem. Educ.* 45:511-513.

Newbry, B. W., Jones, R. A., and Lee, G. F., 1979, "Application of the OECD Eutrophication Modeling Approach to TVA Impoundments," Draft Rep. to Tennessee Valley Authority, March.

Parker, F. A., and Krenkel, P. A., 1969, "Thermal Pollution-State of the Art," Natl. Center for Research and Training in the Hydrologic and Hydraulic Aspects of Water Pollution Control, Rep. No. 3, Vanderbilt Univ., Nashville, Tennessee, December.

Peltier, W. H., 1970, Relationship of nutritional and environmental factors to selected rooted aquatic macrophytes: Part II. Influence of climatic factors on the rate of aquatic macrophyte growth, *in:* "TVA Activities Related to Study and Control of Eutrophication in the Tennessee Valley," Natl. Fertilizer Development Center, Muscle Shoals, Alabama.

Pfitzer, D. W., 1954, Investigation of waters below storage reservoirs in Tennessee, *Trans. N. Am. Wildl. Conf.*, 19:271-282.

Pfitzer, D. W., 1962, "Investigation of Waters Below Large Storage Reservoirs in Tennessee," Tennessee Game and Fish Commission, Nashville, 233 p.

Rast, W., and Lee, G. F., 1978, "Summary Analysis of the North American (U.S. Portion) OECD Eutrophication Project: Nutrient Loading-Lake Response Relationships and Trophic State Indices," U.S. EPA 600/3-78-008.

Tarzwell, C. M., 1930, Changing the Clinch River into a trout stream, *Trans. Am. Fish. Soc.*, 68:228-233.

Taylor, M. P., 1971, Photoplankton productivity response to nutrients correlated with certain environmental factors in six TVA reservoirs, *in:* "Reservoir Fisheries and Limnology," G. F. Hall ed., Am. Fish. Soc., Special Publ. No 8., Washington D.C.

Vollenweider, R. A., 1976, Advances in defining critical loading levels for phosphorus in lake eutrophication, *Mem. Inst. Ital. Idrobiol.*, 33:53-83.

Weibe, A. H., 1958, "The Effects of Impoundments upon the Biota of the Tennessee River System," Intl. Union for the Conservation of Nature and Natural Resources, 6th Annual Assembly, 7th Tech. Session, Athens, Greece.

THE REGULATED STREAM AND SALMON MANAGEMENT

J. H. Mundie

Department of Fisheries and Oceans
Fisheries and Marine Service
Pacific Biological Station, Nanaimo, B.C., Canada
V9R 5K6

INTRODUCTION

The purpose of this paper is to provide a broad review of
the implications, both positive and negative, of regulated stream
discharge for salmon management. Generally, regulation is taken
to mean the intentional alteration of the discharge of a natural
stream for a particular purpose. Usually, the alteration is the
consequence of impounding water behind a dam; the stored water
may be used to generate power, to irrigate crops, or to provide a
recreational lake. To cover the application, however, of regulated
discharge to salmon production it is instructive to extend the
definition to include control of flow in semi-natural or artificial
side-channels to rivers. Comparisons can then be made of the
effects of partial and of total regulation, and new potentials can
be identified.

The basic processes in biological production operate, of
course, vertically. Sunlight falls on the earth's surface; plants
respond by growing upwards; even decomposition is dependent on
leaves falling to the ground. Superimposed on these processes are
horizontal ones. Seeds are dispersed by wind, and currents in
water distribute nutrients and plankton; tides impose their influence; fish, birds and mammals make extensive migrations. Changes
in the scale of these secondary processes affect the primary ones.

The lotic environment, where water moves in one overall direction, is the most obvious example of horizontal environmental
forces, and their extremes--minimum and maximum flows--have a major

influence on the kinds and quantities of plants and animals present. For salmon, regulation may affect the success of migrations of smolts and adults, survival of eggs in gravel, the amounts of drifting food for juveniles, the spatial and territorial requirements of fry, and the susceptibility of young fish to predation from birds. For a review of the major effects of discharge on fish, see Fraser (1972). Regulation in relation to salmon can be viewed from two main aspects: the negative impact of dams, and the positive consequences of flow control for the specific purposes of incubating eggs and rearing juveniles.

In this paper five examples of regulation and its effects on salmon are examined, i.e., regulation by a very large dam, regulation of a river by a small dam for generating hydroelectric power, regulation of a river explicitly to increase salmon production, total regulation in an artificial channel for incubating one species, and total regulation in an artificial channel for rearing one species by drawing on the invertebrates of the parent river to provide food organisms. These examples illustrate the range of possible relationships between regulation and salmon production. For each, some outstanding research questions are identified. All of this will be familiar to salmon biologists, but other biologists and engineers will see links with their own disciplines.

There are five species of Pacific salmon in North America: coho (*Oncorhynchus kisutch*), sockeye (*O. nerka*), chinook (*O. tschawytscha*), chum (*O. keta*), and pink (*O. gorbuscha*). Coho usually spend 1 or 2 years in streams before going to sea; sockeye spawn in lakes or rivers, and their progeny feed for at least a year in a lake before migrating to sea; chinook go to sea after 3 months or a year in a stream; pink and chum salmon go to sea as fry after emerging from stream gravel. The adults of all species die after spawning.

The Atlantic salmon (*Salmo salar*) spends 2 or 3 years in rivers before migrating. The adults return to sea after spawning.

THE EFFECTS OF LARGE DAMS

To the general public the best-known effects of flow control on salmon are the notorious consequences of large dams. Reference may be made to Netboy (1974) for a historical review of the development of hydroelectric power and associated salmon problems in North America. Briefly, these effects are: obstruction of upstream passage of adult fish; delays in the migration of spawners through long reservoirs to reach spawning grounds; flooding of spawning grounds by impounded water; delays in the migration of fry downstream through reservoirs; reluctance of fry to move through stratified water masses; losses of fry in their passage over or through dams; accumulations of predators below dams; changes in discharge below dams that are unacceptable to fish; and changes in water quality, especially temperature and nitrogen content, that cause avoidance or stress.

The solution offered most persistently to these problems is transportation of fish by fish-ways, locks, or trucking. Yet the magnitude of the facilities or operations required prohibits their feasibility, quite apart from the losses of adults and fry incurred.

The penalties associated with large dams have led to much concern within government agencies responsible for management of anadromous fishes. Proposals for new dams are therefore viewed very critically. As an example--admittedly an extreme one--of a proposed major dam and its possible consequences, the dam at Moran Canyon on the Fraser River, B.C., may be cited (reviewed by Geen, 1975). Such a dam, along with its downstream re-regulation dam, which minimizes the short-term effects of discharge from the main dam, could generate 4350 MW and provide flood control for the lower Fraser River. It would have a head of 215 m and would flood 270 km of the river. Current estimated costs are $881 million.

All species of Pacific salmon spawn in the Fraser River and its tributaries, but sockeye and chinook are most abundant above Moran Canyon. These fish support a large commercial fishery in the Fraser River and Georgia Strait, the average annual catch of sockeye originating from these stocks having a value of $8.9 million between 1946 and 1969. In addition, perhaps 100,000 salmon are taken annually for food by native Indians.

If the Moran Canyon dam were built, the passage of water through the reservoir created would take 2 months in early summer, whereas formerly it took 1-2 days. The reservoir would stratify, turbidity would be reduced, and 15 million t of sediment would settle annually. Winter increases, and spring and summer decreases, in discharge would occur below the dam and probably would affect primary and secondary production in the Strait of Georgia itself. As a result, the survival of young salmon during their early sea life would decrease.

Migrant adult fish would be stressed by the altered discharge below the dam and by the supersaturation of the plunging water with nitrogen. Their passage over the dam, however, would present the most intractable problem. As many as 750,000 adults/day could be moving during the peak of the run. Economical and safe methods of transporting adult fish in such numbers do not at present exist. The downstream migrants would be delayed 11-25 days by the reservoir. Predation and residualism (loss of migratory tendency) would

result in losses of up to 85%. Half the fish going through the
turbines likely would die; mortalities also would occur at the
spillway. The conclusion emerges that, even with today's technology, a high-head dam at Moran Canyon would eliminate upstream
salmon and reduce downstream stocks. Further research is not
likely to alter this conclusion.

One further aspect of large dams in relation to salmon deserves
mention here. The large-scale alteration of wet-lands by dams may
have far-reaching effects on juvenile salmon that depend on marshes
in the course of their progress to sea (see Kistriz, 1978). Further
negative aspects of dams are reviewed by Baxter (1977).

SALMON PRODUCTION DOWNSTREAM OF SMALL HYDROELECTRIC POWER DAMS

As an example of the co-existence of salmon and small power
dams, the John Hart Dam of the B.C. Hydro and Power Authority on
Campbell River, east central Vancouver Island, may be cited. The
main stem of the river is about 80 km long and consists largely of
natural and man-made lakes. Because of a natural obstruction, however, only about 5.6 km of river upstream from the mouth have supported, historically, runs of anadromous fish. The construction in
1947 of a 30.4-m head spillway dam further reduced this distance to
4.9 km. Nevertheless, the system supports an important fishery of
chinook salmon noted for their large Tyee strain (chinook over
13.6 kg in weight). The other species of Pacific salmon and
steelhead trout also are present. In 1972, 53,000 salmon were
taken in a sport fishery off the mouth of the river. The sport
fishery attracts fishermen from throughout western North America,
and tourist expenditures in the area exceed $5 million per season.

The lower Campbell River is 45-106 m wide. The substrate
varies from cobbles and boulders in the upper reaches, through
medium gravel in the mid-stretch, to mixed gravel, sand, and silt
in the tidewater region. The large chinook salmon favour spawning
gravel in the 5- to 15-cm range. The amounts of this type of
gravel are not extensive; moreover, little or no recruitment of
gravel can occur below the dam, and small and medium gravel can be
scoured from the river by rapid increases in discharge. Rearing
habitat for chinook fry, which spend 3 months in the river before
migrating, and for coho fry, which spend 14 months, is limited.

The lower river is maintained, under normal operating conditions, at an average discharge of 112 m^3/sec, and the range is
31-122 m^3/sec. (The highest discharge recorded since the building
of the dam is 835 m^3/sec; this was exceptional.) This regime has
maintained chinook production, with escapements to the river of
4,000-5,000 adults.

For its average discharge the lower Campbell River is swift
and shallow. The scarcity of rearing habitat is made evident by
the occurrence of a large out-migration of fry of chinook and coho
after their emergence from the gravel. It is unlikely that these
migrants survive. The resident fry of both species initially take
up position close to the bank (see Mundie, 1969). Here they feed
in shallow water, especially in slow back eddies created by out-
cropping rock. This habitat, in addition to having suitable depth
and velocity, appears to offer visual points of reference for the
fry. In consequence, much of their early diet consists of terrestrial food, e.g., Collembola; zooplankton, which is abundant in the
drift, is not of major consequence at this time. As the fish grow
larger they leave this locality and acquire territories in swifter,
deeper water.

In 1970 an expansion of the hydroelectric potential of the
Campbell River system was proposed. One way would be to increase
discharge periodically by 141 m^3/sec. Resulting daily discharges
would range between 28 and 263 m^3/sec. The consequences of this
proposed regime were examined in detail (Hamilton and Buell, 1976),
and it was concluded that the increased discharge and the abrupt
changes associated with fluctuating flow would seriously reduce
salmon and trout production. The causes of loss were identified:
increases in water depth and velocity would exceed limits tolerated
by adults and therefore would result in reduction in spawning area;
abrupt changes in flow would displace spawners and inhibit spawning
behaviour; the higher discharges would transport juveniles downstream and reduce suitable low-velocity feeding areas; changes in
discharge alternately would inundate marginal habitat or leave
juveniles stranded. Finally, it was recommended that the proposal
be dropped and an alternative be pursued involving diversion of the
additional 141 m^3/sec.

The study identified discharges appropriate for the life
stages of the fish in different months. A minimum discharge of 70
m^3/sec was recommended, and flows greater than 99 m^3/sec were to be
avoided during the rearing period of chinook (March 1-July 1) to
minimize loss of rearing habitat and dislodgement of benthos.
Flows up to 122 m^3/sec from December 1 to March 1 were considered
acceptable as they should not affect buried eggs. Special attention was given to rates of change, for sudden changes in flow,
especially in darkness, could be disastrous to fry. It was
recommended that rates of change should not exceed 0.25 m^3/min.

This study included observations (Mundie and Mounce, 1976) on
transport of invertebrates and leaf litter at high rates of discharge, but it omitted both fish diets and the basic features of
the benthos, e.g., species composition, diversity and standing
crop. (See Ward, 1976, for a review of benthos below dams.)

Clearly, some understanding of fish diets and benthos is required
for optimizing flow regime for fish production.

The Campbell River system is an example of compatibility
between peak discharges, with fairly rapid changes in river levels,
and fish production. More commonly, regulation follows intense
demand for water for various uses, and therefore raises the question: What is the minimum acceptable flow for fish in a stream?
Much attention has been paid to ways of answering this (e.g.,
Stalnaker and Arnett, 1976; Neuman and Newcombe, 1977), and there
has been growing recognition that recommendations cannot be made
from the properties of the fish alone; responses from the total
system must be assessed. The most instructive approach, where
possible, seems to be field experimentation in which stream communities are subjected to a range of discharge regimes. In this way
the annual dominant discharge could be defined that best sets the
stage, by sorting and transporting bed and nutrient materials, for
community production through the seasons. Again, experimentation
could establish the levels and duration of low flows that result in
the adverse consequences of accumulations of silt and organics,
encroachment of riparian vegetation, and dominance of the stream
community by aquatic angiosperms.

REGULATION OF NATURAL STREAMS FOR SALMON PRODUCTION

The Big Qualicum River on the east coast of Vancouver Island,
B.C., is a pioneering flow-control project explicitly implemented
to increase production of chum, coho, and chinook populations
(Lister and Walker, 1966). The river has about 10.4 km accessible
to salmon and is a major contributor to a chum fishery in Johnstone
Strait off the northeast coast of the island. The spawning populations from 1950 to 1958, before flow regulation, were 10,000-
100,000 chum, 2,000-5,000 coho and 200-2,000 chinook salmon.
Maximum and minimum discharges recorded were 200 and 0.39 m³/sec.

Regulation was introduced primarily to increase egg-to-fry
survival of chum by stabilizing winter discharge. The river lent
itself to control because it flows mainly from one source, Horne
Lake, which could be used as a storage reservoir, and because its
single tributary could be diverted. Control was obtained by
impounding fall and winter inflows in Horne Lake by a dam at the
natural outlet. The required storage capacity (175 million m³)
requires a fluctuation in the lake surface of 24 m. Drawdown is
accomplished by a vertical shaft, 30 m deep, which has three separate intakes; these are 5.3, 11.7, and 24.9 m below the lake
surface. Water drawn into the shaft passes down a tunnel 518 m
long and is discharged to the river bed. The three intakes allow
water to be drawn simultaneously from different depths, so that

when the lake is stratified some temperature control of the river is possible.

Egg-to-fry survival of chum salmon before regulation of the river averaged 13% and after regulation 29% (E. A. Perry, pers. comm.). An inverse relation was found between survival rate and the dominant discharge during incubation (November–May), with highest survival occurring at values below 22.6 m^3/sec. Coho smolt output from the river increased from about 28,000 to 36,000 after control. No marked increase in survival of chinook fingerlings could be demonstrated; the reasons for this are obscure.

An important research question in this type of regulation is: What dominant discharge is required each fall to dislodge and transport silt from the gravel so that egg-to-fry survival is maintained? Another is: Given the higher egg-to-fry survival following from regulation, what are the factors (e.g., territory or food) that now operate to set limits on production of coho smolts or chinook fingerlings?

A corollary of flow regulation at the Big Qualicum River was that it facilitated the further expansion (1967) of chum production by means of an artificial spawning channel (see below). This channel gives an egg-to-fry survival of about 65% and yields over 20 million fry annually (Paine et al., 1975).

TOTAL REGULATION FOR INCUBATING ONE SPECIES

The outstanding application of regulated flow to salmon management is the completely artificial spawning channel, for this can provide both stable flow and optimum size of gravel for the eggs of the species being raised. Spawning channels can give two to nine times the egg-to-fry survival of a natural river. In general (see Clay, 1961 for an engineering account), the intake should permit diversion of water from a river at several levels to avoid flooding of the channel. Exclusion of silt, and fine bed-load materials, from the channel is necessary. Intakes, therefore, should be sited at the concave bends of rivers (see Vanoni, 1975), and a settling pool should be provided at the top of the channel. Fine perforated screens, preferably of slotted aluminum, make the intake fry-proof in both directions; i.e., fry must not escape from the channel nor must they enter from the river. Gravel size, and water depth and velocity, can be chosen in relation to the known requirements of the species, and roughness is obtained with surface boulders. Cleaning is necessary either by flushing or, more effectively, with expensive equipment that dislodges silt (Anon., 1973). Riparian trees are needed to shade the channel, so that spawning fish are not disturbed, and to give protection from insolation and freezing

winds. Pools are included; adults wait in these until spawning
territory is available.

Weaver Creek Spawning Channel, which flows into Harrison
River, B.C., and is operated by the International Pacific Salmon
Fisheries Commission (Cooper, 1977), may be taken as an example.
This was the first channel for sockeye salmon. It is 2,930 m long,
6 m wide, and provides 17,429 m^2 of gravel in a serpentine arrange-
ment. The slope is 0.0006 between 27 drop structures 15 or 30 cm
high. The gravel is 40 cm deep and ranges in size from 1.2 to 7.6
cm. The discharge is 0.56 m^3/sec, giving a velocity of 0.36 m/sec
and a depth of 0.24 m. The sides of the channel are on a 1–1.5
slope.

In recent years 18,000–27,000 spawners, more than half of
which are female, have been allowed into the channel at a spawning
density of 0.90 females/m^2. Egg deposition amounts to 41–65
million, and egg-to-fry survival is 61–81%.

The channel is nine times more efficient, per unit area, at
producing returning adults than natural spawning grounds. The
returning sockeye runs from brood years 1971 and 1972 were 148,952
and 194,744 adults, and the landed values of the commercial catches
were $665,352 and $997,838. The capital cost of the channel was
$280,725 in 1964–65, and the overall benefit/cost ratio, on the
basis of 1975 dollars, is 9.5 (Cooper, 1977).

Spawning channels, therefore, can be highly successful. There
are major aspects, however, that require precise definition. The
most important are the relationships of slope of bed, gravel composition, and permeability of gravel (Cooper, 1965). Again, channels
require periodic cleaning. Whatever method is used has to do the
work of a river in freshet; this might amount, of course, to a 20-
fold increase in flow over minimum discharge, sustained for days or
weeks. Artificial cleaning, therefore, is a substantial and costly
undertaking. Research is needed on the development of effective
and inexpensive methods that can dislodge silt and transfer it to
land.

Various troubles can arise with channels. Adults of unwanted
species may enter and spawn. The timing of runs and the density of
spawners affect survival of eggs; successive waves of females may
enter and disturb previously deposited eggs. Frazil ice can cause
egg mortality in winter. Algal production in the spring, at the
time of fry output, can be so high that screens become blocked.
Although many of the dead salmon are lifted out and buried, channels
become enriched by decaying carcasses. It would be helpful, therefore, to have a biological indicator that would give some measure
of the conditions in the gravel. This could be derived from the

species composition and abundance of the benthos, and could supplement data on oxygen and ammonia. Some warning might then be
obtained of when a channel should be cleaned before egg survival
declined. Finally, increased fry output, above certain limits, may
not result in higher numbers of returning adults. At certain
levels of production, density-dependent mortality may operate in
the freshwater or marine phase of the fish.

Channels are usually drained and allowed to stand dry in
summer after the fry have emerged and migrated. There may be
opportunity, however, for using them as rearing channels for some
species (see next section). Atlantic salmon or steelhead trout,
for example, after incubation in a spawning channel, might feed in
the channel on a combination of natural and artificial foods for 2
or 3 months. Some reduction of discharge would probably be
necessary. The fingerlings produced could be planted in depleted
rivers in late summer.

TOTAL REGULATION FOR REARING ONE SPECIES

Artificial channels exist in which juvenile salmon or trout
are raised to the smolt stage. They usually have sluggish flows
and are stocked at high density. Commercial fish food is supplied
to the fish at frequent intervals from automatic feeders. It is
possible, however, to raise salmonids semi-naturally in regulated
channels alongside rivers and to draw upon the invertebrate drift
of the river for natural food for the fish. To justify costs, large
numbers of fish are reared, and the bulk of the food must be artificial but, because of the metabolic properties of streams, fish
wastes and uneaten commercial food are assimilated by the benthos
and production of fish-food organisms is increased.

The concept of optimizing salmonid production in this way
(Mundie, 1974) adopts the stream features of riffle, pool, and
overhead cover as the basic essentials. The first produces and
transports invertebrate drift, the second provides a habitat for
the fish, and the third gives protection from predators. Regulation of discharge in the channel is necessary for maintaining
stability of the system.

The installation of such a channel alongside a river is
greatly facilitated, of course, if the river itself is already
regulated, for this can guarantee the modest necessary flow in the
channel during the dry season. This is a fundamental requirement.

A semi-natural channel consisting of a continuous series of
riffles and pools has been constructed alongside the Big Qualicum
River on Vancouver Island, B.C. (Mundie and Mounce, 1978). The

objective of this pilot project is to reduce the costs of traditional hatchery-rearing methods, which use concrete ponds or raceways, and to raise healthy salmon smolts having wild-type physiological and behavioural characteristics.

Construction of this channel began with an excavation 396 m long and 4.5 m wide at water level. The overall slope was 0.004. There are 25 riffles, each 6 m long and 15 cm deep, alternating with 25 pools 10 m long and 0.9 m deep. The velocity of water over the riffles is 0.6 m/sec at the surface, and stop-logs are set in the gravel at the tail of each riffle to prevent movement of the gravel. The pools have a surface velocity of 0.1 m/sec. Complete exchange of water in the channel takes 23 min. The gravel is 30 cm deep, and its size ranges from 1.3 to 3.8 cm. Boulders on the surface create turbulence. The intake is provided with a protective trash-rack. Downstream from the trash-rack, in the river itself, is a shallow weir of removable stop-logs; this ensures that sufficient river water can be diverted to the channel at low flows in summer. The outlet consists of a concrete housing with inclined screens that can be lowered to let the smolts out when they migrate to the estuary.

A stainless-steel inclined screen is installed in a concrete pad in the fifth pool from the intake. This screen divides the channel into a short upstream section used for incubating eggs and for rearing the young fry from April to July. The remainder of the channel holds fish from the previous brood. They are reared here from the previous July until their outmigration in May or June. The upstream fry are then moved down into the main rearing section.

The aim is to raise 0.5 million coho smolts annually. The fish are fed from April until May or June of the following year. They are fed commercial fish food by hand three times per day every second day when the water is above 5°C. Throughout this feeding period they obtain drifting aquatic insects derived from the riffles. Feeding takes place in both the riffles and pools. The benthos responds to enrichment from faeces and uneaten food; increases mainly are found in Chironomidae and naiid Oligochaeta (Williams et al., 1977). In consequence, the amount of commercial food required by the fish is less than 50% of standard hatchery rations.

In the first 2 years of operation, losses of fry amounting to 40% occurred from predation by herons, mergansers, and mink. The channel therefore was covered with coarse netting to keep out the larger birds, and floating plywood cover was provided for the fish in each riffle and pool. There is evidence that this has reduced predation.

After the smolts leave, the channel is washed with a fire hose to dislodge sand, organic material, and algae. The gravel becomes re-colonized by drifting insects in about 2 weeks. The screen in pool 5 is partly removed to allow the new brood to occupy the length of the channel. A floating fish shocker is used to expedite this redistribution of fry.

The smolts released from the channel are indistinguishable from wild smolts, being low in fat content and having the natural pathogens and immune responses of wild fish. The first returns of adults, which have spent 18 months at sea, are currently being assessed. Numerous fish that were tagged before their migration have returned to the channel outlet.

The advantages of rearing salmonids in semi-natural streams appear to be the gravity feed (pumps are not used), the contribution of natural food (so that automatic feeders are not required and the fish are therefore widely dispersed), the re-cycling of wastes, and the low costs of food and labour. The disadvantages lie in losses of fry, high demand on space, the impracticability of grading fish to different sizes, and the difficulty of transferring all fish from the upper section to the lower. This approach to rearing, however, is so new that the scope may widen. For example, controlled organic enrichment of the riffles may promote substantially greater production of food organisms. Again, additions to the channel of early fry from the parent river would broaden the genetic spectrum of the reared stock and so reduce one of the negative aspects of all large-scale fish-culture procedures.

CONCLUSION

The five examples of flow regulation in relation to salmon have been chosen to illustrate the range of major possibilities. The first--very large dams--has a wholly negative effect on salmon production. The second--small dams operated within a limited range of discharges and their rates of change--allows downstream production of salmon, although it may be uncertain whether the average annual output is greater or less than that before dam construction. The other three examples are of planned attempts to use flow regulation to increase the efficiency of incubation, or rearing, or both. These planned projects have been based on the known ecological requirements of eggs and juvenile stages, but their implementation has necessarily involved judgements where precise information has been deficient. It follows that the efficiency of design and management of flow-controlled production facilities could be improved if research were directed at carefully identified questions. These questions fall into two main categories. The first covers the physical relationships of depth and slope, gravel-size

distribution, and permeability of gravel; the second covers benthic
production and invertebrate drift in relation to the physical
properties of gravel and its organic content.

It may be assumed that the ecological significance of regulated
discharge will become better understood from further experience
under widely varied circumstances. New applications, at present
unforeseen, will emerge. Close liaison between engineers and
biologists will hasten this process.

ACKNOWLEDGMENTS

The author's best thanks go to Dr. D. F. Alderdice and
Mr. J. M. B. Hume for critical comments on the manuscript.

REFERENCES

Anon., 1973, "International Pacific Salmon Fisheries Commission
    Annual Report, 1972," New Westminster, B.C., Canada, 36 p.
Baxter, R. M., 1977, Environmental effects of dams and impound-
    ments, *Annu. Rev. Ecol. Syst.*, 8:255-283.
Clay, C. H., 1961, "Design of Fishways and Other Fish Facilities,"
    Dep. Fisheries of Canada, Queen's Printer, Ottawa, 301 p.
Cooper, A. C., 1965, "The Effect of Transported Stream Sediments on
    the Survival of Sockeye and Pink Salmon Eggs and Alevin," Int.
    Pac. Salmon Fish. Comm. Bull. 18, 71 p.
Cooper, A. C., 1977, "Evaluation of the Production of Sockeye and
    Pink Salmon at Spawning and Incubation Channels in the Fraser
    River System," Int. Pac. Salmon Fish. Comm. Prog. Rep. 36,
    80 p.
Fraser, J. C., 1972, Regulated discharge and the stream environ-
    ment, *in*: "River Ecology and Man," R. T. Oglesby, C. A.
    Carlson, and J. A. McCann, eds., Academic Press, N.Y.
Geen, G. H., 1975, Ecological consequences of the proposed Moran
    Dam on the Fraser River, *J. Fish. Res. Board Can.* 32, 126-135.
Hamilton, R., and Buell, J. W., 1976, "Effects of Modified Hydrol-
    ogy on Campbell River Salmonids," Environment Canada, Fish.
    Mar. Serv. Tech. Rep. Ser. PAC/T-76-20, 156 p. + 21 p.
Kistriz, R. U., 1978, "An Ecological Evaluation of Fraser Estuary
    Tidal Marshes: The Role of Detritus and the Cycling of Ele-
    ments," Westwater Research Centre Tech. Rep. 15, 59 p., Univ.
    British Columbia, Vancouver, B.C.
Lister, D. B., and Walker, C. E., 1966, The effect of flow control
    on freshwater survival of chum, coho and chinook salmon in the
    Big Qualicum River. *Can. Fish. Cult.*, 37:3-25.

Mundie, J. H., 1969, Ecological implications of the diet of juvenile coho in streams, p. 135-152, *in*: "Symposium on Salmon and Trout in Streams," T. G. Northcote, ed., H. R. MacMillan Lectures in Fisheries, Univ. British Columbia, Vancouver, B.C.

Mundie, J. H., 1974, Optimization of the salmonid nursery stream, *J. Fish. Res. Board Can.*, 31:1827-1837.

Mundie, J. H., and Mounce, D. E., 1976, Effects of changes in discharge in the Lower Campbell River on the transport of food organisms of juvenile salmon, Appendix A, 21 p., *in*: "Effects of Modified Hydrology on Campbell River Salmonids," R. Hamilton and J. W. Buell, eds., Environment Canada, Fish. Mar. Serv. Tech. Rep. Ser. PAC/T-76-20.

Mundie, J. H., and Mounce, D. E., 1978, Application of stream ecology to raising salmon smolts in high density, *Verh. Int. Verein. Limnol.*, 20:2013-2018.

Netboy, A., 1974, "The salmon: Their Fight for Survival," Houghton Mifflin Co., Boston, 613 p.

Neuman, H. R., and Newcombe, C. P., 1977, "Minimum Acceptable Stream Flows in British Columbia: A Review," Fisheries Management Rep. 70, B.C. Fish and Wildlife Branch, Victoria, B.C., 49 p.

Paine, J. R., Sandercock, F. K., and Minaker, B. A., 1975, "Big Qualicum River Project 1972-3," Environment Canada, Fish. Mar. Serv. Tech. Rep. PAC/T-75-15, 126 p.

Stalnaker, C. B., and Arnette, J. L., eds., 1976, "Methodologies for the Determination of Stream Resource Flow Requirements: An Assessment," Utah State University, Logan, Utah, 199 p.

Vanoni, V. A., ed., 1975, "Sedimentation engineering," Am. Soc. Civil Engineers, N.Y., 745 p.

Ward, J. V., 1976, Effects of flow patterns below large dams on stream benthos: A review, p. 235-253, *in*: "Instream Flow Needs Symposium, Vol. 2," J. F. Orsborn and C. H. Allman, eds., Am. Fish. Soc. and Am. Soc. Civil Engineers, Bethesda, Maryland.

Williams, D. D., Mundie, J. H., and Mounce, D. E., 1977, Some aspects of benthic production in a salmonid rearing channel, *J. Fish. Res. Board Can.*, 34:2133-2141.

THE USE OF HABITAT STRUCTURE PREFERENDA FOR ESTABLISHING FLOW

REGIMES NECESSARY FOR MAINTENANCE OF FISH HABITAT

Clair B. Stalnaker

Cooperative Instream Flow Service Group
USFWS
2625 Redwing Road
Fort Collins, Colorado   80526

INTRODUCTION

The pursuit of simplicity in the complex management of instream and out-of-stream water uses very early led to the *minimum*, or "base," flow concept and the myth that a consistent methodology could be used to establish the desired minimum flow. Such a single flow value has not proven to be practical. As water becomes fully appropriated to upstream use and storage, the minimum flow, if not violated, tends to become the average condition. This situation is manifest in a flat or stepped but fixed hydrograph throughout the stream reach of concern for much of the year.

This paper focuses on the instream physical habitat conditions that influence the carrying capacity of the target fish species. Emphasis is placed on the dynamic aspects of the physical stream habitat conditions over seasons and years under changing water and nutrient supplies, which suggests that these must be understood and treated as stochastic values. The approach being developed by the Cooperative Instream Flow Service Group (IFG) for evaluating the biological integrity of stream systems is outlined.

HABITAT/STANDING CROP MODELS

Many studies have documented the importance of the physical parameters of temperature, depth, velocity, and substrate in the microhabitat specialization of stream fishes (Hynes, 1970; Giger, 1973; Hooper, 1973; Hunter, 1973; Bovee, 1974; Gorman and Karr,

1978; Paragamian, 1978). Under conditions of suitable habitat structure and temperature, the distributions of depth and velocity are the dominant aspects of the flow regime that dictate fish species distributions and territorial behavior (Gorman and Karr, 1978). In marginal habitats temperature may interact with other factors to determine distribution and abundance of stream fishes (Dettman, 1978). Matthews and Hill (1979) found that temperature, velocity, and depth had great influence on redshiner (*Notropis lutrensis*) habitat selection and stated that "the strong influence of current speed and depth suggested that the influence of all other variables could have been overridden by those two factors."

Considerable effort has been expended over the past two decades toward better understanding the relationship between discharge and fish populations, although the methodologies available to date examine only the relationship between streamflow and fish habitat (Stalnaker and Arnette, 1976; Ott and Tarbox, 1977; Wesche, 1978). These methods generally fall into three categories: (1) those related to a percentile of historic flows as recorded by stream gauging; (2) "threshold" methods keyed upon critical cross sections of streams (Choice of these "critical" sites is subjective, but the methods theoretically examine the habitat features most limiting to the fish.); and (3) multiple-transect approaches that attempt to define the spatial aspects of the micro-habitat throughout a sampled stream reach.

Attempts to develop predictive fish standing crop models appear encouraging in that regression models relating standing crop of fishes to "habitat quality" have emerged that account for a high percentage of the observed variability. Binns (1976, 1977) developed a regression model with 10 attributes (critical period stream flow, annual stream flow variation, maximum summer stream temperature, mean water velocity, cover, stream width, stream-bank stability, food abundance, food diversity, and nitrates) of a stream regressed against "trout" standing crop. Biomass and habitat measurements were highly correlated (r = 0.95). Gorman and Karr (1978) developed a "habitat diversity" index comprised of three variables, depth, velocity, and substrate type. Significant correlations were found between habitat diversity and species diversity for two midwestern streams and one Panama stream.

Nickelson and Hafele (1978) developed regression models for coho salmon (*Oncorhynchus kisutch*) juveniles, cutthroat trout (*Salmo clarki*), and steelhead (*Salmo gairdneri*) juveniles. Pool volume, cover, depth, and velocity were the model variables (Nickelson and Beidler, 1979). Paragamian (1978) measured density (standing crop) of smallmouth bass (*Micropterus dolomieui*) simultaneously with depth and substrate size at 12 different reaches in an Iowa stream.

Regression of bass abundance on proportion of gravel and cobble was significant ($P < 0.05$).

IFG INCREMENTAL APPROACH TO ASSESSING FISH HABITAT STRUCTURE

Four major components of a stream system determine the productivity of the fishery (Karr and Dudley, 1978a,b): (1) flow regime; (2) physical habitat structure (channel form, substrate distribution, and riparian vegetation); (3) water quality; and (4) energy (watershed inputs in the form of sediments, particulate organic matter, and nutrients). Each of the components is interrelated with the others, and a detailed description of each is necessary before fish production can be explained.

An incremental simulation approach to assessing the relationship between stream flow and fish habitat structure has been developed by the Cooperative Instream Flow Service Group (IFG). The assessment process uses a hierarchical and modular approach and computer simulation techniques. The modules and various models available constitute the "building blocks" for the simulation and analyses warranted by a given instream-flow investigation. The quality of the physical habitat, being a deterministic function of flows, is described as a stochastic variable.

A modular approach with indicated information flow is proposed (Fig. 1). The IFG is pursuing this course of action by providing a role in model construction, which is to be periodically refined by a continual flow of knowledge and theory from applied research.

A hierarchical approach based upon different levels of knowledge, data, and analyses is necessary.

In terms of the degree of sophistication of analysis the following levels of present management application are identified. Components and linkages are as illustrated in Fig. 1.

Level 1 - Best professional judgement and use of simple calculations and nomographs, both within the system components and in linkages.

Level 2 - Mathematical models within each component are brought to bear. However, the linkage or information flow from one component to another rests with the judgment of the investigator.

Level 3 - Fairly sophisticated mathematical approaches are used, and linkages among various components are structured in a system-simulation manner.

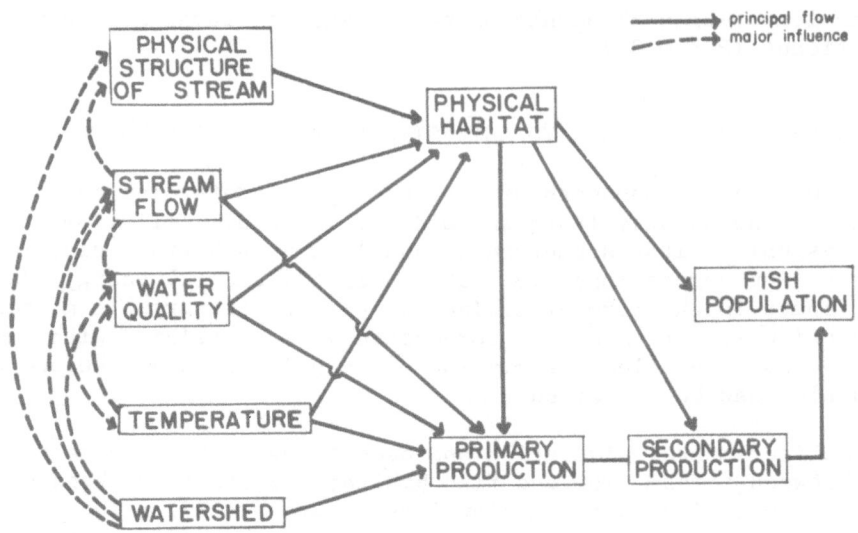

Fig. 1.   Major components and linkages necessary for analysis of
          change in stream systems.

Level 4 - This represents the truly holistic approach to
ecosystems and the frontier of knowledge.

     From the land and water management perspective, Level 3 would
be the application of the state-of-the art at its highest level of
operational feasibility. However, most management decisions are
presently made at Levels 1 and 2 and often with a mixture of Levels
1, 2, and 3 among various components. It is the goal of the IFG to
bring all components to Level 3 sophistication with dynamic
simulation techniques.

     IFG's incremental approach, as presently applied, is composed
of five steps: (1) field measurement of channel, physical, and
chemical characteristics; (2) hydraulic simulation of the spatial
distribution of combinations of depths, velocities, substrate, and
cover objects; (3) simulation of the temporal distribution of
temperature and chemical constituents; (4) application of habitat
evaluation criteria for species and life stages, for each flow
regime and channel condition under investigation; and (5) display of
the changing habitat usability over time.

     Fig. 2 illustrates information flow in the computerized system
as presently used. To be valid, parallel analyses at Level 1 or
above must indicate that watershed nutrient input and channel form
will not change significantly. If these components are assumed to
change, a site-specific analysis of this change is mandatory.

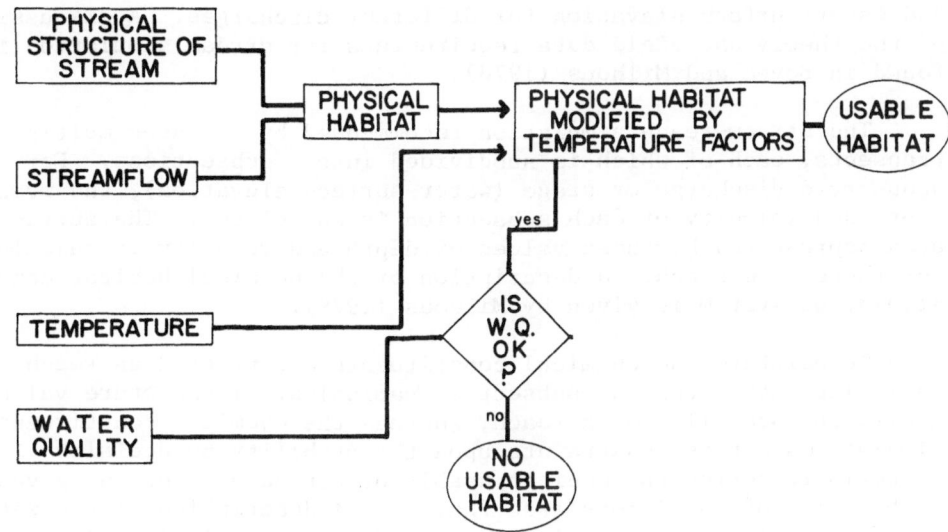

Fig. 2. Information flow in the IFG incremental approach as presently operating.

This methodology is intended to be used as a decision-making tool and is specifically tailored to demonstrate the impact of alternative flow regimes upon fishery habitat potential. The methodology was developed to provide a quantitative assessment of the following three fundamental types of instream flow problems:

1.   Quantification of instream flow requirements:
      A.   Area-wide planning activities
      B.   Protection of stream flow under state reservation or licensing provisions

2.   Negotiation of water delivery schedules:
      A.   Monthly flow regimes for average, wet, and dry years
      B.   Minimum releases

3.   Impact analyses:
      A.   Flow depletion
      B.   Flow augmentation
      C.   Channel alteration

Simulations

Several computer programs are available that can predict the hydraulic parameters of depth, velocity, width, wetted perimeter,

and water-surface elevation for different discharges. A discussion
of the theory and field data requirements for stream simulation is
found in Bovee and Milhous (1978).

The stream-reach simulation recommended by IFG uses multiple
transects, each of which is subdivided into n subsections. For any
unobserved discharge or stage (water-surface elevation), the mean
depth and velocity of each subsection is calculated. The surface
area represented by these values of depth and velocity is calculated
for the entire reach. A description of the physical habitat computer
simulation system is given by Milhous (1979).

Temperature and chemical constituents are treated as reach
variables rather than as subsection variables. Temperature values
modify the usability of a reach, whereas the chemical constituent
simulations act as constraints upon the usability by decision
criteria rendering the habitat usable or not usable for any given
combination of conditions (see Fig. 2). A description of the water-
quality simulation system is given by Grenney et al. (1975).

The stream-reach simulation takes the form of a multi-
dimensional matrix of the calculated surface areas of a stream
having different combinations of hydraulic parameters (i.e., depth,
velocity, substrate, and cover, when applicable). This is the total
summation of surface areas within the stream reach with a given
combination of depth and velocities. These areas are not necessarily
contiguous (Stalnaker, 1979).

Habitat Evaluation Criteria

To evaluate the magnitude of impacts caused by changes in
stream hydraulics and temperature, it is necessary to develop an
information base for each species or group of species of interest.
This information base is in the form of habitat evaluation criteria.

Biological criteria are primarily aimed at those parameters
affecting fish distribution that are most directly related to stream
flow and channel morphology, namely, depth, velocity, temperature,
and substrate. Cover, a habitat parameter of paramount importance
to many species, is also indirectly related to stream flow. Cover
may be incorporated into an assessment by evaluating the usability
of available cover objects in reference to the flow parameters
around them.

A computer file (FISHFIL) of species criteria for these physical
parameters is being developed and maintained by the IFG (Bovee and
Cochnauer, 1977). The type of criteria developed are illustrated in

Fig. 3. Water-quality criteria, as promulgated by USEPA (1976), are used with the water quality constraint model.

The expressed assumption is that the distribution and abundance of any species is not primarily influenced by any single parameter of stream flow, but related by varying degrees to all stream-flow parameters. Furthermore, these criteria are based on the assumption that individuals of a species tend to select the most favorable · conditions in a stream, but will also use less favorable conditions, with the preference for use decreasing where conditions are less favorable.

Most flow-assessment methodologies used to date address only one, or occasionally two, life stages (Stalnaker and Arnette, 1976). Frequently, a particular life-history stage, or a certain period,

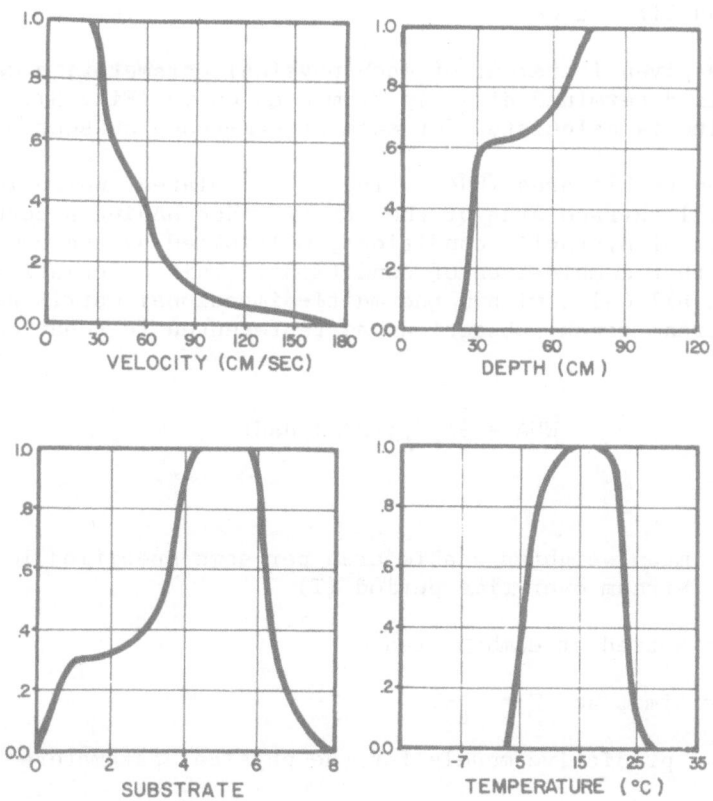

Fig. 3. Habitat evaluation criteria for adult brown trout (Y axis is normalized with optimum conditions set at 1.0) [from Bovee (1978)].

is singled out as being critical for the continued well-being of a
fish population. For example, spawning success is commonly considered
a critical factor in the maintenance of a fish population, but
habitat evaluations for fry and juvenile fish are almost universally
neglected. With the incremental, simulation approach, criteria are
used for all life stages.

Habitat Suitability

The quality of the physical habitat for a specific instream use
is a deterministic function of the flows. The IFG incremental
methodology is specifically designed to allow a stochastic, or time
series, approach to displaying this function.

Within acceptable water-quality conditions, a composite suitability of any defined combination of hydraulic and temperature
conditions encountered in a stream reach is determined for each
species and life stage.

For a given increment of each physical parameter, a weighting
function is determined directly from each curve (Fig. 3). A composite value is calculated for each stream-reach subsection.

A mean usable area (WUA) value is calculated, which is defined
as the total surface area of the stream reach having a certain
combination of hydraulic conditions, multiplied by the composite
value for that combination of conditions. This calculation is
applied to all cells within the multi-dimensional matrix and summed
for all stream-reach subsections as represented in Equation (1).

$$\overline{\text{WUA}} = \frac{1}{T}\int_t \int_A \psi(H(\ell))\,dAdt \tag{1}$$

where,

$\overline{\text{WUA}}$ = mean weighted usable area per some specified length of
stream over time period (T)

A = wetted streambed area

t = time, and T = $\int_t dT$

H(ℓ) = predictive models for the physical parameters

ℓ = space-time variables

ψ = habitat suitability function (weighting function)

The weighted usable area (WUA) computation roughly equates an area of suboptimal habitat to an equivalent area of preferred habitat.

For each species and life stage, weighted usable area is calculated for the range of discharge of interest (Fig. 4). The shape of this curve is a function of the channel form and distribution of hydraulic conditions throughout the sampled reach.

From such habitat-discharge relationships, and historical flow records or stochastic projections of flow events, one can generate

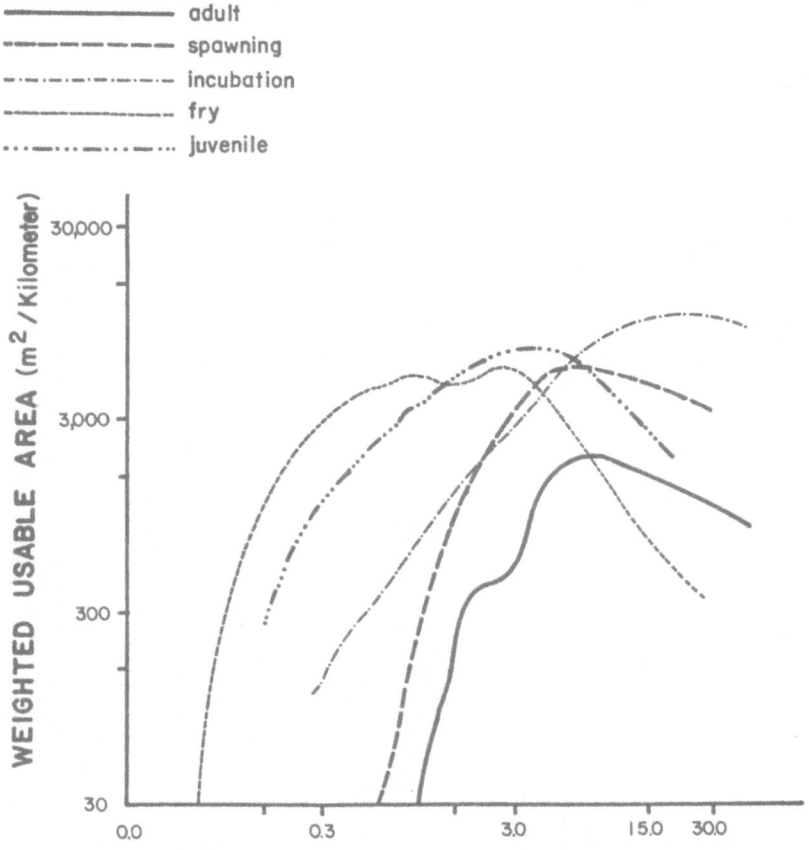

Fig. 4. Habitat/discharge relationship for five life stages of
         steelhead (*Salmo gairdneri*) in a coastal California
         stream.

plots of physical habitat suitability over time (Fig. 5). Such
plots can assist in identifying critical time periods for a given
life stage, limiting habitat availability for each life stage (i.e.,
physical carrying capacity), and differential habitat availability
for species. Because changes in streamflow characteristics will
initiate differential species reactions, the approach is particularly
useful in evaluating potential changes in species dominance.

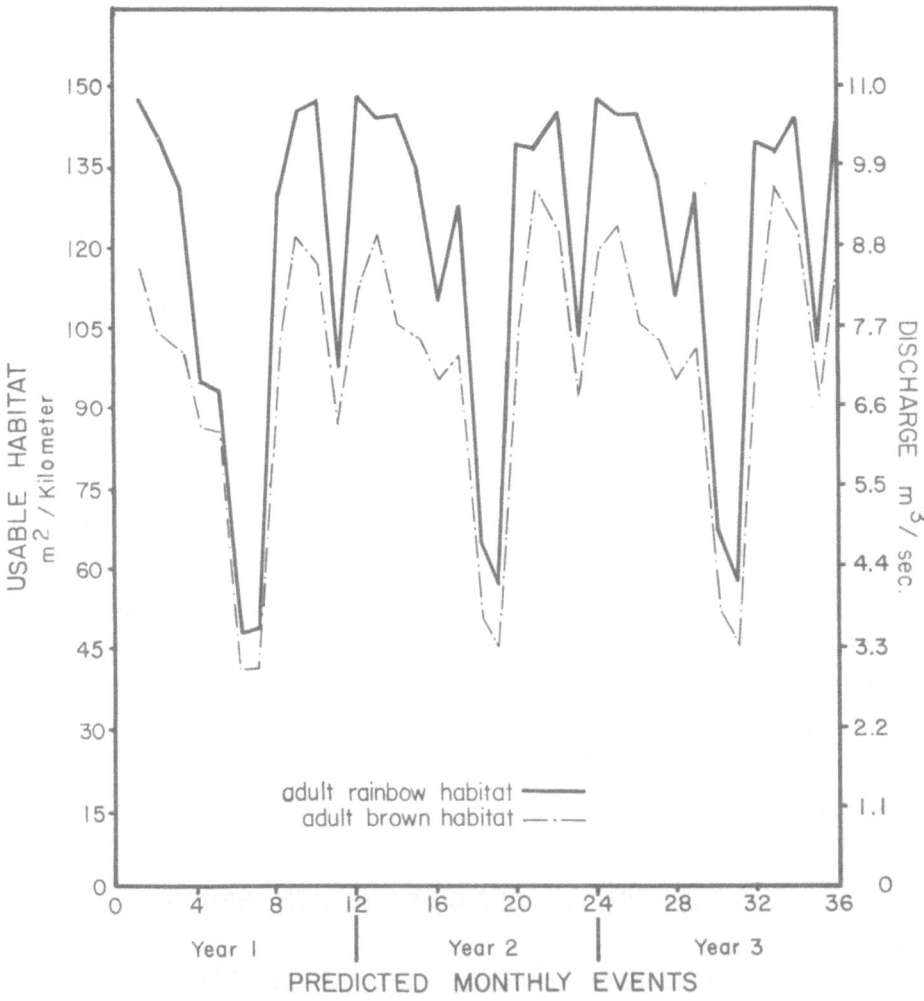

Fig. 5.   Example output relating usable habitat to monthly time
          series of discharges over a three-year period.

The IFG incremental methodology has been specifically designed as a communication link among fishery biologists, hydrologists, and hydraulic engineers. Engineers have been working for several years with hydraulic simulation models of streams. Hydrologists have been displaying water supply information as flow-duration and frequency-of-occurrence graphs. Similarly, fishery biologists have been collecting data on stream fishes. The IFG incremental methodology arranges the fishery data in such a way as to be directly analyzed in conjunction with the stream simulation and monthly time series of flow events. As a result, hydrologists and fishery biologists have a common means to display and discuss the interrelationships of their professions. The incremental methodology thus allows system-operating criteria to be developed considering the stochastic variation of the production of benefits from use of water instream.

Application To Instream-Flow Quantification

The following example demonstrates how a fishery manager might formulate an instream-flow request for presentation to a state water licensing authority.

Data requirements consist of stream-discharge records, species-periodicity chart (Table 1) and weighted usable area (WUA) computations for an array of discharges that describe all flow conditions encountered during an average year (1 in 2 year monthly events). The approach is to:

Table 1. Species Periodicity Chart--Steelhead Trout (o = Limited Occurrence and x = Primary Occurrence)

Life Stage	J	F	M	A	M	J	J	A	S	O	N	D
						Month						
Adult immigration	x										o	o
Spawning	o	o	x									
Incubation	o	o	o	x	x							
Fry emergence			o	o	x							
Juvenile rearing	x	x	x	x	x	x	x	x	x	x	x	x

1. Identify what life stage of the target fish species is to
    be given preferential consideration during each month of
    the year;

2. Obtain or construct the average annual hydrograph;

3. For each life stage, choose from Fig. 4 the lowest dis-
    charge that will maintain the same amount of usable
    habitat as the given average monthly flow;

4. Identify the average monthly flow that is limiting the
    life stage during the months of concern;

5. Build a flow recommendation around these limiting flows
    (Table 2).

The more historical information there is available on the fish
population and the streamflows, the better the analysis. Recon-
structed habitat usability plotted on a monthly basis along with
monthly population or density estimates can be very valuable in
associating limiting time of year and years of poor year-class-
strength with streamflow.

Testing

The Cooperative Instream Flow Service Group is cooperating with
local, state, and federal agency biologists in application and tests
of the incremental simulation approach on over 300 stream reaches
throughout the United States.

Research on behavior of fish and invertebrates is needed to
develop evaluation criteria broadly applicable to the nation's
streams. As these data become available, statistical analyses can
be continually conducted to examine the degree of interdependence
among the streamflow parameters as they influence species preferences. Regionalized differences in habitat evaluation criteria
should emerge and, when established, will become part of the species
data base being accumulated by the Project Impact Evaluation Group
of the U.S. Fish and Wildlife Service. Information available
through 1978 has come primarily from the western U.S. and is best
for the salmonids.

A test of the correlation between WUA and standing crop of
brown trout (*Salmo trutta*) was conducted with data provided by
Wesche (1976). All observations were made during the late summer
months, when habitat for the adult brown trout was assumed to be
most limited. Temperatures and water quality were similar among all
streams, and food was assumed to be non-limiting to the fish

Table 2. Example Instream-Flow Determination for Coastal Steelhead Trout [Monthly Instantaneous Flow (Q) Values in M^3/Sec]

Month	Life Stage of Overriding Concern	Median Flow Q_{1-2}	Instream Q Needed to Maintain WUA	Q Available for Other Uses
January	Adult	2.97	2.97[a]	0
February	Spawning	6.80	5.66	1.13
March	Spawning	7.36	5.66[a]	1.70
April	Incubation	9.91	5.10	7.65
May	Incubation	5.10	5.10[a]	0
June	Juvenile	2.32	0.57	1.76
July	Juvenile	1.13	0.57	0.57
August	Juvenile	0.74	0.57	0.17
September	Juvenile	0.57	0.57	0
October	Juvenile	0.62	0.51	0.06
November	Adults	1.08	1.08[a]	0
December	Adults	1.59	1.59[a]	0

[a]Monthly flow which is limiting the life stage.

populations. Fishing pressure was considered insignificant, and therefore populations were assumed to be at or near carrying capacity.

Criteria illustrated in Fig. 3 were used in applying the model. These criteria are from Bovee (1978) and developed independently from the data of Wesche (1976). A total of 21 data sets were available. Only two sets obviously violated the above assumptions in that fishing pressure was known to be very high and the observed standing crop was very low. The other 19 sets were analyzed

by computing WUA for each stream reach.   Absolute values for each
reach were regressed, revealing a high correlation between the
standing crop and habitat suitability (r = 0.85) and between biomass
and habitat suitability (r = 0.9) (Fig. 6).   Other similar analyses
are planned.

DISCUSSION

Evidence to date supports the validity and utility of the
incremental simulation approach to decision making on "trout and
salmon" streams.   The basic concepts are applicable to all stream
types, if used with caution.   In all cases, the degree of change
within each of the components displayed in Fig. 1 must be understood
before this approach can be placed in context.   The most useful role

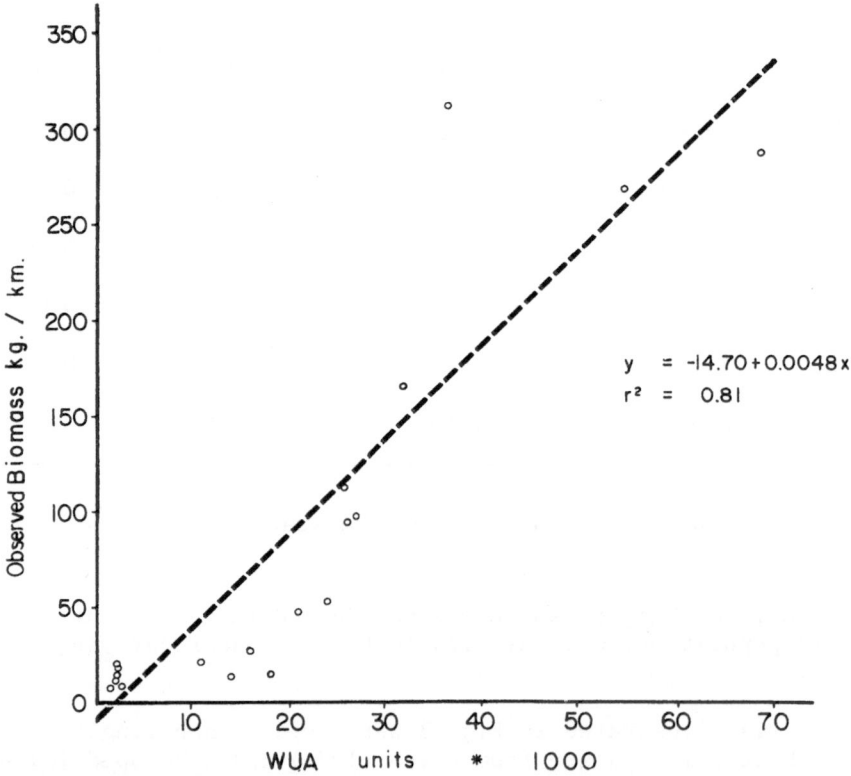

Fig. 6.   Correlation between standing crop of brown trout, *Salmo
          trutta*, and weighted usable area in eight Wyoming streams
          [computed from data provided by Wesche (1976)].

of simulation is to examine the dynamics of the usability of specific, but representative, stream reaches for specified species in order to determine *if* or *when* streamflow may be limiting.

The flow requirements for maintaining any desired level of the stream channel fish habitat structure must be dynamic and can only be protected by establishing *instream-flow regimes* for wet years (sediment and bed load transport); average years (establishes the base level of fish production); and dry years (provides minimal survival conditions for "seed" stock necessary for replenishing the stream reach).

Continual research and development should ultimately provide the aquatic biologist with useful tools to evaluate all components and linkages in a structured, consistent, and cost-efficient manner. The relationship between the usable habitat (WUA) and the standing crop when at carrying capacity for various fish and invertebrate species and species associations is the subject of needed research.

Simultaneous observations of standing crop and habitat are needed when the habitat is at "carrying capacity," before such tests can be made.

Flow requirements may differ among fish species and life stages as well as other instream uses, thus forcing the management agency and the public to define the management objectives for the stream reach in question.

The environmental management agencies will have to display the instream flow requirements in a manner conducive to *negotiating* with competing uses. Many mistakes are likely to be made. However, the greatest mistake is for the aquatic ecologist and fishery manager to offer no recommendations for establishing and reserving an instream-flow regime. To do so is tantamount to allowing the out-of-stream users to dictate the direction of the water allocation process. This could result in complete dewatering of certain river reaches and eventual depletions down to the 7-day Q10 in those reaches to which water quality standards are rigorously applied.

The first and most important rule for establishing an instream-flow regime must be the description and bounds of the stream reach to which the flow applies, i.e., from point A on a river downstream to point B. This definition of the reach is necessary in order to maintain some degree of certainty in the management and allocation of water among the many water uses of a river.

Second, all uses for which flows are to be protected must be identified. These may include navigation, hydropower generation, fish and wildlife, recreation, aesthetics, waste assimilation, and

estuarine inflow. Most often overlooked are periodic high flows
necessary to move bed load, flush sediments, and generally maintain
the desired stream channel characteristics. These collectively have
often been lumped together under the label *stream resource maintenance
flows*.

Water management must recognize the dynamic nature of the
instream uses and develop contingency plans for various water supply
conditions that equitably distribute the losses among all instream
and out-of-stream uses during low water years. This recognition
must come before any degree of certainty can be brought to the water
allocation and management process.

REFERENCES

Binns, N. A., 1976, Evaluation of habitat quality in Wyoming trout
    stream, Unpubl. ms. of presented paper, 33 p.
Binns, N. A., 1977, "Evaluation of Trout Habitat that Would Be
    Impacted by Cheyenne's Proposed Phase II Water Development in
    the North Fork Little Snake River Drainage," Wyo. Game and
    Fish Dep., Fish Div., Admin. Rep. Proj. 5076-09-6002.
Bovee, K. D., 1974, "The Determination, Assessment and Design of
    'Instream Value' Studies for the Northern Great Plains Region,"
    M.S. Thesis, Univ. Missouri, Columbia, 129 p.
Bovee, K. D., 1978, "Probability of Use Criteria for the Family
    Salmonidae," Instream Flow Info. Pap. No. 4, FWS/OBS-78/07,
    Coop. Instream Flow Serv. Group, Fort Collins, Colorado, 80 p.
Bovee, K. D., and Cochnaur, T., 1977, "Development and Evaluation of
    Weighted Criteria, Probability-of-Use Curves for Instream Flow
    Assessments: Fisheries," Instream Flow Info. Pap. No. 3,
    FWS/OBS-77/63, Coop. Instream Flow Serv. Group, Fort Collins,
    Colorado, 38 p.
Bovee, K. D., and Milhous, R. T., 1978, "Hydraulic Simulation in
    Instream Flow Studies: Theory and Techniques" Instream Flow
    Info. Pap. No. 5, Coop. Instream Flow Serv. Group, Fort Collins,
    Colorado, 131 p.
Dettman, D. H., 1978, "Distribution, Abundance and Microhabitat
    Segregation of Rainbow Trout and Sacramento Squawfish in Deer
    Creek, California," M.S. Thesis, Univ. California, Davis, 40 p.
Giger, R. D., 1973, "Streamflow Requirements of Salmonids," Anadro-
    mous Fish. Proj. 14-16-0001-4150, Oreg. Wildl. Comm., 117 p.
Gorman, O. T., and Karr, J. R., 1978, Habitat structure and stream
    fish communities, *Ecology*, 59:507-515.
Grenney, W. J., Dison, L. S., and Teuscher, M. C., 1975, "Assessment
    of Proposed River Management and Planning Alternatives by Water
    Quality Simulation Modeling," Utah Water Res. Lab., Logan,
    FRWA20-3.

Hooper, D. R.. 1973, "Evaluation of the Effects of Flows on Trout
    Stream Ecology," Pacific Gas Elec. Co., Emeryville, California,
    97 p.
Hunter, J. W., 1973, "A Discussion of Game Fish in the State of
    Washington as Related to Water Requirements," Wash. Dep. Game,
    Unpubl. ms., 66 p.
Hynes, H. B. N., 1970, "The Ecology of Running Waters," Univ.
    Toronto Press, Toronto, 555 p.
Karr, J. R., and Dudley, D. R., 1978a, Biological integrity of a
    headwater stream, *BioScience* (submitted).
Karr, J. R., and Dudley, D. R., 1978b, A primer on the biological
    integrity of running waters, Unpubl. ms., 12 p.
Matthews, W. J., and Hill, L. G., 1979, Influence of physiochemical
    factors on habitat selection by redshiners, *Notropus lutrensis*
    (Pisces:Cyprinidae), *Copeia*, 1979:70-81.
Milhous, R. T., In press, The PHABSIM system for instream flow
    studies, *Proc. Summer Computer Simulation Conf.*, Toronto,
    Canada.
Nickelson, T. E., and Beidler, W. M., 1979, "Willamette Basin
    Streamflow Studies. 1977," Oreg. Dep. Fish and Wildl., Fish
    Res. Proj. 2-4-8-20-02, Final Job Rep., 30 p.
Nickelson, T. E., and Hafele, R. E., 1978, "Streamflow Requirements
    of Salmonids," Oreg. Dep. Fish and Wildl., Federal Air Proj.
    AFS-62, Annu. Prog. Rep., 26 p.
Ott, A. G., and Tarbox, K. E., 1977, "'Instream flow' Applicability
    of Existing Methodologies for Alaskan Waters," Prepared for
    Alaska Dep. Fish and Game and Alaska Dep. Nat. Resources,
    Woodward-Clyde Consultants, Anchorage, Alaska, 70 p.
Paragamian, V. L., 1978, "Population Dynamics of Smallmouth Bass in
    the Maquoketa River and other Iowa Streams - Physical and
    Chemical Characteristics of the Maquoketa River," Iowa Conserv.
    Comm., Federal Aid Proj. No. F-89--2, Annu. Prog. Rep., 56 p.
Stalnaker, C. B., 1979. Methodologies for preserving instream
    flows, the incremental method, *Proc. Instream Flow Management:
    State-of-the-Art*, Upper Mississippi River Basin Comm.,
    Twin Cities, Minnesota.
Stalnaker, C. B., and Arnette, J. L., 1976, "Methodologies for the
    Determination of Stream Resource Flow Requirements: An Assess-
    ment," U.S. Fish and Wildl. Serv., Off. Biol. Serv., Washington,
    D.C., 199 p.
U.S. Environmental Protection Agency, 1976, "Quality Criteria for
    Water," USEPA, Washington, D.C., 256 p.
Wesche, T. A., 1976, Development and application of a trout cover
    rating system for IFN determinations, p. 224-234, *in*:
    "Instream Flow Needs, Vol. II," J. F. Orsborn and D. H. Allman,
    eds., West. Div. Am. Fish. Soc., Bethesda, Maryland.
Wesche, T. A., 1978, "Determining Instream Flows for Management of
    Aquatic and Riparian Ecosystems," Rep. U.S. For. Serv. prepared
    by Water Resources Res. Inst., Laramie, Wyoming, 158 p.

MODEL PREDICTIONS OF EFFECTS OF IMPOUNDMENT ON PARTICULATE

ORGANIC MATTER TRANSPORT IN A RIVER SYSTEM

J. R. Webster, E. F. Benfield, and J. Cairns, Jr.

Biology Department and Center for Environmental Studies
Virginia Polytechnic Institute and State University
Blacksburg, Virginia 24061

INTRODUCTION

Seston may be defined as all organic and inorganic matter
suspended in water (Ruttner, 1963). The organic fraction of seston,
including both living and nonliving material, is generally known as
particulate organic matter (POM). A substantial number of studies
over the last quarter-century have demonstrated that POM is extremely
important in the energetics of freshwater, marine, and estuarine
ecosystems. The concern in this paper is with the dynamics of POM
in stream ecosystems and the effects of impoundment on POM dynamics
in a river system.

A general model of biological factors affecting POM in streams
is illustrated in Fig. 1. There are two broad categories of POM:
coarse particulate organic matter (CPOM), usually considered to be
material greater than 1 mm diameter, and fine particulate organic
matter (FPOM), particles between 1 mm and 0.45 μm in diameter.
Stream POM is derived from several possible sources. In low-order,
woodland streams, the major source of POM is allochthonous material,
primarily leaves, falling directly or blowing and sliding into the
stream from the stream margin (e.g., Cummins, 1974). In nonwoodland
streams and in woodland streams sufficiently wide to allow light
penetration through the canopy, periphytic algae are a major source
of fixed carbon (e.g., Minshall, 1978) and algal sloughing contributes to POM (e.g., Swanson and Bachmann, 1976). In larger streams,
macrophytes may provide substantial portions of autochthonous
production (e.g., Cummins, 1975) and, on death, contribute to POM.
Additionally, though not shown in Fig. 1, periphyton and possibly
macrophytes may contribute indirectly to the seston through herbivore

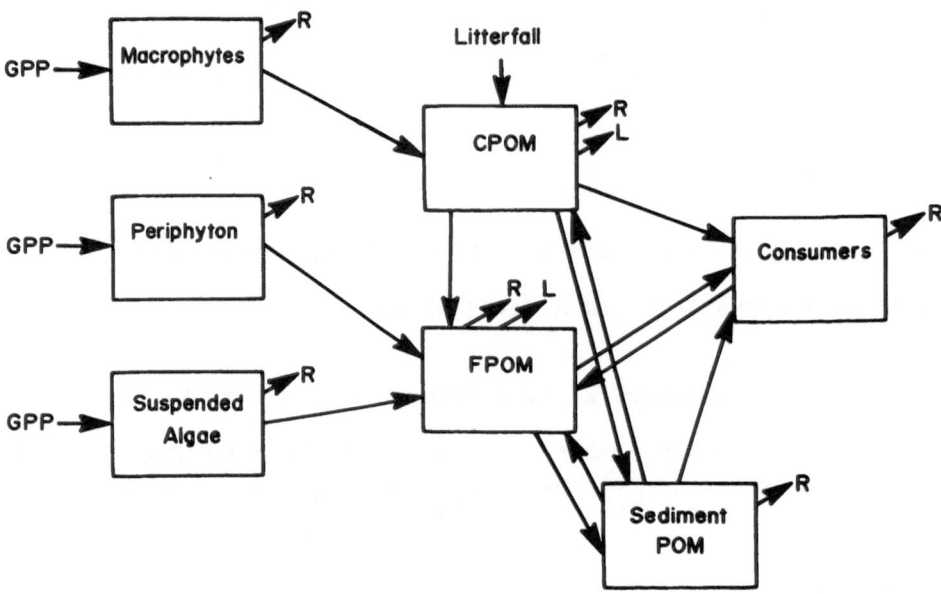

Fig. 1. General model of particulate organic matter (POM) in a
         stream ecosystem; GPP = gross primary production, CPOM
         = coarse POM, FPOM = fine POM, R = respiration, and L
         = leaching.

consumption and through flocculation of leachates. In large rivers,
and especially in reservoirs, suspended algae are the major source
of primary production and POM. Also, floodplain capture of allochthonous material may be an important source of POM. Once in the
stream, POM is processed or degraded via respiration by associated
microflora, leaching of dissolved organic material, capture and
consumption by consumers, and sedimentation. POM is returned from
sediments by increased flows, and a large fraction of the POM
ingested by collectors is returned by egestion (Wallace et al.,
1977).

    A realization that POM is important in the energetics of stream
ecosystems has led to a proliferation of investigations of POM
dynamics in a variety of streams. General results of some of the
more recent studies are listed in Table 1. Stream orders have been
combined for the sake of convenience and because of uncertainty of
the precise order of certain streams. Several difficulties encountered in attempts to derive broadly applicable models for POM
dynamics in streams from the literature are immediately apparent in
Table 1. Methods for sample collection varied somewhat, but that
factor is probably less bothersome than variation in methods used
for sample analysis.

In a comparison of POM transport in streams of various orders, both within and between geographic regions, Sedell et al. (1978) found that there was little variation in POM concentration between stream orders within specific regional basins; however, they determined that concentrations varied between basins by as much as an order of magnitude. In addition to finding no strong correlation between stream order and POM transport characteristics, Sedell et al. (1978) were unable to demonstrate strong relationships between either unit stream power or gradient and POM transport, such as has been demonstrated for inorganic particles. They suggested that differences in behavior of organic and inorganic particles can be attributed to lower specific gravities and higher surface-to-volume ratios of organic particles.

Flow is one stream variable that often has been correlated positively with POM concentration (Nelson and Scott, 1962; Maciolek, 1966; Bormann et al., 1974; Wetzel and Manny, 1977; Malmqvist et al., 1978; Naiman and Sedell, 1979). Maximum concentrations of POM are usually associated with episodic high discharges related directly or indirectly to precipitation (Bormann et al., 1969; Fisher and Likens, 1973; Fisher and Minckley, 1978). Seasonal timing of high discharges has varying effects on POM concentration (Liaw and MacCrimmon, 1977; Wetzel and Manny, 1977; Bilby and Likens, 1979). For example, a June storm produced a twofold greater organic load in Catahoula Creek (Mississippi) than at a similar discharge in February (de la Cruz and Post, 1977). A peak discharge in May, equaling about half the February discharge, produced an organic loading almost equal to the June load.

One general conclusion to be derived from the various studies of POM transport in all types of streams is that most of the material is transported as extremely small particles (e.g., Maciolek, 1966; Maciolek and Tunzi, 1968; Fisher and Likens, 1973; Sedell et al., 1978; Naiman and Sedell, 1979). CPOM entering headwater streams tends to move only short distances before becoming trapped by obstructions and is usually processed in place (e.g., Peterson and Cummins, 1974; Malmqvist et al., 1978; Naiman and Sedell, 1979a). Rising discharges suspend smaller and lighter particles and transport them downstream. Receding discharges allow fine particles to be trapped by obstructions (Bilby and Likens, 1979) or to settle in low gradient reaches of the stream (Malmqvist et al., 1978). Subsequent discharge peaks repeat the process. Continued mechanical action, coupled with a variety of biological processing factors (Boling et al., 1975), reduces the particle size of POM. Naiman and Sedell (1979a) summarized the phenomenon concisely. Small streams, by virtue of their resistance to the movement of large particles, retain CPOM; thus, more FPOM is transported than other size fractions. Larger streams transport FPOM because that is the

Table 1. POM Concentrations in Streams

Location	Method[a]	POM Conc Range mg l⁻¹	POM Conc mean annual mg l⁻¹	Duration	Authors
Stream Order 1-3					
Augusta Creek, Mich. (1st Order)	3	0.6-6.4[b]		weekly, 24 mo	Wetzel and Manny (1977)
Augusta Creek, Mich. (3rd Order)	3	0.6-13.96[b]		weekly, 24 mo	Wetzel and Manny (1977)
Augusta Creek, Mich. (2nd Order)	1	2.4-15.0	7.13	1-2d/qtr	Sedell et al. (1978)
Augusta Creek, Mich. (3rd Order)	1	3.01-7.02	4.77	1-2d/qtr	Sedell et al. (1978)
Bear Brook, N.H.	3		0.31	monthly? 19 mo	Fisher and Likens (1973)
Camp Creek, Idaho	1	0.82-7.60	2.62	2-3d/qtr	Sedell et al. (1978)
Catahoula Creek, Miss.	1	4.80-11.85	7.0	Feb.-Aug.	de la Cruz and Post (1977)
Convict Creek, Calif.	3	0.30-2.1	0.67	weekly, 12 mo	Maciolek (1966)
Devils Creek, Ore.	1	0.54-1.35	0.97	2-3d/qtr	Naiman and Sedell (1979a)
Hubbard Brook (W-6), N.H.	1	<0.1-10.0		variable 4 yr	Bormann et al. (1974)
Hubbard Brook (W-2), N.H.	1	<0.1-10.0		variable 4 yr	Bormann et al. (1974)
Laurel Creek, Calif.	3	0.5-3.0		July, Aug.,Oct.	Maciolek and Tunzi (1968)
Mack Creek, Ore.	1	0.212-1.270	0.59	2-3d/qtr	Naiman and Sedell (1979a)
Panther Creek, Ga.	1	0.7-3.3		quarterly 12 mo	Malas and Wallace (1977)
Rhode River Watershed, Md.	1		3.7-22.7	weekly, 3 yr[c]	Pierce and Dulong (1977)
Roaring Brook, Mass.	1	0.04-3.2		bimonthly, 3 mo	McDowell and Fisher (1976)
Smith Creek, Mich.	1	2.62-10.4	5.89	1-2d/qtr	Sedell et al. (1978)
Stampen (stream), Sweden	3	4.0-22.0		monthly, 13 mo[d]	Malmqvist et al. (1978)
White Clay Creek, Penn.	1	0.97-3.65	1.36-2.29	2-3d/qtr	Sedell et al. (1978)
Stream Order 4-5					
Buck and Doe Run, Penn.	1	0.89-6.91	3.1	1-2d/qtr	Sedell et al. (1978)
Kalamazoo River, Mich.	1	2.15-6.91	4.76	weekly, 12 mo	Sedell et al. (1978)
Little Miami River, Ohio	1	1.0-31.0		2-3d/qtr	Weber and Moore (1967)
Lookout Creek, Ore.	1	0.31-1.10	0.58	monthly, 12 mo	Naiman and Sedell (1979a)
Middle Oconee, Ga.	4	8.4-47.0	10.0-20.0 (Ave)	weekly, 3 mo[e]	Nelson and Scott (1962)
Palouse River, Idaho	4	2.2-3.5		monthly, 12 mo	Buscemi (1969)
River Pilica, Poland	4	2.4-8.5		5d/wk 10 mo	Penczak, et al. (1976)
River Ricklcon, Sweden	1	0.2-5.0	1.0-1.5	2-3d/qtr	Karlström and Backlund (1977)
Salmon River, Idaho	1	0.72-9.4	3.68	2-3d/qtr	Sedell et al. (1978)
Shetucket River Conn.	5	0-3.0		3 h thundstm., Oct.	Klotz and Matson (1978)
Tallulah River, Ga./N.C.	1	1.5-4.4		Apr.,Oct.,Nov.	Wallace et al. (1977)
W. Br. Patuxent River, Md.	2	0.39-34.4[b]		monthly, 24 mo	Keefe et al. (1976)

		Stream Order 6-7			
Altamaha River, Ga.	1	0.66-1.2		Apr.,Oct.,Nov.	Wallace et al. (1977)
Brazos River, Tex.	2	2.76-11.04[b]	7.2[b]	monthly, 12 mo	Malcolm and Durum (1976)
Grand River, Ont.	3	1.0-26.2	6.9	monthly, 24 mo	Liaw and MacCrimmon (1977)
McKenzie River, Ore.	1	0.616-0.919	0.742	2-3d/qtr	Naiman and Sedell (1979a)
Nanaimo River, B.C.	2	0.28-0.68[b]		quarterly, 12 mo	Seki et al. (1969)
Nanaimo River, B.C.	1	0.10-1.2[b]		bimonthly, 12 mo	Naiman and Sibert (1978)
Neuse River, N.C.	2	0.396-9.8[b]	5.64[b]	monthly, 12 mo	Malcolm and Durum (1976)
Patuxent River, Md.	2	0.67-10.3[b]		monthly, 24 mo	Keefe et al. (1976)
River Thames, England	1	2.09-19.83	5.47	Apr.-Mar.	Berrie (1972)
Salmon River, Idaho	1	0.51-1.98	1.24	2-3d/qtr[f]	Sedell et al. (1978)
Sopchoppy River, Fla.	2	0-8.04[b]	3.2[b]	monthly, 3 mo	Malcolm and Durum (1976)
South Platte River, Colo.	1	0-40.7		monthly, 12 mo[g]	Ward (1974)
South Platte River, Colo.	1	0.8-5.7		monthly, 12 mo[h]	Ward (1976)
		Stream Order >7			
Ohio River, Ill.	2	1.56-8.76[b]	3.60[b]	monthly, bimonthly, 11 mo	Malcolm and Durum (1976)
Mississippi River, La.	2	2.16-12.24[b]	7.56[b]	monthly, 6 mo[i]	Malcolm and Durum (1976)
Missouri River, Neb.	2	0.96-380.0[b]	18.36[b]	monthly, 16 mo	Malcolm and Durum (1976)

[a]1-filtration; loss on ignition
2-filtration; carbon analysis via machine
3-filtration; dichromate oxidation
4-centrifugation; loss on ignition
5-other

[b]POM computed as $\frac{POC}{0.5}$

[c]Data summarized for 9 small streams in watershed

[d]Data summarized for 17 sites

[e]Data summarized for 4 sites

[f]Data summarized for 2 sites

[g]Data summarized for 4 sites

[h]Data summarized for 8 sites

[i]Data summarized for 3 sites

predominant fraction they receive from tributaries. These authors
conclude that FPOM is probably generated continuously and rapidly.

Goldman and Kimmel (1978) state that dams are barriers to
natural drainage and result in interruptions of organic matter
transport that may profoundly affect downstream food webs. The few
studies available concerning reservoir effects on POM show that
different reservoirs have different effects. Some papers point to
serious disruption in the flow of organic matter, while others point
to limnetic production in reservoirs as a positive influence downstream. Perhaps the central difficulty is that few studies have
been sufficiently broad to encompass large reaches of stream in
which reservoirs are integral components. Our objective in this
study was to construct a model of POM dynamics in a river system
that would reflect, as much as possible, known mechanisms of POM
transport and utilization. We then placed an impoundment on the
model river and observed the effects on POM dynamics.

MODEL DEVELOPMENT

General Modeling Approach

In building the model to assess effects of impoundment, we
considered the entire river system above an impoundment and for some
distance downstream. The interdependence of reservoirs and their
parental rivers clearly indicates that they should be studied as a
single functional unit, a river-reservoir ecosystem. Vannote
(unpublished manuscript) and Cummins (1975) point out that the
general framework within which stream ecosystems should be viewed is
as a continuum from headwaters to mouth.

Present biologically oriented models of stream ecosystems are
essentially point models (e.g., Hall, 1972; McIntire, 1973; Boling
et al., 1975; Webster et al., 1975; McIntire and Colby, 1978). The
fundamental difference between a stream and a lake is the unidirec-
tional flow of water, which, for the most part, precludes feedback
from downstream components (Webster and Patten, 1979). This essential feature is lacking in point models. Point models, in effect,
treat streams as lakes. O'Neill et al. (unpublished manuscript) and
Boling (personal communication) are attempting to avoid this error
by modeling stream ecosystems as series of point models. This
approach is certainly an improvement, but still treats streams as
linearly connected lakes. In our model development we used the
civil engineering approach, pioneered by Streeter and Phelps (1925)
and further developed by O'Connor (1962), Dobbins (1964), and
Thomann (1972), in which stream distance is treated as an independent
variable. The model equation for POM is a partial differential
equation describing the rates of change of POM with respect to both

time and distance. We combined concepts from current biological and civil engineering models and placed them on a framework of geomorphological models that describe the physical conditions for biological functions. In addition, we used hydraulic engineering principles in developing several parts of the model.

We attempted to develop a general model of POM transport in streams. All inputs were functions, rather than empirical values, in order to minimize the number of site specific parameters. Where it was necessary to numerically parameterize the model, we used data from the New River and Claytor Lake. The section of river used extends from the headwater near Boone, North Carolina, north 400 km to Glen Lyn, Virginia, near the West Virginia border. Over this reach the river goes from first to sixth order. Human impact on the New River includes impoundment, industrial and urban effluents, and agricultural runoff. Upper reaches of the river, though far from pristine, have received minimal impact, and one section has been designated for protection under the Scenic Rivers Act. Approximately 290 km below the headwaters, the New River is impounded by a hydroelectric power dam, which forms Claytor Lake. We emphasize that our model is not meant to be a simulation of the effects of Claytor Lake on POM transport in the New River. Rather, at this time, we intend the model to be a means of summarizing current knowledge of POM dynamics in a river-reservoir ecosystem and a means of identifying areas where essential information is lacking.

Geomorphic Foundation

The geomorphic foundations of our model rest on three empirical equations relating elevation, stream width, and mean annual stream flow to distance.

Elevation and slope. The longitudinal profile of a river (a plot of elevation versus distance) typically has the shape of a negative exponential; that is, it is usually concave (e.g., Leopold et al., 1964; Morisawa, 1968). Data from the New River very closely approximate this relationship ($r^2 = 0.99$, $N = 46$). Slope (or gradient) of the stream was calculated as the derivative of the elevation curve:

$$E = 966.9 \, e^{-0.00197x} \tag{1}$$

and

$$G = -1.907 \, e^{-0.00197x}, \tag{2}$$

where E is the elevation (m), x is distance measured from the head-waters (km), and G is channel gradient (m km^{-1}).

Streamflow. Mean annual streamflow was assumed to be directly proportional to stream distance in the analysis by Leopold and Maddock (1953). Using data from seven gaging stations on the New River [Kanawha River Basin Coordinating Committee (KRBCC), 1971], we found a better approximation ($r^2 = 0.98$, N = 7) by using a power function:

$$Q = 0.0054 \ x^{1.718}, \tag{3}$$

where Q is mean annual discharge (m^{3}s^{-1}). We introduced seasonal fluctuation in flow, using a sine function with a wavelength of one year, a mean equal to the mean annual flow, an amplitude of 1.2 times the mean, and a peak in mid-March. These parameters are based on data from the New River (KRBCC, 1971). In addition, an annual flood occurring in mid-March, lasting approximately 3 days, and with a flow of 20 times the mean annual flow was included in the model.

Stream width. Stream width approximates a power function of discharge both for temporal variation of streamflow at one point and spatial variation of mean annual flow (Leopold and Maddock, 1953). Because mean annual flow is a power function of distance (Equation 3), width can be treated as a power function of distance. Using measurements taken from Geologic Survey 15 minute series topographic maps, we found a significant power function relationship between width and distance ($r^2 = 0.89$, N = 22):

$$W = 2.18 \ x^{0.740}, \tag{4}$$

where W is stream width (m). Temporal variability in stream width was not incorporated into our model.

Hydraulic Parameters

Derivations of other model hydrodynamic parameters are based on the Manning equation:

$$V = \frac{1}{n} \ R^{2/3}G^{1/2}, \tag{5}$$

where V is stream velocity (m s^{-1}), n is Manning's roughness coefficient, and R is hydraulic radius (m); and on the flow continuity equation:

$$Q = WDV, \qquad (6)$$

where D is mean depth (m). In the Manning equation (Equation 5), we have approximated hydraulic radius with mean depth, a good approximation for natural river channels (Leopold and Maddock, 1953). We used n = 0.04 for the entire river.

Solving Equations 5 and 6 simultaneously, we obtained an equation for mean depth as a function of empirically derived parameters:

$$D = \left(\frac{Qn}{WG^{1/2}}\right)^{3/5}. \qquad (7)$$

We then calculated velocity and cross-sectional area (A) from these parameters:

$$V = \frac{Q}{WD} \qquad (8)$$

and

$$A = WD. \qquad (9)$$

Reservoir Parameters

Our model reservoir began at a dam 300 km below the river headwater and extended upstream 25 km. From Equation 1, the surface elevation of the reservoir was 562.3 m. Reservoir depth was calculated as

$$D = 562.3 \ -E, \ 275 < x \leq 300, \qquad (10)$$

where E was calculated from Equation 1. Width of the reservoir was calculated from a linear equation derived from regression of map-measured widths of Claytor Lake on distance:

$$W = 268.0 + 18.4 \ (x - 275), \ 275 < x \leq 300. \qquad (11)$$

The fit of this equation was not strong but statistically significant ($r^2 = 0.48$, N = 25). After calculating width and depth, we calculated velocity and cross-sectional area from Equations 8 and 9, as before.

Biological Parameters

In our model development we have so far used only a much simplified version of Fig. 1. As shown in Fig. 2, we considered a single, combined category of POM with a single input from litterfall and an output due to breakdown. Suspended POM interacted with sediment POM through deposition and erosion.

Litterfall dynamics were based on the model of litterfall into a lake, developed by Gasith and Hassler (1976), in which litterfall decreased linearly with distance from the shoreline to zero at 10 m. We assumed a forest literfall rate of 300 g m^{-2} y^{-1}, which is about average for deciduous forests (Bray and Gorham, 1964). Our litterfall equation was:

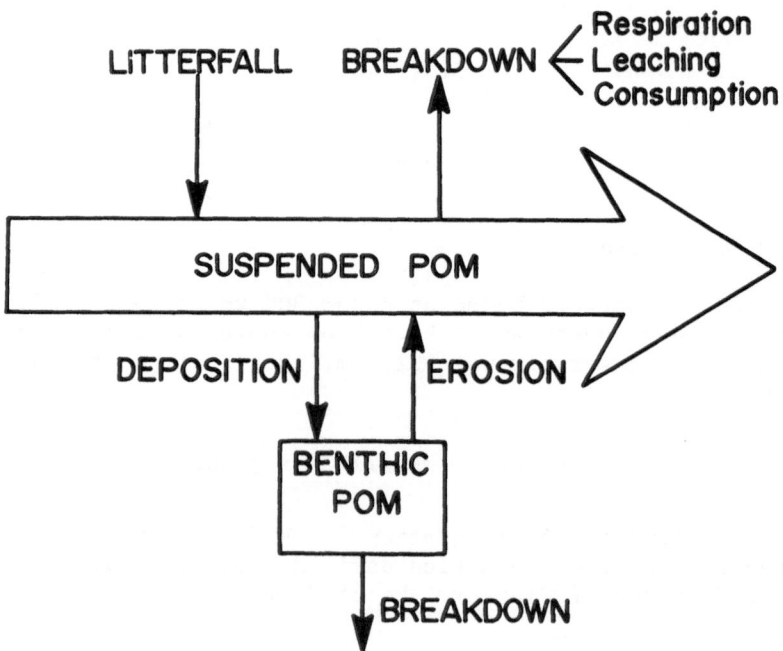

Fig. 2. Simplified model of particulate organic matter (POM) in a
         stream ecosystem.

$$Z = \begin{cases} 300\ W - 15\ W^2, & W \leq 10 \\ 1500, & W > 10, \end{cases} \qquad (12)$$

where Z is litterfall (g per linear meter). According to this equation, litterfall input increased rapidly with stream distance to the point where stream width exceeded 10 m (7.8 km downstream). From there downstream, litterfall input was constant at 1500 g m^{-1}. To simulate seasonal litterfall, we modeled all input to occur as a pulse in October and November.

Paul (1978) measured leaf breakdown rates in the New River, using mesh bags of box elder, sycamore, sugar maple, and dogwood. For these species, which represent a broad spectrum of breakdown rates, we estimated an average breakdown rate (Jenny et al., 1949; Olson, 1963) of 2.74 y^{-1}. This constant was used for breakdown of both suspended and sediment POM.

Sedimentation and Erosion

Because information concerning sedimentation and resuspension of organic particles is lacking, our model is based on the stream power approach developed by Bagnold (1966) for inorganic particles. In the equation

$$I_s = \Omega(a\ V/V_f), \qquad (13)$$

I_s is the suspended sediment load a stream can carry, Ω is available stream power, a is a constant, and V_f is fall velocity of the suspended particles. Available stream power is given by

$$\Omega = \rho g Q G, \qquad (14)$$

where ρ is the density of water and g is the force of gravity. The suspended sediment load can also be written as

$$I_s = \frac{\sigma - \rho}{\sigma}\ mgVW, \qquad (15)$$

where σ is the density and m is the mass of the suspended particles, so that $\frac{\sigma - \rho}{\sigma}$ is the immersed weight of the suspended sediment. Substituting Equation 14 into Equation 13 and solving 13 and 15 simultaneously for m yields

$$m = \frac{c\,Q\,G}{W}, \qquad\qquad (16)$$

where c is a constant combining a, V_f, ρ, and σ. The units of m are mass per unit area. Converting m to concentration by dividing by depth yields

$$S = c\,V\,G, \qquad\qquad (17)$$

where S is suspended sediment concentration (g m^{-3}).

Equation 17 can be compared to empirically derived equations. Combining Equations 7 and 8 gives

$$V \propto Q^{2/5}. \qquad\qquad (18)$$

Converting S, sediment concentration, to sediment load, L = SQ, we obtained

$$L \propto Q^{1.4}. \qquad\qquad (19)$$

Leopold and Miller (1956) state that the exponent of this equation is between 1.5 and 2.0 for natural river channels [citing Leopold and Maddock (1953)]. However, Leopold and Maddock (1953) and Leopold et al. (1964) report that the exponent is between 2 and 3. Müller and Förstner (1968) found values of the exponent from 0 to 2.5.

We used Equation 17 to calculate maximum POM load in the river. We evaluated c with data from Newbern (1978) for the New River. In late fall 1976 the New River at Galax, Virginia, was at approximately mean annual flow (50 m^3 s^{-1}). POM concentrations during this period ranged between 2 and 3 mg l^{-1}. We calculated c = 3.44, based on the assumption that POM was maximal at that time of year.

In our model, if the POM concentration was above the maximum (S_{max}) calculated from Equation 17, POM was transfered from suspension to sediment at the rate

$$DR = 4335\,\frac{S - S_{max}}{S_{max}}, \qquad\qquad (20)$$

which gives a deposition rate (y^{-1}) of 50% of the excess per hour when the POM concentration is twice the maximum. Actual deposition (DP) was then calculated as

$$DP = DR(S - S_{max}). \tag{21}$$

Similarly, resuspension occurred when the POM concentration was less than the maximum and POM was available in the sediment according to the equation

$$E = 90(S_{max} - S), \tag{22}$$

where E is resuspension input ($g\ m^{-3}\ y^{-1}$). This equation is based on the assumption of 25% uptake of the deficit per day.

Simulation Equations

Our simulation equation for the rate of change of POM concentration (S) with respect to time (t) and distance (x) came from Thomann (1972) and was based on mass balance:

$$\frac{\partial S}{\partial t} = \frac{1}{A}\frac{\partial(QS)}{\partial x} - KS + \frac{Z}{A} + \frac{S_T}{A}(\frac{\partial Q}{\partial x}) + E - DP, \tag{23}$$

where K is breakdown rate, S_T is the POM concentration of tributary inputs, and other symbols are as above. For tributary inputs we used:

$$S_T = \begin{cases} S(x), & x \leq 275 \\ S(275), & x > 275. \end{cases} \tag{24}$$

That is, above the point where the lake starts, all tributary POM concentrations equaled the main river POM concentrations. From 275 km on downstream, tributary inputs remained constant.

This equation was solved by the method of characteristics (e.g., Chester, 1971). By changing to another independent variable, ξ, we broke down Equation 23 into a set of three ordinary differential equations:

$$\frac{dt}{d\xi} = 1$$

$$\frac{dx}{d\xi} = \frac{Q}{A} \tag{25}$$

$$\frac{dS}{d\xi} = \frac{S_T - S}{A} \frac{\xi Q}{\xi x} - KS + \frac{Z}{A} + E - DP.$$

The first equation was solved analytically: $t = \xi + t_O$, where t_O is the initial time. The rate of change of discharge with respect to distance was calculated by differentiating Equation 3. The other two equations were then solved numerically by the fourth-order Runge-Kutta integration technique, with initial conditions t_O = variable, $x_O = 0.001$, and $S_O = 0$. The generated solution was a downstream series of POM concentrations, with time also increasing with downstream distance.

Because sediment POM does not move downstream, the same solution technique could not be used. We found it necessary to treat sediment POM as a series of compartments. We used a total of 70 compartments, distributed at decreasing intervals downstream. Between downstream simulation runs, each sediment POM compartment was updated according to the equation

$$\frac{dB_i}{dt} = (DP_i - E_i) D - K B_i, \tag{26}$$

where B_i is the sediment POM concentration (g m^{-2}) in compartment i, and DP_i and E_i are the deposition inputs and erosion outputs to compartment i. DP_i and E_i were calculated as the average values of DP (Equation 21) and E (Equation 22) over the distance represented by compartment i. Multiplication by depth (D) converted from concentrations to area units.

RESULTS AND DISCUSSION OF SIMULATION

Some of the output from a 14-month simulation produced by our model is shown in Figs. 3 and 4. With respect to the river above the reservoir, the model produced the following results:

1. POM concentration was highest during leaf fall and
    decreased through the rest of the year. It is difficult
    to separate seasonal effects from flood effects in pub-
    lished data; however, there are studies with which the
    model results can be compared. In studies of low-order

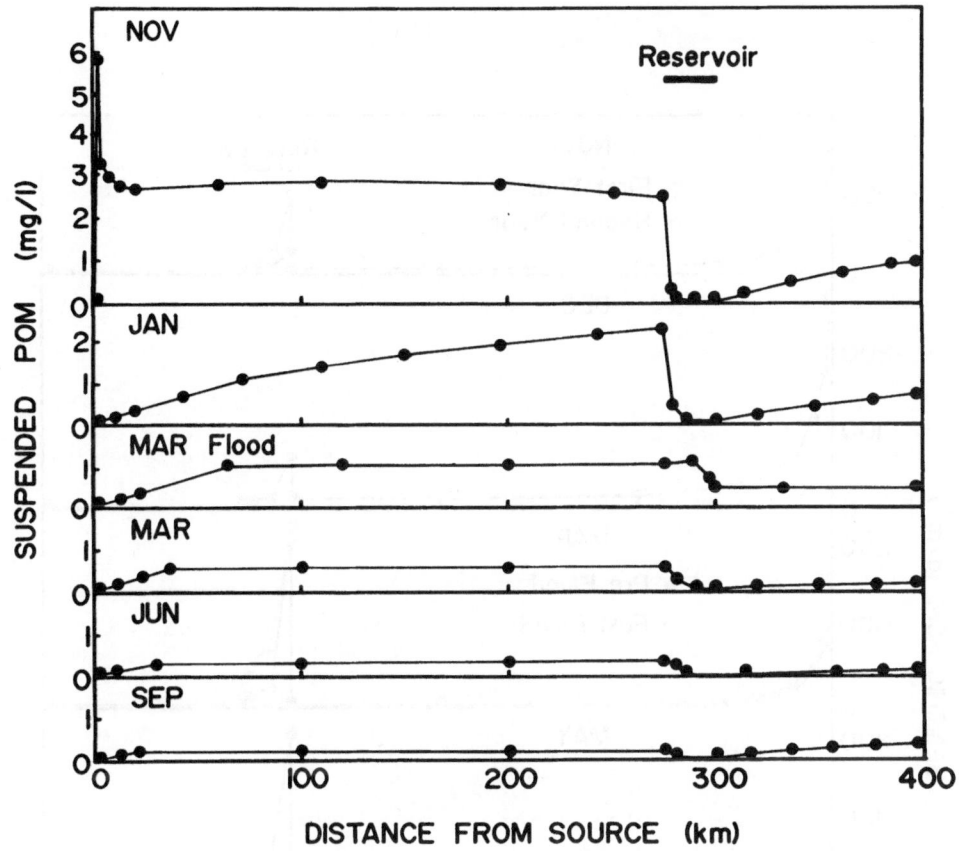

Fig. 3. Model-predicted suspended particulate organic matter (POM) as a function of distance from the river source.

woodland streams Naiman and Sedell (1979) found a winter POM concentration peak. Using a continuous sampler in a woodland stream, Malmqvist et al. (1978) found a POM peak in winter associated with high flows. FPOC concentrations in small streams at Hubbard Brook were highest during summer, when flows were highest (Hobbie and Likens, 1973). Wetzel and Manny (1977) found a summer peak, associated with rapid vegetative growth, and an early winter peak associated with high flows. Minimum POC concentration occurred during leaf fall. Data from small forested watersheds at Coweeta Hydrologic Laboratory (Wallace et al., unpublished) show that base flow POM concentrations are consistently lower in winter than summer. In general, these studies suggest that POM is retained in low-order streams much longer than predicted by our model.

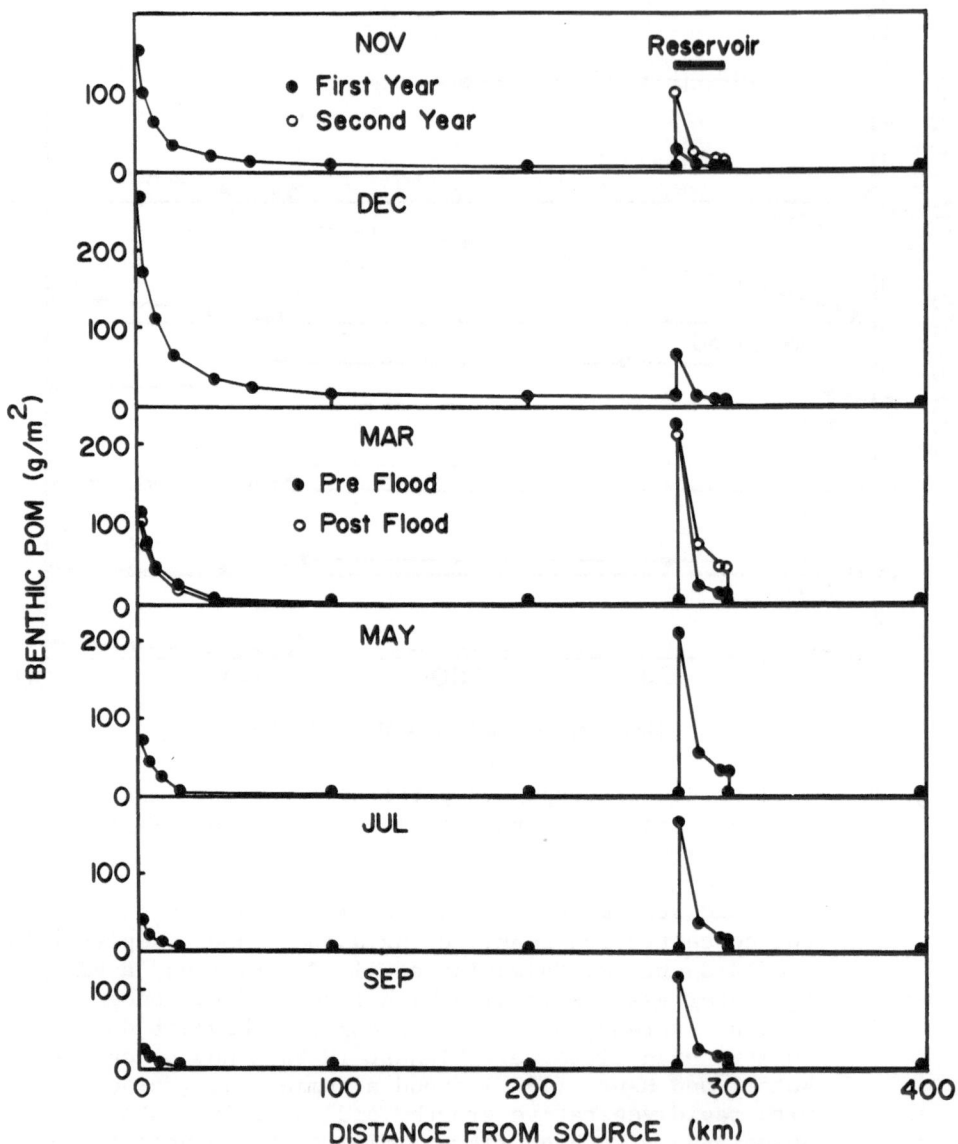

Fig. 4. Model-predicted benthic particulate organic matter (POM)
         as a function of distance from the river source.

Also, as a number of studies have shown, the input of
allochthonous leaf material to headwater streams continues
beyond leaf fall because of blow-in (e.g., Fisher and
Likens, 1973; McDowell and Fisher, 1976; Comiskey et al.,
1977; Webster, 1977). In studies of larger streams not
immediately affected by lakes, high POM concentrations
during low-flow periods are often reported and are usually
associated with autochthonous production (Weber and Moore,
1967; Berrie, 1972; Naiman and Sedell, 1979).

2. During leaf fall, POM concentration peaked a short distance
downstream from the headwaters, then decreased downstream.
At other times of the year POM concentration increased or
remained constant downstream. Data from first- to fourth-
order streams at Coweeta Hydrologic Laboratory (Wallace,
unpublished) show peak concentrations within the first
2 km for all seasons except spring, when POM concentration
increased slightly from the headwaters to a point 6 km
downstream. Malmqvist et al. (1978) found that POM
concentration in a small woodland stream in Sweden
increased to a peak 5-6 km downstream from the headwaters,
then decreased downstream. Data from the River Continuum
Study (Sedell et al., 1978; Naiman and Sedell, 1979) show
a variable pattern. At each of four sites in Oregon,
Pennsylvania, Michigan, and Idaho, samples were collected
from four streams spanning first to seventh order. In
Oregon the peak annual average POM concentration was in
the lowest-order stream; in Michigan and Idaho it was in
the next-to-lowest order stream. In Pennsylvania the peak
was in the highest-order stream. In a more thorough
analysis of the Oregon data, Naiman and Sedell (1979) found
that POM concentration was always higher in the first-
order stream than in the third- and fifth-order streams,
which were generally similar. POM concentrations in the
seventh-order streams were higher than in the third- and
fifth-order streams in spring, summer, and autumn, but
lower in winter. In spring, POM concentration in the
seventh-order stream was higher than in the first-order
stream. It is evident from these studies that an understanding of the longitudinal distribution of POM in stream
basins is not yet possible. Such factors as autochthonous
production and increased downstream inputs from nonforested
areas produce a more complex system than currently
depicted in our model.

3. Benthic POM was greatest just after leaf fall and was
gradually depleted over the rest of the year. This
pattern is generally consistent with studies of low-order
streams, where inputs are primarily allochthonous (e.g.,

Minshall, 1967), though some studies also show a spring
peak (e.g., Nelson and Scott, 1962; Tilly, 1968; Webster
and Patten, 1979).

4. There was a consistent downstream decrease in POM standing
    crop. This is consistent with a few available studies.
    Naiman and Sedell (in press) showed a downstream decrease
    in POM standing crop in streams from first to seventh
    order in the Oregon Cascades. Webster (1977) found a
    similar pattern for first- to fourth-order streams in the
    southern Appalachians. Our model shows a complete deple-
    tion of benthic POM in all but the upper reaches of the
    river by late spring. This suggests a needed refinement
    in the part of the model concerned with resuspension of
    deposited POM.

The following results of the simulation refer to changes
occurring within and below the impoundment.

5. POM concentrations decreased as the river flowed through
    the reservoir. There is little question that substantial
    amounts of suspended organic material settle from reser-
    voir waters. Armitage (1977) estimated that up to 91% of
    the total solids transported by the River Tees (England)
    was retained by the Cow Green Reservoir. Lind (1971)
    reported that Lake Waco (Texas) trapped 71% of the partic-
    ulate organics entering from the Bosque River catchment
    area, and Maciolek (1966) reported that microseston in the
    Convict Creek (California) catchment was seriously
    disrupted by Convict Lake. This phenomenon is associated
    with the loss of stream power and ability to transport
    even small low-density organic particles. One exception
    to this generalization occurred in our simulation during
    the spring flood, when material was picked up in the
    shallow end of the lake and deposited farther down the
    lake.

6. Deposition in the reservoir increased POM standing crop
    primarily during high winter-spring flows, especially the
    early spring flood. During summer and early fall, benthic
    POM was depleted by decomposition. Our 14-month simula-
    tion showed a realistic year-to-year increase in reservoir
    benthic POM.

7. There was no deposition of benthic POM below the reservoir.
    In a review of downstream reservoir effects, Neel (1963)
    noted that reservoir releases to smaller streams may
    remove small sediment particles, leaving a bottom composed
    of larger particles. One problem of the Aswan Dam has

been downstream erosion caused by silt-free, high-velocity
flows (Sterling, 1971). Those observations suggest that,
in some cases, POM would not be deposited below a reservoir. In other cases, where flow below the reservoir is
less than reservoir input because of water diversion or
where a more constant flow is maintained, the opposite
effects occur. Gravels may become compacted and the
interstices filled with fines (Fraser, 1972).

8.   POM concentration increased with distance downstream from
the dam. This effect was caused by tributary inputs and,
in autumn, was accentuated by inputs from riparian vegetation. This increase has been documented in several
studies conducted below reservoirs with hypolimnial
releases. Ward (1974, 1976) found that water discharged
from the hypolimnion of Cheesman Lake on the South Platte
River (Colorado) was very low in organic particles. POM
gradually increased downstream due to a combination of
allochthonous and autochthonous sources. Ward also noted
that the rich community of filter-feeding invertebrates
frequently reported downstream from reservoirs was substantially reduced downstream from Cheesman Dam. Gore (1977)
found a similar situation downstream from the Tongue River
Reservoir (Montana), which has a hypolimnial release.
Although low temperature of water released from these dams
was implicated as contributing to depressed macroinvertebrate communities, absence of a suitable food source was
cited as another factor.

     Dams with releases nearer the surface may have
essentially opposite effects downstream to those with deep
releases. Maciolek and Tunzi (1968) reported that microseston in the outlet of Lurel Lake (California) was 85%
greater than at the inlet and that the bulk was limnoplankton generated in the lake. In Belwood Lake on the
Grand River (Ontario), Spence and Hynes (1971) reported
that more suspended organic matter was entering than
leaving the lake in summer. However, lake level fell in
late summer, and there was a large die-off of limnoplankton, which were washed into the river through the dam.
Suspended organic matter downstream increased to several
times that carried by the river into the reservoir.
Cushing (1963), Simmons and Voshell (1978), and Merkley
(1978) reported that, although the reservoirs they studied
had a clarifying effect on water as it passed through the
lakes, there were sufficient exports of both living and
moribund limnoplankters in the discharges to support large
populations of filter-feeding macroinvertebrates. Lind
(1971) reported that POM entering Lake Waco (Texas) from

the Bosque River catchment was chiefly "sestonic detritus," while that leaving the system was living and dead limnoplankton.

9. The reservoir increased overall processing efficiency of the river. We defined processing efficiency as the difference between total particulate input to the river and total particulate output (Webster, 1977; Webster and Patten, 1979). Total input to the river estimated from the model was 1.84×10^5 T y^{-1}. Output without the reservoir was 6.60×10^3 T y^{-1}, a processing efficiency of 96.4%. With the reservoir in place, the output was 3.07×10^3 T y^{-1}, a 98.3% processing efficiency. Output from the river system was reduced to less than half. Only 240 T y^{-1} of the extra 3.53×10^3 T y^{-1} processed was stored permanently on the reservoir bottom. Because of the ability of reservoirs to trap and decompose POM, total processing of organic material in a river with a reservoir is greater than in an undammed river. This effect is not evident in published studies, because of increased autochthonous (limnoplankton) and anthropogenic (sewage) inputs. Lind (1971) estimated that Lake Waco had little effect on total organic matter transport in the Bosque River. Sedimented POM was equaled by limnoplankton production. From a study of the New River, Newbern (1978) found that annual organic matter transport at a point 90 km upstream from Claytor Lake was approximately equal to transport at a point 90 km downstream from the dam. However, in addition to the reservoir, between these two sites were a number of towns with sewage effluents to the river, so that total organic-matter processing within the reach was considerable.

CONCLUSIONS

Through our modeling effort, we identified several areas where our model does not accurately reflect current knowledge of POM behavior in river-reservoir ecosystems. These include: appropriate seasonality of allochthonous inputs, autochthonous sources of POM, filter-feeder utilization of POM, and deposition and resuspension characteristics of POM. Despite the inadequacies of our model, we feel it is an advance over previous stream-ecosystem models because it includes the longitudinal or continuum aspect. By including a reservoir as part of this system, we have been able to analyze reservoir effects on POM dynamics within this larger framework.

ACKNOWLEDGMENTS

   Dr. M. Williams of the Mathematics Department of Virginia
Polytechnic Institute and State University helped solve the partial
differential equations. J. Kirby, Center for Environmental Studies,
Virginia Polytechnic Institue and State University, helped assemble
the morphological data on the New River. We also acknowledge Darla
Donald for editorial assistance in preparation of this manuscript
and Betty Higginbotham for typing the original manuscript.

REFERENCES

Armitage, P. D., 1977, Invertebrate drift in the regulated River
    Tees and an unregulated tributary Maize Beck, below Cow Green
    Dam, *Freshwater Biol.*, 7:167–183.
Bagnold, R. A., 1966, "An Approach to the Sediment Transport Problem
    of General Physics," USGS Prof. Pap. 422-I.
Berrie, A. D., 1972. The occurrence and composition of seston in the
    River Thames and the role of detritus as an energy source for
    secondary production in the river, *Mem. Inst. Ital. Idrobiol.*,
    29(Suppl.):473–483.
Bilby, R. E., and Likens, G. E., 1979, Effect of hydrologic fluctua-
    tions on the transport of fine particulate organic carbon in a
    small stream, *Limnol. Oceanogr.*, 24:69–75.
Boling, R. H., Jr., Goodman, E. D., Van Sickle, J. A., Zimmer, J. O.,
    Cummins, K. W., Reice, S. R., and Peterson, R. C., 1975, Toward
    a model of detritus processing in a woodland stream, *Ecology*,
    56:141–151.
Bormann, F. H., Likens, G. E., and Eaton, J. S., 1969, Biotic
    regulation of particulate and solution losses from a forest
    ecosystem, *BioScience*, 19:600–611.
Bormann, F. H., Likens, G. E., Siccama, T. G., Pierce, R. S., and
    Eaton, J. S., 1974, The export of nutrients and recovery of
    stable conditions following deforestation at Hubbard Brook,
    Ecol. Monogr., 44:255–277.
Bray, J. R., and Gorham, E., 1964, Litter production in forests of
    the world, *Adv. Ecol. Res.*, 2:101–157.
Buscemi, P. A., 1969, Chemical and detrital features of Palouse
    River, Idaho, runoff flowage, *Oikos*, 20:119–127.
Chester, C. R., 1971, "Techniques in Partial Differential Equations,"
    McGraw Hill Book Co., New York.
Collier, B. D., Cox, G. W., Johnson, A. W., and Miller, P. C., 1973,
    "Dynamic Ecology," Prentice-Hall, Englewood Cliffs, New Jersey.
Comiskey, C. E., Henderson, G. S., Gardner, R. H., and Woods, F. W.,
    1977, Patterns of organic matter transport on Walker Branch
    Watershed, *in*: "Watershed Research in Eastern North America,"
    Vol. I, D. L. Correll, ed., Chesapeake Bay Center for Environ.
    Stud., Smithsonian Inst., Edgewater, Maryland.

Cummins, K. W., 1974, Structure and function of stream ecosystems, *BioScience*, 24:631-641.

Cummins, K. W., 1975, The ecology of runnings waters: Theory and practice, *in*: "Proceedings Sandusky River Basin Symposium," Tiffin, Ohio.

Cushing, C. E., Jr., 1963, Filter-feeding insect distribution and planktonic food in the Montreal River, *Trans. Am. Fish. Soc.*, 92:216-219.

de la Cruz, A. A., and Post, H. A., 1977, Production and transport of organic matter in a woodland stream, *Arch. Hydrobiol.*, 80:227-238.

Dobbins, W. E., 1964, BOD and oxygen relationship in streams, *ASCE J. Sanitary Eng. Div.*, 90:53-78.

Fisher, S. G., and Likens, G. E., 1973, Energy flow in Bear Brook, New Hampshire: An integrative approach to stream ecosystem metabolism, *Ecol. Monogr.*, 43:421-439.

Fisher, S. G., and Minckley, W. L., 1978, Chemical characteristics of a desert stream in flash flood, *J. Arid Environ.*, 1:25-33.

Fraser, J. C., 1972, Regulated discharge and the stream environment, *in*: "River Ecology and Man," R. T. Oglesby, C. A. Carlson, and J. A. McCann, eds., Acad. Press, New York.

Gasith, A., and Hassler, A. D., 1976, Airborne litterfall as a source of organic matter in lakes, *Limnol. Oceanogr.*, 21:253-258.

Goldman, C. R., and Kimmel, B. L., 1978, Biological processes associated with suspended lake sediment and detritus in lakes and reservoirs, *in*: "Current Prospectives on River-Reservoir Ecosystems," J. Cairns, Jr., E. F. Benfield, and J. R. Webster, eds., N. Am. Benthol. Soc., Springfield, Illinois.

Gore, J. A., 1977, Reservoir manipulations and benthic macroinvertebrates in a prairie river, *Hydrobiologia*, 55:113-123.

Hall, C. A. S., 1972, Migration and metabolism in a temperate stream ecosystem, *Ecology*, 53:585-604.

Hobbie, J. E., and Likens, G. E., 1973, Output of phosphorus, dissolved organic matter, and fine particulate carbon from Hubbard Brook watersheds, *Limnol. Oceanogr.*, 18:734-742.

Jenny, H., Gessel, S. P., and Bingham, F. T., 1949, Comparative study of decomposition rates of organic matter in temperate and tropical regions, *Soil Sci.*, 68:417-432.

Kanawha River Basin Coordinating Committee, 1971, "Kanawha River Comprehensive Basin Study," Vol. V, U.S. Dep. Interior, Bureau Outdoors Reclamation, N.E. Reg., Philadelphia.

Karlström, U., and Backlund, S., 1977, Relationship between algal cell number, chlorophyll a, and fine particulate organic matter in a river in northern Sweden, *Arch. Hydrobiol.*, 80:192-199.

Keefe, C. W., Flemer, D. A., and Hamilton, D. H., 1976, Seston distribution in the Patuxent River estuary, *Chesapeake Sci.*, 17:56-59.

Klotz, R. L., and Matson, E. A., 1978, Dissolved organic carbon
    fluxes in the Shetucket River of eastern Connecticut, U.S.A.,
    Freshwater Biol., 8:347-355.
Leopold, L. B., and Maddock, T., Jr., 1953, "The Hydraulic Geometry
    of Stream Channels and Some Physiographic Implications," USGS
    Prof. Pap. No. 252.
Leopold, L. B., and Miller, J. P., 1956, "Ephemeral Streams -
    Hydraulic Factors and Their Relation to the Drainage Net,"
    USGS Prof. Pap. 282-A.
Leopold, L. B., Wolman, M. C., and Miller, J. P., 1964, "Fluvial
    Processes in Geomorphology," W. H. Freeman and Co.,
    San Francisco.
Liaw, W. K., and MacCrimmon, H. R., 1977, Assessment of particulate
    organic matter in river water, *Int. Rev. Gesamt. Hydrobiol.*,
    62:445-463.
Lind, O. T., 1971, Organic matter budget of a central Texas reser-
    voir, *in*: "Reservoir Fisheries and Limnology," G. E. Hall,
    ed., Special Publ. No. 8, Am. Fish. Soc., Washington, D.C.
Maciolek, J. A., 1966, Abundance and character of microseston in a
    California mountain stream, *Verh. Int. Verein. Limnol.*,
    16:639-645.
Maciolek, J. A., and Tunzi, M. G., 1968, Microseston dynamics in a
    simple Sierra Nevada lake--stream system, *Ecology*,
    49:60-75.
Malas, D., and Wallace, J. B., 1977, Strategies for coexistence in
    three species of net-spinning caddisflies (Trichoptera) in
    second order Appalachian streams, *Can. J. Zool.*, 55:1829-1840.
Malcolm, R. L., and Durum, W. H., 1976, "Organic Carbon and Nitrogen
    Concentrations and Annual Organic Carbon Load of Selected
    Rivers of the United States," U.S. Geol. Surv. Water Supply
    Pap. 1817-F.
Malmqvist, B., Nilsson, L. M., and Svensson, B. S., 1978, Dynamics
    of detritus in a small stream in southern Sweden and its
    influence on the distribution of the bottom animal communities,
    Oikos, 31:3-16.
McDowell, W. H., and Fisher, S. G., 1976, Autumnal processing of
    dissolved organic matter in a small woodland stream ecosystem,
    Ecology, 57:561-569.
McIntire, C. D., 1973, Periphyton dynamics in laboratory streams:
    A simulation model and its implications, *Ecol. Monogr.*,
    43:399-420.
McIntire, C. D., and Colby, J. A., 1978, A hierarchical model of
    lotic ecosystems, *Ecol. Monogr.*, 48:167-190.
Merkley, W. B., 1978, Impact of Red Rock Reservoir on the Des Moines
    River, *in*: "Current Prospectives on River-Reservoir Ecosystems,"
    J. Cairns, Jr., E. F. Benfield, and J. R. Webster, eds., N. Am.
    Benthol. Soc., Springfield, Illinois.

Minshall, G. W., 1967, Role of allochthonous detritus in the trophic structure of a woodland springbrook community, *Ecology*, 48:139-149.

Minshall, G. W., 1978, Autotrophy in stream ecosystems, *BioScience*, 28:767-771.

Morisawa, M., 1968, "Streams, Their Dynamics and Morphology," McGraw-Hill Book Co., New York.

Müller, G., and Förstner, U., 1968, General relationship between suspended sediment concentration and water discharge in the Alpenrhein and some other rivers, *Nature*, 217:244-245.

Naiman, R. J., and Sedell, J. R., 1979, Characterization of particulate organic matter transported by some Cascade Mountain streams, *J. Fish. Res. Board Can.*, 36:17-31.

Naiman, R. J., and Sedell, J. R., In press, The river continuum: Benthic organic matter as a function of stream order, *Hydrobiologia*.

Naiman, R. J., and Sibert, J. R., 1978, Transport of nutrients and carbon from the Nanaimo River to its estuary, *Limnol. Oceanogr.*, 23:1183-1193.

Neel, J. K., 1963, Impact of reservoirs, *in*: "Limnology in North America," D. G. Frey, ed., Univ. Wisconsin Press, Madison.

Nelson, D. J., and Scott, D. C., 1962, Role of detritus in the productivity of a rock-outcrop community in a piedmont stream, *Limnol. Oceanogr.*, 7:396-413.

Newbern, L., 1978, "Detritus Transport in the New River," M.S. Thesis, Virginia Polytechnic Inst. and State Univ., Blacksburg.

O'Connor, D. J., 1962, "The Effect of Stream Flow on Waste Assimilation Capacity," Proc. 17th Purdue Industrial Waste Conf., Lafayette, Indiana.

Olson, J. S., 1963, Energy storage and the balance of producers and decomposers in ecological systems, *Ecology*, 44:322-332.

Paul, R. W., Jr., 1978, "Leaf Processing and the Effects of Thermal Perturbation on Leaf Degradation in the New River, Va," Ph.D. Thesis, Virginia Polytechnic Inst. and State Univ., Blacksburg.

Penctak, T., Molinski, M., and Zalewski, M., 1976, The contribution of autochthonous and allochthonous matter to the trophy of a river in the Barbel region, *Ecol. Pol.*, 24:113-121.

Peterson, R. C., and Cummins, K. W., 1974, Leaf processing in a woodland stream, *Freshwater Biol.*, 4:343-368.

Pierce, J. W., and Dulong, F. T., 1977, Discharge of suspended particulates from Rhode River subwaters heads, *in*: "Watershed Research in Eastern North America," Vol. II, D. L. Correll, ed., Chesapeake Bay Center for Environ. Stud., Smithsonian Inst., Edgewater, Maryland.

Ruttner, F., 1963, "Fundamentals of Limnology," 3rd ed., Univ. Toronto Press, Toronto.

Sedell, J. R., Naiman, R. J., Cummins, K. W., Minshall, G. W., and
    Vannote, R. L., 1978, Transport of particulate organic material
    in streams as a function of physical processes, *Verh. Int.
    Verein. Limnol.*, 20:1366-1375.
Seki, H., Stephens, K. V., and Parsons, T. R., 1969, The contribution
    of allochthonous bacteria and organic materials from a small
    river into a semi-enclosed sea, *Arch. Hydrobiol.*, 66:37-47.
Simmons, G. M., Jr., and Voshell, J. R., Jr., 1978, Pre- and post-
    impoundment benthic macroinvertebrate communities of the North
    Anna River, *in*: "Current Prospectives on River-Reservoir
    Ecosystems," J. Cairns, Jr., E. F. Benfield, and J. R. Webster,
    eds., N. Am. Benthol. Soc., Springfield, Illinois.
Spence, J. A., and Hynes, H. B. N., 1971, Differences in benthos
    upstream and downstream of an impoundment, *J. Fish. Res. Board
    Can.*, 28:35-43.
Sterling, C., 1971, Aswan Dam looses a flood of problems, *Life*,
    70:46-47 (as cited in Collier et al., 1963).
Streeter, H. W., and Phelps, E. B., 1925, "A Study of the Pollution
    and Natural Purification of the Ohio River," Public Health
    Bull. 146, USPHS, Washington, D.C.
Swanson, C. D., and Bachmann, R. W., 1976, A model of algal exports
    in some Iowa streams, *Ecology*, 57:1076-1080.
Thomann, R. V., 1972, "Systems Analysis and Water Quality Manage-
    ment," McGraw-Hill Book Co., New York.
Tilly, L. J., 1968, The structure and dynamics of Cone Spring, *Ecol.
    Monogr.*, 38:169-197.
Wallace, J. B., Webster, J. R., and Woodall, W. R., 1977, The role
    of filter feeders in flowing waters, *Arch. Hydrobiol.*,
    79:506-532.
Ward, J. V., 1974, A temperature-stressed stream ecosystem below a
    hypolimnial release mountain reservoir, *Arch. Hydrobiol.*,
    74:247-275.
Ward, J. V., 1976, Comparative limnology of differentially regulated
    sections of a Colorado mountain river, *Arch. Hydrobiol.*,
    78:319-342.
Weber, C. I., and Moore, D. R., 1967, Phytoplankton, seston and
    dissolved organic carbon in the Little Miami River at
    Cincinnati, Ohio, *Limnol. Oceanogr.*, 12:311-318.
Webster, J. R., 1977, Large particulate organic matter processing in
    stream ecosystems, *in*: "Watershed Research in Eastern North
    America," Vol. II, D. L. Correll, ed., Chesapeake Bay Center
    for Environ. Stud., Smithsonian Inst., Edgewater, Maryland.
Webster, J. R., and Patten, B. C., 1979, Effects of watershed
    perturbation on stream potassium and calcium dynamics, *Ecol.
    Monogr.*, 49:51-72.
Webster, J. R., Waide, J. B., and Patten, B. C., 1975, Nutrient
    recycling and the stability of ecosystems, *in*: "Mineral
    Cycling in Southeastern Ecosystems," F. G. Howell, J. B. Gentry,
    and M. H. Smith, eds., USERDA Conf.-740513.

Wetzel, R. G., and Manny, B. A., 1977, Seasonal changes in particu-
    late and dissolved organic carbon and nitrogen in a hardwater
    stream, *Arch. Hydrobiol.*, 80:29-39.

MACROINVERTEBRATE RESPONSE TO FLOW MANIPULATION IN THE
STRAWBERRY RIVER, UTAH (U.S.A.)

Robert D. Williams and Robert N. Winget

Department of Zoology
Brigham Young University
Provo, Utah 84602

INTRODUCTION

The effects of flow regulation on stream communities immediately below reservoirs have been reported by several authors (Radford and Hartland-Rowe, 1971; Trotsky and Gregory, 1974; Ward, 1976a; Gore, 1977). Ward (1976a) reviewed the effects of reduced flow below dams on the physical and biological characteristics of streams. Attempts to provide criteria for recommending adequate stream flows have resulted in several survey methods using physical stream parameters, such as width, depth and velocity, pool-to-riffle ratio, and wetted perimeter (Stalnaker and Arnett, 1976). Recommended flows have also been based on a percentage of the historical stream flow (Tennant, 1975) or "adequate" water coverage for aquatic insect production and fish spawning (Hooper, 1973; Gore, 1978). Recommended minimum flows do not, however, take into account the long-term implications of altering the natural flow regime of a lotic system.

Soldier Creek Dam was completed in July 1973 to enlarge a reservoir located 7 km upstream from the new dam. In 1975 a study of the macroinvertebrates of the Strawberry River was initiated to determine effects of (1) short-term flow fluctuations and (2) constant reduced flows over an extended period. An additional objective was to recommend a flow regime that would maintain desired habitat and macroinvertebrate community structure under conditions of reduced discharge.

DESCRIPTION OF STUDY AREA

The study was conducted on the 2.2-km stretch of the Strawberry River from 2.0 km below Soldier Creek Dam to the confluence of Willow Creek. The dominant substrate is composed of rounded boulders, rubble, and gravel. The study area has a mean gradient of 1.8%. Stream-bank stability is high (75-100% stability), with a dense riparian vegetation of shrubs, grasses, and conifers along most of the stream.

The river water is moderately hard (200-230 mg/liter $CaCO_3$); pH fluctuated between 7.9 and 8.5; dissolved oxygen ranged from 8 to 12 mg/liter. Orthophosphate (0.04-0.13 mg/liter) and nitrate nitrogen levels (<0.05 mg/liter) are representative of mesotrophic conditions. These water quality values are similar to those before completion of Soldier Creek Dam. Because of the small size of the river, the roughness of the substrate, and a moderate gradient, water temperature rapidly equilibrates with prevailing atmospheric conditions. Water temperatures in the study section, measured by Ryan Thermographs, were similar to those of unregulated streams having diurnal summer fluctuations near 12°C, with no obvious difference following completion of Soldier Creek Dam.

METHODS AND PROCEDURES

Benthic macroinvertebrates were collected from two gravel-rubble riffles with a Surber sampler (280 μm mesh), which enclosed 929 cm^2 of substrate.

During the summer and autumn of 1975, three Surber samples were taken from each riffle at each treatment flow (0.11, 0.34, 0.71, and 0.42 m^3/sec). The treatment flows were of 8-10 days duration. In August and October of 1976, a total of nine samples were collected on each date during a reservoir release flow of 0.34 m^3/sec.

Based upon measurements of width, depth, velocity, and pool-to-riffle ratios at various flows, a minimum discharge regime was initiated in 1977. It was determined that minimum flows of 0.22 m^3/sec were necessary during periods of high irrigation and recreation needs, but that only 0.11 m^3/sec was required during low need periods. Eight samples were taken in August (0.22 m^3/sec) and eight were collected in October (0.11 m^3/sec) during 1977.

RESULTS

The macroinvertebrate populations sampled at approximately two-week intervals from 27 July to 12 September 1977 showed a twofold increase in the number of organisms/m^2 for each successive period (Table 1). While numbers/m^2 increased, the mean individual weight of the organism decreased. The decrease in weight per organism reflects the natural life-cycle changes, with the appearance of numerous small larvae.

As in 1975 the number of organisms/m^2 in 1976 and 1977 increased between the August and September-October samples (Table 2). The mean weight per organisms decreased from August to September in 1975 and from August to October in 1976, but in 1977 the values were nearly identical, although lower than in previous years. Of special interest was the decrease in weight per organism between years. The yearly increase in numbers/m^2 accompanied by the reduction in weight per organism represented a community composition change, with an increase in the number of smaller larvae. Diversity, represented by the Shannon-Weaver Index ($\bar{d}$), remained fairly constant during all three years of the study, but the taxonomic composition changed.

Although there were no changes in the community structure related to short-term (8-10 days) flow fluctuations in 1975, there were changes in the trophic structure of the community from 1975 to 1977 (Table 3). Most notable was an increase in algal scrapers (*Baetis* and *Micrasema*) and filter feeders (Simuliidae). Other community changes included increased numbers of low-flow tolerant species, such as *Paraleptophlebia* (Trotzky and Gregory, 1974), chironomids, and oligochaetes. Detrital feeders/shredders (*Ephemerella grandis, Pteronarcella badia*, and *Hexatoma*) were either greatly reduced in numbers or eliminated from 1975 to 1977. Elmid riffle beetles were also reduced in density. There was an increase in macrophytes and algae from 1975 to 1977.

DISCUSSION

Because water quality and temperature in the study section were similar before and after completion of Soldier Creek Dam, it is highly probable that the changes in the aquatic macroinvertebrate community of the Strawberry River during 1973-1977 are attributable, directly or indirectly, to the reduced flows. Zoobenthic samples collected from Strawberry River below the confluence of Willow Creek, in a section with a more normal discharge pattern, did not reflect the community changes seen in the regulated study section (Table 4).

Macroinvertebrate data collected in 1975 failed to show that altered stream flow over short periods (8-10 days) had any major effect on the macroinvertebrate community. Flow-related change was masked by emergence, egg laying, and larval hatch. However, in association with the reduced and more uniform flows from Soldier

Table 1. Summary of Macroinvertebrate Data at Four Test Flows
          During the Summer of 1975 in the Strawberry River Below
          Soldier Creek Dam

Date	Q m³/sec	n	No. of Taxa	x̄ No. per m² (Standard Deviation)	x̄ Wt. g per m² (Standard Deviation)	x̄ Wt. per Organism (g x 10⁻³)
27 July 75	0.34	6	26	2,750 (73)	8.33 (5.38)	3.00
12 Aug 75	0.71	6	27	5,833 (269)	5.42 (2.95)	0.93
24 Aug 75	1.42	6	32	10,632 (304)	6.95 (3.80)	0.65
12 Sep 75	0.11	6	26	19,142 (760)	6.69 (1.62)	0.35

Table 2. Summary of Macroinvertebrate Numbers, Weights, and
          Shannon-Weaver Index Values (d̄) in the Strawberry River
          During August and September-October Over a Three-Year
          Period

Date	Q m³/sec	n	No. of Taxa	x̄ No. per m² (Standard Deviation)	x̄ Wt. g per m² (Standard Deviation)	x̄ Wt. per Organism (g x 10⁻³)	d̄
Aug 1975	1.42	6	32	10,632 (304)	6.95 (3.80)	0.65	2.41
Sep 1975	0.11	6	26	19,142 (759)	6.69 (1.62)	0.35	1.87
Aug 1976	0.34	9	25	22,292 (901)	3.72 (0.11)	0.20	2.47
Oct 1976	0.34	9	26	60,385 (3,761)	3.21 (0.84)	0.06	1.70
Aug 1977	0.22	8	26	52,207 (1,740)	4.92 (0.19)	0.09	2.05
Oct 1977	0.11	8	30	61,782 (1,943)	4.77 (0.24)	0.08	2.05

Table 3.   Macroinvertebrate Community Response to Flow Manipulation
as Shown by Changes in Mean Density (Organisms/m^2) of
Selected Taxa in the Strawberry River.   Functional
Grouping According to Merritt and Cummins (1978)

Taxa		1975	1976	1977	Functional Group*
Gastropoda	Aug	13	13	38	F
	Oct	0	0	187	D/H
Oligochaeta	Aug	872	787	1,872	D/H
	Oct	465	1,965	3,931	
Hydracarina	Aug	120	251	1,076	Pa
	Oct	72	438	452	
Ephemeroptera					
Baetis spp.	Aug	2,376	7,879	31,043	C/G, Sc
	Oct	10,652	42,394	41,146	H
Paraleptophlebia sp.	Aug	7	17	30	S/D, C/G
	Oct	4	38	14	
Ephemerella grandis	Aug	185	10	1	C/G, S
	Oct	323	14	2	
Ephemerella inermis	Aug	48	326	351	C/G, S
	Oct	16	579	416	
Plecoptera					
Pteronarcella badia	Aug	2	0	0	Sc/C, E
	Oct	4	0	0	
Trichoptera					
Brachycentrus americanus	Aug	14	461	825	C/F, Sc
	Oct	25	454	1,083	
Micrasema sp.	Aug	66	861	667	S/G, H
	Oct	102	799	803	
Hydropsyche sp.	Aug	4	50	196	C/F
	Oct	45	148	72	
Rhyacophila sp.	Aug	13	40	44	E
	Oct	38	43	61	
Coleoptera					
Elmidae	Aug	899	324	222	C/G, Sc
	Oct	708	320	366	
Diptera					
Antocha monticola	Aug	7	68	114	C/G
	Oct	93	148	380	
Hexatoma sp.	Aug	13	0	0	E
	Oct	2	0	0	
Simuliidae	Aug	565	5,254	9,082	F, C
	Oct	638	5,222	5,767	
Chironomidae	Aug	5,094	5,830	6,352	H, E, D,
	Oct	5,619	8,029	6,320	G, C

*Collectors (C), Gatherers (G), Shredders (S), Scrapers (Sc), Engulfers (E),
Detritus (D), Filterers (F), Herbivores (H), Piercers (P), Parasitic (Pa)

Table 4. Comparison of Mean Density (Organisms/m^2) of Selected
Macroinvertebrate Taxa from Two Sections of the Strawberry
River

Taxa	Regulated Section (Above Willow Creek) Oct. 1977	Section Influenced by Seasonal Flows From Willow Creek Oct. 1977
Gastropoda	187	0
Oligochaeta	3,391	829
Hydracarina	452	430
Ephemeroptera		
Baetis spp.	41,146	4,196
Cinygmula sp.	0	32
Heptagenia sp.	0	22
Paraleptophlebia sp.	14	43
Ephemerella grandis	2	22
Ephemerella inermis	416	22
Plecoptera		
Pteronarcella badia	0	18
Isoperla patricia	0	21
Isoperla fulva	0	24
Megarcys sp.	0	8
Trichoptera		
Brachycentrus sp.	1,083	323
Micrasema sp.	803	43
Glossosoma sp.	0	1,367
Hydropsyche sp.	72	377
Hydroptila sp.	0	11
Rhyacophila sp.	61	65
Coleoptera		
Elmidae	366	1,119
Diptera		
Antocha monticola	380	75
Holorusia sp.	0	22
Psychodidae	0	65
Simuliidae	5,767	32
Chironomidae	6,320	3,637
Empididae	0	43
Stratiomyidae	0	11

Creek Dam from 1973 to 1977, there have been changes in the zoobenthic community. These changes have resulted from changes in the
physical and biological characteristics of the stream.

Mean monthly discharges for the Strawberry River from 1963 to
July 1973, before the completion and subsequent filling of Soldier
Creek Reservoir, ranged from 0.34 m^3/sec to over 7.0 m^3/sec, with
a mean of 2.08 m^3/sec for April and of 2.97 m^3/sec for May (Fig. 1).
A relatively natural flow regime was maintained, despite the smaller
reservoir upstream, because of unregulated tributaries that entered
the Strawberry River in the section below the old dam. With the
closing of Soldier Creek Dam in July 1973, the runoff pattern has
been greatly changed (Fig. 2). Except for a high treatment flow of
1.42 m^3/sec in August 1975, the mean monthly flows have been considerably lower than preimpoundment levels. Normal high runoff during
the months of April, May, and June are completely eliminated.

The closing of Soldier Creek Dam and the resulting low uniform
release flows have changed the availability of stream habitat
(Table 5). Mean width and depth between 0.11 m^3/sec and 1.42 m^3/sec
increased little. Water width and depth may be only slightly
affected by increased flow (Curtis, 1956; Minshall and Winger,
1968). Mean velocity was affected more than depth or width by a
change in discharge. Mean depths and velocities reached reported
optimum levels for insect production and fish spawning between
discharges of 0.34 m^3/sec and 0.71 m^3/sec (Hooper, 1973; Gore,
1978). However, adequate point velocities and depth measurements in
riffle areas could be attained at lower discharge, since mean values
include higher and lower measurements bracketing optimum conditions
for insect and fish production.

At 0.23 m3/sec there was over twice as much riffle as pool
habitat. With each increased flow level over 0.23 m^3/sec there was
a more balanced pool-to-riffle ratio. Decreases in current velocities from 32.0 to 15.2 cm/sec, at flows reduced from 0.34 to 0.11
m^3/sec greatly decreased the microhabitat available to the
benthic organisms in regions with heterogeneous substrates. The
elimination of high seasonal runoff, which annually flushed the
stream, also decreased habitat heterogeneity. The elimination in
1976 and 1977 of many of the species present in 1975, such as the
plecopteran *Pteronarcella badia*, can be related to changes in flow
regime and losses of habitat heterogeneity.

Another habitat change observed during the three-year study
that greatly affected water velocities was the increase in the
density of aquatic macrophytes and epilithic algae. Large mats of
algae formed on the rubble substrate in the riffle habitat. Barber
and Kevern (1973) found that increases in macrophyte density
decreased water current, with a corresponding deposition of detrital

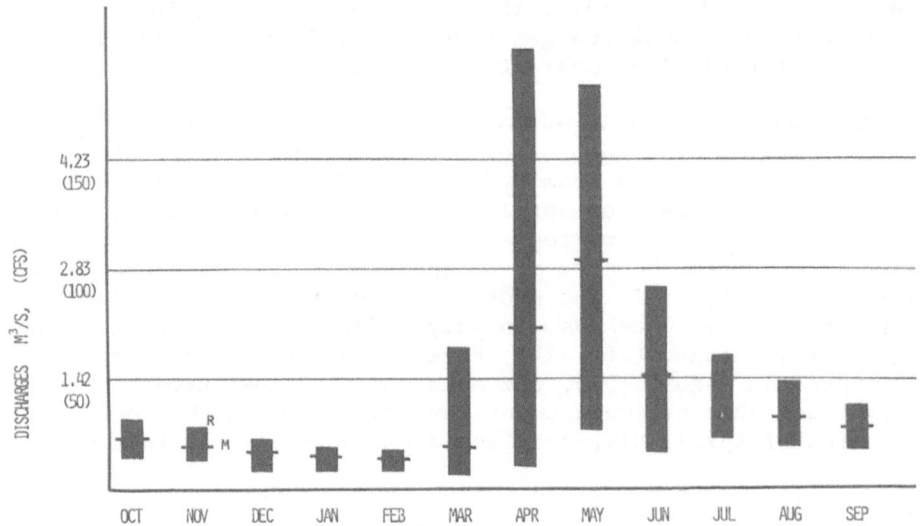

Fig. 1.   Strawberry River range of daily mean (R) and mean of
          monthly mean discharges (M) from 1963 to the closing
          of Soldier Creek Dam in July 1973.

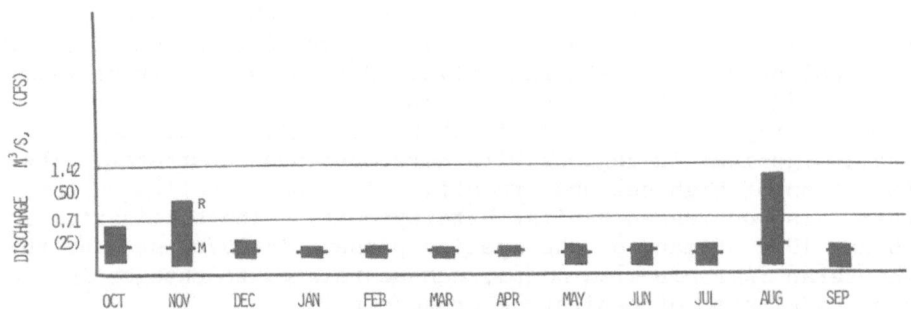

Fig. 2.   Strawberry River range of daily mean (R) and mean of
          monthly mean discharges (M) below Soldier Creek Dam
          from July 1973 to September 1977.

Table 5. Summary of Mean Changes in the Physical Characteristics
          of the Strawberry River at Five Discharges

Q m³/sec (cfs)	x̄ Water Width (M)	x̄ Depth (M)	x̄ Velocity (cm/sec)	% Stream in Riffle	% Stream in Pool
0.11 (4)	8.32	0.09	15.2	86.9	13.2
0.23 (8)	8.75	0.13	29.9	74.1	25.2
0.34 (12)	9.24	0.14	32.0	62.1	37.6
0.71 (25)	9.36	0.19	45.7	60.9	39.1
1.42 (50)	10.03	0.25	57.9	53.5	46.5

materials. Ward (1976b) stated that increased epilithic algae below
impoundments result from the clarifying effects of the reservoir,
the more stable substrate, increased nutrients, and higher winter
temperatures. These large mats of algae act as additional habitat,
trapping detrital material and forming false bottoms (Pfitzer, 1954;
Spence and Hynes, 1971), which offer considerable shelter from the
current and create an abundant food supply.

Heavy algae growth allows establishment of species otherwise
incapable of maintaining themselves in the stream. It also elimi-
nates those species specialized to exist on clean rock surfaces.
The increased algal growth in the Strawberry River, acting as an
additional food source and increasing niche diversification for
herbivorous organisms, allowed substantial increases in the numbers
of *Baetis* spp. The increase in numbers of other taxa, such as
Ephemerella inermis, Chironomidae, and *Antocha monticola*, may also
be attributed to the increased algal growth. Moore (1978) reported
that during summer months some oligochaetes utilize epilithic algae
as their major food source.

The reduction in current velocity associated with reduced flows
greatly limits many organisms specialized to exist in moderately
fast to torrential waters (Radford and Hartland-Rowe, 1971; Haddock,
1977). Many stream species are confined to running waters because
of respiratory requirements (Ward, 1976a). In the Strawberry River,
filter-feeding organisms, dependent on the current to carry food to
them (Hilsenhoff, 1971; Spence and Hynes, 1971), were not negatively
affected by the reduced flows. The simuliids as well as *Hydropsyche*,
Brachycentrus, and *Micrasema* all increased during the low-flow
period, possibly because of increased drift from enriched epilithic
algae, which would sustain their increased numbers. Increased

numbers of nematodes, oligochaetes, and gastropods in the Strawberry
River reflected the buildup of organic sediment in the stream in 'the
absence of seasonal flushes. Pearson et al. (1968) and Ward (1976a)
also noted an increase in gastropods under reduced flow conditions.
Gore (1977) reported that a reduction in water flow allows dominant
forms to increase, while rare species may be eliminated.

      Although the number of species has remained approximately the
same over the three-year study period, the species composition has
changed markedly. The macroinvertebrate community in 1975 was
comprised of ephemeropterans, plecopterans, and other species
normally associated with rocky streams. However, an extended period
of low, uniform flow changed the community by increasing the relative
abundance of species more tolerant of the flow conditions. This
change in community structure appears to be confined only to the
section of stream above the confluence of Willow Creek. Samples
taken below Willow Creek during 1976 and 1977 yielded data similar
to those during 1975 (R. Black, U.S'. Bureau of Reclamation, personal
communication).

      An additional problem created by reduced flows in the study
section was an increase in beaver activity over the three-year
period. Normally, seasonal runoff would wash out beaver dams.
Without a high seasonal runoff, beaver dams were built and sustained
in the study section. However, below Willow Creek all dams built
since 1975 were eliminated by the 1978 runoff. The new dams in the
study site caused inundation and siltation of previously high
production riffles, including the sampling sites of 1975. This
siltation fills the interstices of substrates, sealing off the
hyporheic zones and greatly reducing habitat heterogeneity (Ward,
1976a). The hyporheic zones have been shown to be important sources
of community repopulation following periods of environmental stress
(Hynes, 1974).

CONCLUSIONS

      The short term flow fluctuations below Soldier Creek Dam on the
Strawberry River had negligible effects on the stream benthic
community. Uniformly low flows over extended periods appeared to be
the major cause of changes in the community. The elimination of
high seasonal discharges allowed enhanced algal growth, beaver dams,
and increased sedimentation, which reduced or changed habitat
heterogeneity, including reducing the availability of the hyporheic
zone. The annual release of flows simulating normal runoff would do
much to reduce the adverse effects of reduced flows on the habitat
and biota of the Strawberry River.

ACKNOWLEDGMENTS

The authors wish to thank the U.S. Bureau of Reclamation,
Central Utah Project, Provo, Utah, for financial assistance and
support data used in this paper. We also wish to acknowledge the
assistance of Drs. James Barnes and Duane Smith of Brigham Young
University.

REFERENCES

Barber, W. E., and Kevern, N. R., 1973, Ecological factors influ-
    encing macroinvertebrate standing crop distribution,
    Hydrobiologia, 43:53–75.
Curtis, B., 1959, Changes in a river's physical characteristics
    under substantial reductions in flow due to hydroelectric
    diversion, *Calif. Fish Game*, 45:181–188.
Gore, J. A., 1977, Reservoir manipulations and benthic macroinverte-
    brates in a prairie river, *Hydrobiologia*, 55:113–123.
Gore, J. A., 1978, A technique for predicting in-stream flow require-
    ments of benthic macroinvertebrates, *Freshwater Biol.*,
    8:141–151.
Haddock, J. D., 1977, The effect of stream current velocity on the
    habitat preference of a net-spinning caddisfly larva
    Hydropsyche oslari Banks, *Pan-Pacific Entomol.*, 53:169–174.
Hilsenhoff, W. L., 1971, Changes in the downstream insect and
    amphipod fauna caused by an impoundment with a hypolimnion
    drain, *Ann. Entomol. Soc. Am.*, 64:743–746.
Hooper, D. R., 1973, "Evaluation of the Effects of Flows on Trout
    Stream Ecology," Pacific Gas and Elec. Co., Dept. Energy,
    Emeryville, California.
Hynes, H. B. N., 1974, Further studies on the distribution of stream
    animals within the substratum, *Limnol. Oceanogr.*, 19:92–99.
Kraft, M. E., 1972, Effects of controlled flow reduction on a trout
    stream, *J. Fish. Res. Board Can.*, 29:1405–1411.
Merritt, R. W., and Cummins, K. W., 1978, "An Introduction to the
    Aquatic Insects of North America," Kendall Hunt Publ. Co.,
    Dubuque, Iowa.
Minshall, W. G., and Winger, P. V., 1968, The effect of reduction in
    stream flow on invertebrate drift, *Ecology*, 49:580–582.
Moore, J. W., 1978, Importance of algae in the diet of the
    oligochaetes *Lumbriculus variegatus* (Müller) and *Rhyacodrilus
    sodalis* (Eisen), *Oecologia*, 35:357–363.
Pearson, W. D., Kramer, R. H., and Franklin, D. R., 1968, Macro-
    invertebrates in the Green River below Flaming Gorge Dam, 1964–
    1965 and 1967, *Proc. Utah Acad. Sci., Arts, Lett.*, 45:148–167.
Pfitzer, D. W., 1954, Investigations of waters below storage reser-
    voirs in Tennessee, *Trans. N. Am. Wildl. Conf.*, 19:271–282.

Radford, D. S., and Hartland-Rowe, R., 1971, A preliminary investi-
    gation of bottom fauna and invertebrate drift in an unregulated
    and a regulated stream in Alberta, *J. Appl. Ecol.*, 8:883-903.
Spence, J. A., and Hynes, H. B. N., 1971, Differences in benthos
    upstream and downstream of an impoundment, *J. Fish. Res. Board
    Can.*, 28:35-43.
Stalnaker, G. B., and Arnette, J. L., eds., 1976, "Methodologies for
    the Determination of Stream Resource Flow Requirements: An
    Assessment " U.S. Fish Wildl. Serv., Office Biol. Serv.
Tennant, D. L., 1975, "Instream Flow Regimes for Fish, Wildlife,
    Recreation, and Related Environmental Resources," U.S. Fish
    Wildl. Serv. Unpubl. Rep., Billings, Montana.
Trotsky, H. M., and Gregory, R. W., 1974, The effects of water flow
    manipulation below a hydroelectric power dam on the bottom
    fauna of the upper Kennebec River, Maine, *Trans. Am. Fish.
    Soc.*, 103:318-324.
Ward, J. V., 1976a, Effects of flow patterns below large dams on
    stream benthos: A review, *in:* "Instream Flow Needs, Vol. II,"
    J. F. Orsborn and C. H. Allman, eds., Am. Fish. Soc.,
    Bethesda, Maryland.
Ward, J. V., 1976b, Comparative limnology of differentially regu-
    lated sections of a Colorado mountain river, *Arch. Hydrobiol.*,
    78:319-342.

SYMPOSIUM SUMMARY AND CONCLUSIONS

James V. Ward and Jack A. Stanford

The primary purpose of these proceedings is to document the extent of stream regulation and resulting ecological ramifications on lotic systems, and to provide directions for further research. Although stated in the introductory chapter, it should be emphasized that the intention was to avoid an advocacy stance in this volume. Because this is the first symposium dealing specifically with the ecology of stream reaches below dams, it was deemed appropriate to approach the subject with scientific objectivity insofar as possible, which is not to imply that the editors and authors do not have strong personal feelings regarding stream regulation.

The most prevalent riverine environments on earth are indeed regulated lotic systems. That few free-flowing river systems remain was a recurrent message at the symposium. It was in that context which K. W. Cummins stated: "We may never know how large rivers in the temperate zone functioned biologically as the result of hundreds of millions of years of evolution and at least ten to twenty thousand years of acclimatization of resident populations."

In this brief summary, focus is directed toward the major concepts developed by the authors. Following an examination of abiotic and biotic components of regulated stream systems, geographical variations are analyzed. In the final section, major conclusions are presented.

THE NATURAL STREAM ECOSYSTEM

The paper by Cummins on unregulated streams provides a scale against which the ecological significance of stream regulation may

be measured. To fully understand the alterations induced by impoundment it is essential to understand the basic structure and function
of natural stream ecosystems.

Ecological processes in streams are a function of stream order,
geomorphic and vegetational setting and the annual hydrograph pattern.

The "River Continuum Hypothesis," which embodies the structural
and functional changes that occur from the headwaters to river mouths,
emphatically illustrates that damming the headwaters will have quite
different ecological implications on the receiving stream than
impounding a large river.

HYDROLOGICAL AND CHEMICAL MODIFICATION

Deposition of sediments causes general aggradation of the
channel entering the reservoir and its tributaries. Increased river
bed elevation in the upstream channel may raise the water table,
with effects on riparian vegetation. Large reservoirs trap nearly
100 percent of inflowing sediment, although under certain circumstances density currents may transport finer particles through the
reservoir. The receiving stream below the dam is subjected to
degradation. The sediment-free water discharged from the reservoir
has a propensity for sediment uptake. Bed armoring by large sediment particles may reduce or limit the degradation process. Elimination of silts and clays from the stream may decrease the moisture-
holding capacity of newly formed banks, with effects on riparian
vegetation. Degradation of the main channel increases the gradient
of tributaries, which increases bank erosion and may induce head
cutting. This new source of sediment may, at least partially. compensate for the loss of fines in the main channel.

The flow regime of the receiving stream is variously modified
by impoundment. Often a dampened flow pattern results from storage
during high flow periods and discharge supplement during periods of
normally low flow. Diversion to another drainage basin will result
in a lower annual discharge, whereas the river system receiving the
abstracted water will have a total discharge greater than the historical flow. Other operational schemes, most notably hydroelectric
generation, result in short-term flow fluctuation, although the
seasonal flow pattern may exhibit increased constancy.

The chemical properties of water discharged from an impoundment
reflect (1) incoming water quality, (2) biotic and abiotic phenomena
within the reservoir, and (3) operational variables. The manager
has, at most, only two operational variables, release depth and flow

regime, immediately and directly under his control. However, these
variables directly and indirectly influence a myriad of factors in
the receiving stream. Hannan indicates that the "biological influence on chemical conditions within the reservoir is often greater
than the influence of other factors combined." Even if an upreservoir site has a temperature profile similar to that of a downreservoir site, chemical conditions may be considerably different, a
fact often overlooked in predicting chemical conditions at the
outflow. Whether concentrations of ions are higher or lower in
outflowing water than in water entering the reservoir depends on the
ion in question and a variety of other factors. Surface-release
impoundments generally act as nutrient sinks, whereas water drawn
from the bottom of a reservoir will have higher nutrient levels
during thermal stratification. Deep-release reservoirs may or may
not release anoxic waters and associated reduced compounds, such
as H_2S.

Webster, Benfield, and Cairns present a model of particulate
organic matter (POM) dynamics for a river-reservoir system. The
ecological significance of disruption of the stream continuum by
impoundment and resultant effects on sedimentary and transport POM
will vary as a function of stream order and the degree of autotrophy/
heterotrophy of the natural stream.

THERMAL MODIFICATION

Because of groundwater inputs and shading, temperatures of
headwater streams are characteristically cool and relatively constant
thermally. In temperate latitudes, streams in middle reaches may
exhibit annual ranges of more than 20°C; diel fluctuations of around
6°C are fairly common in summer. Large rivers have greater thermal
stability because of their greater heat capacity. Streams below
hydroelectric dams may exhibit great temperature fluctuations several times daily in summer due to great variations in discharge.
Shifting discharge depth in stratified reservoirs may induce thermal
shock in organisms of the receiving stream. The same effect may
result from a fixed outlet port as the water level falls in the
reservoir. Fraley describes how a large, shallow reservoir warmed
the waters of the receiving stream with significant effects on a
cold water river.

Streams below deep-release storage reservoirs which stratify
generally exhibit (1) increased seasonal and diel thermal constancy,
(2) winter warm and summer cool conditions, and (3) a delayed seasonal maximum, all of which have significant effects on the biota.

PHYTOBENTHOS

Several factors which may occur in streams below dams enhance conditions for macrophytes and algae. These include increased nutrient levels, decreased turbidity, decreased bank and bed erosion associated with flow constancy, and the absence of ice. In contrast, rapid flow fluctuations may decimate aquatic plants.

Filamentous green algae, such as *Cladophora*, are often enhanced in regulated streams. Submerged angiosperms may occur in regulated portions of streams which otherwise lack vascular plants. Cold stenotherms (e.g., *Hydrurus foetidus*) are often favored below dams. Aside from these generalizations, little is known regarding the phytobenthos of regulated streams. As Lowe pointed out, most of our meager knowledge is derived from stream zoologists incidently examining the flora.

ZOOBENTHOS

The standing crop of zoobenthos below dams may be enhanced or reduced compared with unregulated reaches. Enhancement is normally associated with increased flow constancy; reductions, with rapid flow fluctuations or oxygen depletion. Species diversity, with few exceptions, is reduced and species composition is greatly modified. Plecopterans may be reduced or eliminated; ephemeropterans, although reduced in diversity, may be numerically abundant; dipterans are often enhanced; filter-feeding trichopterans may be the predominant organisms below surface-release dams; amphipods, certain gastropods and oligochaetes are often favored.

The greatly altered macroinvertebrate community structure of the receiving stream has been attributed to a variety of factors. Thermal conditions may eliminate species which depend on a natural temperature regime to cue life cycle phenomena. Temperature and flow alterations have a multitude of direct and indirect ramifications on stream conditions.

Boon reports on the strategies adopted by a filter-feeding trichopteran downstream from an impoundment in Java. By colonizing vesicular rock the species is able to obtain a feeding advantage (seston produced in the reservoir), yet maintain position during extremely high discharge and avoid desiccation during low water.

FISHES

In reference to effects of stream regulation on salmon, Mundie discusses the negative impact of dams and the positive consequences

of flow regulation. In reference to the latter, examples are drawn
from (1) a pioneering flow-control project explicitly implemented
to increase production of salmon in a river, and (2) total regulation
in off-stream incubation and rearing channels.

According to Holden, "approximately 60 per cent of the fishes
presently listed as endangered or threatened in the U.S. are obligate
riverine species." Stream regulation by dams is one of the major
causes of their decline. The greatest impact is on species which are
fairly specialized, especially those with narrow thermal tolerance.

Stalnaker describes the approach being developed by the U.S.
Fish and Wildlife Service's Instream Flow Service Group (IFG). An
incremental approach using simulation modeling is used to assess the
relationship between stream flow and fish habitat structure. Habitat
structure suitability is plotted against various flow regimes for
the life stages of each fish species. The methodology is intended
as a decision-making tool specifically designed to demonstrate the
impact of alternative flow regimes on fishery habitat potential.

GEOGRAPHICAL REVIEWS

The geographical papers reviewed the extent of stream regulation
and resulting ecological effects for four European countries, Canada,
the United States (including a separate paper on the Tennessee Valley),
Africa, and Australia.

In Africa 9,200 km of river are regulated by dams impounding
>100 km^2 surface area. Davies estimates that approximately 652,000
kilometers of stream are regulated if all impoundments >5 km^2 are
considered. The majority of ecological research has dealt with
the reservoir *per se* as inundation proceeds. Some rivers become
seasonal instead of perennial and vice versa. Davies details the
need for quantitative research on effects of regulation on floodplain and coastal ecology, fisheries, zoobenthos, large herbivores,
nutrient dynamics, and salinification.

In Australia, the driest continent, annual runoff averages 45 mm,
one-sixth that of North America. Only a few rivers remain unregulated. Irrigation needs clearly dominate and, therefore, largely
influence stream regulation. Virtually all existing data on ecology
of regulated streams in Australia pertain to the Murray-Darling
drainage division, where 91 percent of the available water is committed for use. Walker's paper logically concentrates on the Murray
River and its tributaries. The outstanding characteristic of
Australian streams, according to Walker, is their flow variability.

Irrigation needs stabilize this natural flow variation, with sup-
posedly profound ecological effects. Even basic ecological data,
however, relating to downstream reaches are lacking.

Although considerably more ecological research has been conducted on regulated streams in Europe, much work remains to be done.
Henricson and Müller report that exploitation of water power during
the last 40 years has caused the disappearance of about 1500 km of
stream in Sweden. Only two large rivers are unregulated. The
environmental effects of regulation on Swedish rivers, however, are
not as severe as those on rivers in other parts of the world. The
authors attribute this to relatively small rivers and water power
projects which are also comparatively small. Typically, rivers have
been converted into "ladders of rather small reservoirs." Although
regulation has important effects on temperature and flow regimes,
water chemistry reverts to normal levels 10-15 years after impoundment. Much basic research should be conducted, especially on the
effects of dam construction and operational variables on stream
biota.

Only Norway has a greater degree of water power utilization
than Sweden. Lillehammer and Saltveit emphasize that nearly all of
the electricity in Norway is derived from water power. Most reservoirs in Norway are formed by damming natural lakes. Small weirs
have been constructed on some regulated streams to retain water in
partly dry river sections. A recent decision has led to a ten-year
protection period for several Norwegian river systems in order to
conduct further studies. Long term studies including periods before
and after regulation are needed to determine the ecological effects
of stream regulation in Norway (or elsewhere).

Décamps, Capblancq, Casanova and Tourenq review previously conducted and ongoing hydrobiological studies on regulated rivers in the
southwest of France. These completely transformed systems are characterized by their instability and fragility. The authors stress the
need for research which combines factors operative in the drainage
basin (e.g., forestry and agricultural practices) with stream limnology.

The demand for water in the United Kingdom is expected to
double by the end of this century. Intensive management of water
resources in England and Wales include reservoir storage, inter-
river transfers, and groundwater augmentation. The major regulatory
schemes in Scotland are associated with hydroelectric power generation. Although a few regulated streams in England have received
relatively intensive study, Armitage concluded that "there is
insufficient data to draw general conclusions about the impact of
flow regulation in Great Britain."

Growth of stream regulation in North America has followed the incessant demand for flood control and hydropower. In the United States (excluding Alaska) only 51 rivers greater than 100 km in length remain free-flowing from headwaters to major confluence. Nearly every feasible high-head dam site has been utilized or is proposed for dam construction.

Hydropower can supply only one per cent or less of the projected energy needs of the U.S. by the year 2000, yet great emphasis on water development projects continues. Low-head hydrogeneration dams and rewind or pumped-storage facilities are being projected as the solution to peaking needs concomitant with the development of numerous continuous-output thermal units (i.e., fossil fuel and nuclear generation plants). Diversion and damming of many headwater streams is ongoing or planned, since lower reaches are already impounded. Recently, major diversions to supply power plant cooling needs and coal slurry have been projected. Stanford and Ward point out that regulation now impacts more on North American rivers than do pollutants. They cite several case histories of highly regulated stream systems that generally characterize the myriad of problems referred to in the subject reviews of this volume.

Krenkel, Lee, and Jones discuss these effects with special reference to the Tennessee Valley, U.S.A. A nearly continuous series of reservoirs impound the mainstream Tennessee River. Tailwater sections typify the effects of stream regulation: increased development of macrophytes and algae, introductions of exotic fishes, and loss of important indigenous faunal components. Such changes have been mediated by controlled flow, altered thermal regima, oxygen depletion, and altered nutrient dynamics.

CONCLUSIONS AND RESEARCH DIRECTION

When the stream continuum is interrupted by a dam, physico-chemical alterations cause profound structural and functional changes at the ecosystem level. The "reset mechanism" (see Cummins) of flooding is lost and, as a result, the regulated stream ecosystem is characterized by loss of spatial and temporal heterogeneity.

Author's texts and discussions with symposium participants indicate several promising areas of research in regulated streams that should contribute significantly toward prediction capabilities and managerial objectives. There is general agreement that too little work has been conducted on too few regulated streams, and a concensus that research should concentrate on cause-effect relationships.

Nutrient loading and concomitant eutrophication response relationships in reservoirs have direct bearing on tailwater ecology,
especially those segments receiving hypolimnial releases (see Krenkel,
Lee, and Jones). Solid data on reservoir mineralization processes
as they relate to availability of particulate and dissolved organic
carbon to tailwater biota are also needed. Nutrient-carbon flux and
other biologically mediated processes in the reservoir are extremely
important deterministic events that influence survival and productivity of specific tailwater biota, but few data are available.

At some point in the regulated stream segment, primary biotic
dependence on allochthonous (i.e., reservoir-derived) nutrients and
carbon may shift to predominantly autochthonous (i.e., instream)
sources. The relationship between autochthonous primary productivity
and import of allochthonous carbon may greatly affect the trophic
scheme in tailwaters. Temporal, quantitative documentation of these
primary processes may elucidate why collectors such as caddisflies
and amphipods often attain high biomass in some tailwaters, while
they are unsuccessful in others. However, as Minshall (1978) has
discussed for natural streams, most of the available particulate
organic carbon is eventually processed irrespective of source.
What is important from the standpoint of community productivity may
be the temporal availability of particulate carbon and the rate at
which it can be processed. Some regulated segments may be thermally
stressed to the point that carbon flux may be limited.

A relatively complete knowledge of the life histories and
experimentally-defined environmental requirements of target species
is necessary. Only then can survival under a defined regulation
scheme be predicted with any certainty. Although many effects are
site-specific, tailwater conditions may best be projected by a
thorough understanding of the ecology of existing tailwater segments
over broad geographical ranges. At present, post-impoundment effects
based upon pre-impoundment conditions cannot accurately be predicted,
nor can useful simulation models be constructed, because of the
limited data base available on the ecology of stream regulation.

It must be emphasised that prior to alteration by man, many
rivers were heterogeneous, but relatively predictable, environments.
Organisms were able to survive and, in fact, maximize diversity, not
in spite of, but because of the temporal and spatial heterogeneity
of their environment. Not only do stream organisms respond in predictive ways, but shoreline and estuarine communities are dependent
upon these pulse events. (These proceedings are perhaps deficient
in that the effects of stream regulation on ecology of estuary ecosystems were purposely omitted. Future symposia should include
such discussions.) Stream regulation invariably reduces enviromental
heterogeneity. Without purposeful introduction of important environmental dynamics (e.g., temperature) into stream regulation, only

those organisms with great genetical plasticity (i.e., adaptive
ability) will be able to appropriately retune life histories. In
a very pertinent paper brought to our attention by Dr. J. H. Mundie,
Birch (1971) proposed that environmental and genetical heterogeneity
enhance long-term survival of species. This idea has since been
further investigated by population geneticists and has been applied
to adaptive strategies in regulated streams (see reference to Zimmerman's work in Stanford and Ward). Birch proposed that species
with great genetical heterogeneity are able to maintain populations
during unfavorable periods in refugia provided by environmental
heterogeneity. Species with such genetical plasticity would be the
most likely organisms to tolerate the regulated stream environment.
However, interspecific competition, changing trophic status of the
upstream reservoir, or redefinition of the discharge schedule may
greatly influence ultimate survival of colonizing or indigenous
tailwater species. Because many variables can be controlled at the
dam by varying the depth or time of release, regulated streams may
be ideal experimental microcosms for delineating such biotic responses.
By controlling these factors it is possible to introduce environmental heterogeneity into the receiving stream.

    Stream regulation imparts profound effects on riverine systems.
The mechanisms enabling organisms to adapt to tailwater conditions
are poorly understood. The types of hypothesis-oriented study that
have been successful in providing understanding of the structure
and, especially, function of unregulated streams are required. The
answers to ecological problems imparted by stream regulation are
already known to lotic organisms--stream ecologists must merely
ask them the right questions!

REFERENCES

Birch, L. C., 1971, The role of environmental heterogeneity and genet-
    ical heterogeneity in determining distribution and abundance,
    p. 109-128, *in*: "Proc. Adv. Study Inst. Dynamics Numbers
    Popul. (Oosterbeck, 1970)," den Boer, P.J. and Gradwell, G. R.
    (eds.), Centre for Agricultural Publication and Documentation,
    Wageningen, The Netherlands.

Minshall, G. W., 1978, Autotrophy in stream ecosystems, *BioScience*
    28:767-771.

CONTRIBUTORS

Patrick D. Armitage
Freshwater Biological Association
River Laboratory
Wareham, Dorset
England BH20 6BB

Ernest F. Benfield
Biology Department and Center for Environmental Studies
Virginia Polytechnic Institute and State University
Blacksburg, Virginia 24061

P. J. Boon
Department of Zoology
The University of Newcastle Upon Tyne
Newcastle Upon Tyne, NE1 7RU
England

John Cairns, Jr.
Biology Department and Center for Environmental Studies
Virginia Polytechnic Institute and State University
Blacksburg, Virginia 24061

J. Capblancq
Laboratoire d'Hydrobiolgie et
Service de la Carte de la Végétation du C.N.R.S.
29, rue Jeanne-Marvig, 31055 Toulouse Cedex
France

H. Casanova
Laboratoire d'Hydrobiologie et
Service de la Carte de la Végétation du C.N.R.S.
29, rue Jeanne-Marvig, 31055 Toulouse Cedex
France

Kenneth W. Cummins
Department of Fisheries and Wildlife
Oregon State University
Corvallis, Oregon 97331

Bryan R. Davies
Institute for Freshwater Studies
Rhodes University
Grahamstown, 6140
Cape Province
Republic of South Africa

Henri Décamps
Laboratoire d'Hydrobiologie et
Service de la Carte de la Végétation du C.N.R.S.
29, rue Jeanne-Marvig, 31055 Toulouse Cedex
France

John J. Fraley
Department of Biology
Montana State University
Bozeman, Montana   59717

Herbert H. Hannan
Aquatic Station
Southwest Texas State University
San Marcos, Texas   78666

Jan Henricson
Fishery Board of Sweden
FÅK
St. Torget 3
S-871 00 Härnösand
Sweden

Paul B. Holden
BioWest, Inc.
P.O. Box 3226
Logan, Utah   84321

R. Anne Jones
Department of Civil Engineering
Colorado State University
Fort Collins, Colorado   80523

Peter A. Krenkel
Water Resources Center
University of Nevada
Reno, Nevada   89506

G. Fred Lee
Department of Civil Engineering
Colorado State University
Fort Collins, Colorado   80523

Albert Lillehammer
Zoologisk Museum
Universitetet i Oslo
Sars Gt. 1
Oslo 5
Norway

Rex L. Lowe
Department of Biological Sciences
Bowling Green University
Bowling Green, Ohio   43403

Karl Müller
Department of Animal Ecology
University of Umeå
S-901 87 Umeå
Sweden

J. Harold Mundie
Department of Fisheries and Oceans
Pacific Biological Station
Nanaimo, British Columbia
Canada V9R 5K6

Svein Jacob Saltveit
Zoologisk Museum
Universitetet i Oslo
Sars Gt. 1
Oslo 5
Norway

Daryl B. Simons
College of Engineering
Colorado State University
Fort Collins, Colorado   80523

Clair B. Stalnaker
Cooperative Instream Flow Service Group
2625 Redwing Road
Fort Collins, Colorado   80526

Jack A. Stanford
Department of Biological Sciences
North Texas State University
Denton, Texas   76203
          and
University of Montana Biological Station
Bigfork, Montana   59911

J. N. Tourenq
Laboratoire d'Hydrobiologie et
Service de la Carte de la Végétation du C.N.R.S.
29, rue Jeanne-Marvig, 31055 Toulouse Cedex
France

Keith F. Walker
Department of Zoology
The University of Adelaide
Adelaide, South Australia
5001 Australia

James V. Ward
Department of Zoology and Entomology
Colorado State University
Fort Collins, Colorado   80523

Jackson R. Webster
Biology Department and Center for Environmental Studies
Virginia Polytechnic Institute and State University
Blacksburg, Virginia   24061

Robert D. Williams
Department of Zoology
Brigham Young University
Provo, Utah   84602

Robert N. Winget
Department of Zoology
Brigham Young University
Provo, Utah   84602

INDEX

The Ecology of Regulated Streams

Edited by James V. Ward
Colorado State University, Fort Collins

and Jack A. Stanford
North Texas State University, Denton

Water development projects have often inspired great controversy over the prophesied ecological effects within the affected riverine segment and in downstream areas. Nearly every major river in the world is now regulated by dams or diversions, yet ecological manifestations of river regulation remain undocumented and management alternatives are not available.

This book is the first comprehensive review of existing knowledge on the ecology of stream regulation. The text focuses specifically on downstream effects, emphasizing the ecological and biological impact of dams. It synthesizes present knowledge on regulated streams, provides a vehicle for scientific resolution of ecological problems resulting from stream regulation, stimulates future research, and provides data for the beneficial modification of future stream regulation projects.

The book begins with a state-of-the-art description of the natural stream ecosystem by K. W. Cummins. Subsequent articles, written by scientists with outstanding research experience in their subject areas, examine all aspects of the limnology of regulated streams. The modifications of physical, chemical, and hydrological gradients in streams below impoundments are examined as causative agents for profound changes in indigenous biotic components, including phytobenthos, zoobenthos, and fishes. Well-known scientists from eight countries discuss problems of stream regulation, offering examples from Scandinavia, Europe, Africa, Australia, and North America. Additionally, the applied problems of instream flow and fish management are discussed in detailed papers.

The editors conclude that regulated streams

ENVIRONMENTAL SCIENCE RESEARCH

In order to deal more effectively with environmental problems, a complete understanding of what is happening to our surroundings is essential. Guided by an editorial board of experts in the field, this dynamic series assesses sources of pollution and detection methods, analyzes damage caused by pollutants, and offers new scientific and engineering solutions to mitigate and control pollution.

THE LANGUAGE OF SCIENCE
Plenum
PUBLISHING CORPORATION

PLENUM PUBLISHING CORPORATION
227 West 17th Street, New York, N.Y. 10011